Exergy Analysis for Energy Conversion Systems

Discover a straightforward and holistic look at energy conversion and conservation processes using the exergy concept with this thorough text. The book explains the fundamental energy conversion processes in numerous and diverse systems, ranging from jet engines and nuclear reactors to human bodies. It provides examples for applications to practical energy conversion processes and systems that use our naturally occurring energy resources, such as fossil fuels, solar energy, wind, geothermal, and nuclear fuels. With more than 100 cases and solved examples, readers will be able to perform optimizations for a cleaner environment, a sustainable energy future, and affordable electricity generation. This text is an essential tool for practicing scientists and engineers who work or do research in the area of energy and exergy, as well as graduate students and faculty in chemical engineering, mechanical engineering, and physics.

Efstathios Michaelides is currently the Tex Moncrief Chair of Engineering at Texas Christian University. He has 40 years of experience teaching and performing research in thermal science and multiphase flow. Among other honors and awards, he has received the Freeman Scholar Award (2002), the ASME Fluids Engineering Award (2014), and the 90th Anniversary ASME-FED Medal (2016).

Exergy Analysis for Energy Conversion Systems

EFSTATHIOS MICHAELIDES
Texas Christian University

CAMBRIDGE
UNIVERSITY PRESS

University Printing House, Cambridge CB2 8BS, United Kingdom

One Liberty Plaza, 20th Floor, New York, NY 10006, USA

477 Williamstown Road, Port Melbourne, VIC 3207, Australia

314–321, 3rd Floor, Plot 3, Splendor Forum, Jasola District Centre, New Delhi – 110025, India

79 Anson Road, #06-04/06, Singapore 079906

Cambridge University Press is part of the University of Cambridge.

It furthers the University's mission by disseminating knowledge in the pursuit of education, learning, and research at the highest international levels of excellence.

www.cambridge.org
Information on this title: www.cambridge.org/9781108480581
DOI: 10.1017/9781108635684

© Efstathios Michaelides 2021

This publication is in copyright. Subject to statutory exception and to the provisions of relevant collective licensing agreements, no reproduction of any part may take place without the written permission of Cambridge University Press.

First published 2021

Printed in the United Kingdom by TJ Books Limited, Padstow Cornwall

A catalogue record for this publication is available from the British Library.

ISBN 978-1-108-48058-1 Hardback

Cambridge University Press has no responsibility for the persistence or accuracy of URLs for external or third-party internet websites referred to in this publication and does not guarantee that any content on such websites is, or will remain, accurate or appropriate.

To Laura

Contents

About the Author		*page* x
Preface		xi
List of Symbols		xiii
List of Abbreviations		xvii
1	**Introduction**	1
	Summary	1
	1.1 Energy – Whither Does It Come? Whence Does It Go?	1
	1.2 Fundamental Concepts of Thermodynamics	4
	1.3 First Law of Thermodynamics	8
	1.4 Second Law of Thermodynamics	12
	1.5 Practical Cycles for Power Production and Refrigeration	16
	1.6 A Note on the Heat Reservoirs	22
	Problems	23
	References	24
2	**Exergy**	25
	Summary	25
	2.1 General Observations on the Capacity of Engines to Perform Work	25
	2.2 The Model Environment	29
	2.3 Maximum Work – Exergy of Closed Systems	30
	2.4 Maximum Power – Exergy of Open Systems	35
	2.5 Exergy of Chemical Resources – Fossil Fuels	40
	2.6 A Note on Semipermeable Membranes	46
	2.7 Exergy of Black Body Radiation	48
	2.8 Exergy of the Water and the Wind	50
	2.9 Exergy of Nuclear Fuel	54
	2.10 Lost Work and Power – Exergy Destruction	57
	2.11 Exergetic Efficiency – Second Law Efficiency	62
	2.12 Characteristics of the Exergy Function	64
	2.13 Models for the Reference Environment	65
	2.14 An Operational Definition of Chemical Exergy	73
	2.15 A Brief Historical Background	74

	Problems		75
	References		76
3	**Energy Conversion Systems and Processes**		**79**
	Summary		79
	3.1	Heat Exchangers	79
	3.2	Vapor Power Plants	92
	3.3	Gas Turbines	95
	3.4	Cogeneration	100
	3.5	Jet Engines	105
	3.6	Geothermal Power Plants	108
	3.7	Fuel Cells	119
	3.8	Photovoltaics Systems	126
	3.9	Solar Thermal Systems	136
	3.10	Wind Turbines	141
	Problems		144
	References		146
4	**Exergy Consumption and Conservation**		**147**
	Summary		147
	4.1	Energy Conservation of Exergy Conservation?	147
	4.2	Maximum Negative Work – The "Minimum Work"	151
	4.3	Refrigeration and Liquefaction	155
	4.4	Drying	161
	4.5	Petroleum Refining	164
	4.6	Water Desalination	169
	4.7	Exergy Use in Buildings	171
	4.8	Exergy Consumption in Transportation	179
	4.9	Energy Storage	187
	Problems		197
	References		199
5	**Exergy in Biological Systems**		**201**
	Summary		201
	5.1	Photosynthesis	201
	5.2	Land Biomass	206
	5.3	Aquatic Biomass	212
	5.4	Animal and Human Systems	214
	5.5	Nonequilibrium Thermodynamics of Biological Systems	238
	5.6	Entropy Production and Exergy Destruction in Humans	242
	Problems		246
	References		247

6 Ecosystems, the Environment, and Sustainability — 251
Summary — 251
6.1 Environmental Effects of Energy Usage — 251
6.2 Ecology and Ecosystems — 252
6.3 The Natural Environment — 258
6.4 Exergy, the Natural Environment, and Ecosystems — 261
6.5 Sustainable Development — 266
Problems — 270
References — 270

7 Optimization and Exergoeconomics — 273
Summary — 273
7.1 Mathematical Optimization Models – Duality — 273
7.2 Definitions of Relevant Economic Variables — 277
7.3 Time Value of Money – Annualized Cost, Net Present Value — 278
7.4 Thermoeconomics and Exergoeconomics — 282
7.5 Uncertainty and Other Limitations — 289
Problems — 293
References — 294

Index — 296

About the Author

Professor Efstathios E. (Stathis) Michaelides is the holder of the Tex Moncrief Chair of Engineering at Texas Christian University (TCU). In the past he was chair of the Department of Mechanical Engineering of the University of Texas at San Antonio, where he also held the Robert F. McDermott Chair in Engineering and was the Founder and Director of the NSF-supported Center on Simulation, Visualization and Real Time Computing (SiViRT); Founding Chair of the Department of Mechanical and Energy Engineering at the University of North Texas (2006–2007); the Leo S. Weil Professor of Mechanical Engineering at Tulane University (1998–2007); Director of the South-Central Center of the National Institute for Global Environmental Change (2002–2007); Associate Dean for Graduate Studies and Research in the School of Engineering at Tulane University (1992–2003); and Head of the Mechanical Engineering Department at Tulane (1990–1992). Between 1980 and 1989 he was on the faculty of the University of Delaware, where he also served as Acting Chair of the Mechanical Engineering Department (1985–1987).

Professor Michaelides holds a BA degree (honors) from Oxford University and MSc and PhD degrees from Brown University. He was awarded an honorary MA degree from Oxford University (1983); the Casberg and Schillizzi Scholarships at St. Johns College, Oxford; the student chapter ASME/Phi,Beta,Tau excellence in teaching award (1991 and 2001); the Lee H. Johnson award for teaching excellence (1995); a Senior Fulbright Fellowship (1997); the ASME Freeman Scholar Award (2002); the Outstanding Researcher Award at Tulane (2003); the ASME Outstanding Service Award (2007); the ASME Fluids Engineering Award (2014); and the ASME 90th Anniversary of FED Medal, 2016.

Professor Michaelides was a member of the executive committee of the Fluids Engineering Division of the ASME (2002–2008) and served as chair of the Division in 2005–2006. He has served as chair (1996–1998) of the Multiphase Flow Technical Committee. He served twice as the President of the ASEE Gulf-South Region (1992–1993 and 2016–2017). He chaired the Fourth International Conference on Multiphase Flows (New Orleans, May 27–June 1, 2001) and was the vice-chair of the Fifth International Conference on Multiphase Flows (Yokohama, Japan, May 2004). He has published 6 other books; more than 150 journal papers; and has contributed more than 250 papers and presentations in national and international conferences.

Preface

When we "conserve energy," do we aim at maximum or minimum mechanical work? Why is it necessary for even the best electric power plants to dissipate and waste more than 50% of the energy in their fuels? What is the maximum amount of useful mechanical work we can get from an energy resource? How can we produce and transport liquefied natural gas with the minimum amount of energy-resource consumption? Exergy – the measure of useful mechanical work that may be extracted from energy resources – offers answers to these and several other questions on energy conversion and utilization.

In the pages of this book we analyze the energy conversion processes starting with two important realizations: first, that the vast majority of energy conversion processes take place in the terrestrial environment, which is the only naturally occurring *reservoir* of heat, work, and mass; second, that all energy conversion processes occur because there are materials, the energy resources, that exist not in thermodynamic equilibrium with the environment. These two realizations and the laws of thermodynamics, lead us to the concept of exergy – the key to the understanding of all energy conversion and conservation processes.

Using the exergy concept, this book provides a new, simple, comprehensive, rigorous, and holistic approach for the analysis of energy conversion and conservation processes. Following a simple and comprehensive exposition of exergy and its relationship to energy resources, the book offers several practical engineering cases and examples on the application of exergy to energy conversion systems that utilize our naturally occurring energy resources – fossil fuels, nuclear fuels, solar, wind, and geothermal. Among the engineering systems that are analyzed are: steam and gas electricity generation units; jet engines; nuclear reactors; heat exchangers; cogeneration; geothermal power plants, including Organic Ranking Cycles (ORCs); biomass as a fuel; photovoltaics; solar thermal systems; wind turbines; and fuel cells.

What are colloquially called *energy conservation systems* are examined under a new perspective, the *maximum work* principle, which emanates from the laws of thermodynamics and is intricately connected to the exergy concept. The exergetic analysis reveals practical methods to reduce the power supplied to engineering systems and provides benchmarks for the consumption of the least amount of resources in actual processes. Examples and cases on the application of the exergy methodology include: natural gas compression and transport; refrigeration; liquefaction; pasteurization; drying of foodstuff; water desalination; lighting; heat, ventilation and air-conditioning; transportation with internal combustion engine vehicles, electric vehicles and fuel-cell vehicles; energy storage; and petroleum refining.

Chapters 1–4 of this book follow the conventional exposition of exergy as a thermodynamic concept and its implications for the operation of engineering systems. These chapters offer a succinct exposition of the basic concepts and the laws of thermodynamics; the thermodynamically rigorous development of the exergy concept and its consequences for the several primary energy sources currently used by the human society; analyses of several power-producing systems including those utilizing renewable energy sources; analyses of power-consuming systems including the establishment of benchmarks for the optimum operation of systems; and analyses of the engineering systems used for transportation. A unique aspect of this book is the inclusion of three chapters on nontraditional thermodynamics subjects of current interest:

1. Biological systems (Chapter 5), a chapter that includes an exergy analysis of the human body as a thermodynamic system and explains in detail and with several examples the exergetic processes of metabolism, thermic effects, the conversion of nutrients to energy, and the production of mechanical power by the muscles.
2. The effects of energy resource utilization on the environment and the ecosystems, including the concept of eco-exergy. Chapter 6 explores the connections and implications of the exergy concept for a cleaner environment and sustainable development.
3. The mathematical optimization of engineering systems and processes, based on the exergy concept. A very important part of Chapter 7 is the uncertainty quantification of the optimization variables and the propagation of uncertainty in the optimum solution.

The inclusion of many cases and solved examples in every chapter further explains the application of exergy to dozens of significant engineering systems and processes. Problems at the end of every chapter offer a challenge and an opportunity for students and professionals to hone their analytical skills and appraise their ability to apply the exergy methodology to a variety of practical engineering systems and processes.

A number of individuals have helped in the writing of this book. First among them are my students in the courses on Thermal Science, Thermodynamics, Sustainable Energy, and Advanced Thermodynamics, which I have taught in five universities during the last 40 years. I have learned from them more than they have learned from me. I am very thankful to my colleagues at these universities as well as to other colleagues I regularly meet during conferences for many fruitful and animated discussions on thermodynamics, energy utilization, and the environment. The arrangements of the WA "Tex" Moncrief Chair of Engineering at TCU have afforded me the opportunity to devote a significant fraction of my time to this book. The Cambridge University Press staff in New York, Steven Elliott and Julia Ford, patiently answered all my inquiries and gave me guidance. I am also very much indebted to my own family, not only for their constant support, but also for lending a hand when this was needed. My wife, Laura, and our three children, Emmanuel, Dimitri, and Eleni, are a constant source of inspiration and were always ready to help. I owe to all my sincere gratitude.

Symbols

Latin

(Capital-case letters of thermodynamic variables denote the total variable. Low-case letters denote the corresponding specific variables)

A	area
A	affinity of reaction
B_j	environmental impact rate of component j
c	specific heat capacity (for solids and liquids)
c_P	specific heat capacity at constant pressure
c_v	specific heat capacity at constant volume
C_D, C_R	drag coefficient, rolling coefficient
D	diameter
e	specific exergy
E	exergy
f	frequency
F	force
F	Faraday constant
F_{HL}	geometric factor (in radiation)
g	gravitational constant
G	Gibbs free energy
ΔG^0	Gibbs free energy of a reaction at ambient conditions
h	convective coefficient (in heat transfer)
h	specific enthalpy
H	enthalpy
ΔH^0	heat of reaction at ambient conditions
I	solar irradiance
I_λ, I_ν	irradiance density
J	flux
k	thermal conductivity
k_B	Boltzmann constant
L	length
L_{ij}	phenomenological coefficients

m	mass
n	polytropic index
N	number of moles
Nu	Nusselt number
P	pressure
$P_{i,a}$	probability of amino acid formation
P_V	present value
Pr	Prandtl number
Q	heat
r	rate of reaction
r_d	discount rate
R	gas constant
R	resistance (electrical or thermal)
Ra	Raleigh number
s	specific entropy
S	entropy
ΔS^0	entropy of reaction at ambient conditions
t	time
T	temperature
U	internal energy
U	overall heat transfer coefficient of heat exchangers
V	volume
V	velocity
V	voltage
x	distance (for work production)
x	dryness fraction
x_i	mole fraction of ith component
z	vertical distance
Z_{env}	environmental exergonomics criterion

Greek

α	amplitude
β	coefficient of performance
Δ	denotes difference
ε	effectiveness (for heat exchangers)
η	efficiency
Θ	entropy production
λ	wavelength
μ	chemical potential
ν	frequency of radiation
ρ	density

List of Symbols

σ	Stefan-Boltzmann constant
τ	time constant
ψ_{ij}	degrees of coupling tensor
ω	moisture content

Superscripts

0	total (for enthalpy, internal energy, and exergy. Includes kinetic, potential, and other forms of energy)
o	reversible (for work and heat).

Subscripts

0	denotes property of the environment
II	exergetic (for efficiency)
a	refers to air
act	actual
B	boiling
C	compressor
e	exit/outlet
f	fuel
f	saturated liquid (for properties)
fg	latent heat
g	saturated vapor (for properties)
g	energy gap (for photovoltaics)
hb	human body
H	high
i, in	inlet
id	ideal (reversible)
lost	refers to exergy destruction (work lost)
L	low
max	maximum
min	minimum
mix	mixture
out	outlet, exhaust
p	primary fluid
P	pump
pp	pinch point
R	resource
s	secondary fluid
sc	solar cell

t	thermal
tr	transmission
T	turbine
w	refers to water
ww	well-to-wheels

Embellishments

A dot (.) above the symbol denotes time rate; a tilde (~) above the symbol denotes a molar quantity (per kmol); an arrow (→) above the symbol denotes a vector.

Abbreviations

ADP	adenosine diphosphate
ATP	adenosine triphosphate
BMR	basal metabolic rate
CAES	compressed air energy storage
CAFE	corporate average fuel economy
CFC	chlorofluorocarbon
COP	coefficient of performance
DEC	direct energy conversion
FWH	feed water heater
GCC	global climate change
GHG	greenhouse gas
HHV	high heating value
HVAC	heat, ventilation, air-conditioning
HX	heat exchanger
IC	internal combustion
LCEA	life cycle exergy analysis
LED	light emitting diode
LHV	low heating value
LMTD	log mean temperature difference
LNG	liquefied natural gas
NADP	nicotinamide adenine dinucleotide phosphate
NOX	nitrogen oxide
NPV	net present value
OP	oxidative phosphorilation
ORC	organic Rankine cycle
PAR	photo-active region
PHS	pumped hydraulic system
PV	photovoltaic
RQ	respiratory quotient
SEER	seasonal energy efficiency ratio
SI	sustainability index
STC	standard terrestrial conditions
TES	thermal energy storage

1 Introduction

Summary

The use of energy has defined our civilization and governs our daily lives. Throughout the day and night modern humans consume enormous quantities of energy resources for the preparation of their food; transportation; lighting, heat, ventilation, and air-conditioning of buildings; entertainment; and a myriad other applications that define modern life. A gigantic global energy industry transports and inconspicuously transforms the energy resources to convenient forms (gasoline, diesel, electricity) that are vital to the functioning of the modern human society. This introductory chapter surveys the types of global primary energy sources, how they are transformed to useful energy forms and how they are used by the consumers. The chapter introduces succinctly the two laws of thermodynamics that govern the conversion of energy from one form to another; it explains the methodology of thermodynamics, which is essential for the understanding of energy conversion processes; and delineates the physical limitations on energy conversion. The thermodynamic cycles for the generation of power and refrigeration are extensively reviewed and the thermodynamic efficiencies of the cycles and energy conversion equipment (turbines, compressors, solar cells, etc.) are defined.

1.1 Energy – Whither Does It Come? Whence Does It Go?

The human civilization has been defined by the utilization of energy forms for space heating, lighting, and food preparation. When the prehistoric humans mastered the use of fire for domestic comfort and cooking in their caves, human civilization began. In the later stages of human civilization, the utilization of animal power (primarily power of horses and oxen) contributed to the development of the agricultural society; helped nourish an increasing human population; and continued with the development of the ancient and medieval social structures. Also, wind-propelled sailboats facilitated trade and brought communities in contact. In the last three centuries the invention of powerful and energy-voracious engines ushered the industrial revolution and the mechanized society. Today's knowledge-based society, where citizens live in relatively great comfort; have more food and natural resources than any other society in the past; use fast modes of transportation to experience far-away lands and social events; and are continuously in contact with electronic devices, necessitates the use of enormous quantities of energy and the production of previously unmatched quantities of power.

Throughout the centuries, the human society has evolved by using the various forms of energy in increasingly larger quantities. Today, the utilization of vast amounts of energy – in thermal, chemical, mechanical, and electrical form – is absolutely necessary for the functioning of the modern society, the prosperity of the nations, and the survival of our civilization.

While the word *energy* is used as a colloquial term for several related concepts – including work, power, heat, and fuel – the human society primarily has the need for mechanical work, and mechanical or electrical power, which are produced by engines that utilize our energy resources. Airplanes and automobiles use liquid hydrocarbon fuels to produce motive power. Tractors and combines are fitted with engines that use hydrocarbons for the production of agricultural goods and food. Households use electricity for lighting, domestic comfort, and entertainment. Computers and wireless networks use electricity to function. The needed power is generated in power plants that convert primary resources – coal, natural gas, nuclear, petroleum, solar, wind, geothermal, and hydroelectric – to electricity.

Because energy is such a vital element of our lives, elaborate networks for the transportation and supply of energy forms to consumers have been developed in the last two centuries. Electricity is fed into cities and households by the transmission lines of the electric grid at very high voltages. Natural gas is transported by complex systems of pipelines, which often transcend national boundaries. Tanker ships crisscross the oceans daily to transport crude oil to refineries, which then supply the consumers with gasoline and diesel fuel via an elaborate system of pipelines, train cars, and trucks. Trainloads of coal are transported daily from coal mines to the electric power plants. In today's world energy enters all aspects of human life, economics, and politics. Actually, the need for increasingly larger quantities of energy resources has defined the lives and geopolitics of humans in the last two centuries.

The global population consumed 392.9 HJ (1 HJ $= 10^{18}$ J) of energy resources in 2015 in several forms [1], which are shown in Table 1.1. It is observed that petroleum products and electricity account for approximately 60% of the energy forms used

Table 1.1 Global energy consumption by form of energy used in 2015.

	Amount (HJ)	Percent (%)
Coal	43.6	11.1
Petroleum products	161.1	41.0
Natural gas	58.5	14.9
Biofuels and waste	44.0	11.2
Electricity	72.7	18.5
Other forms	13.0	3.3
Total	392.9	100

Data from [1].

Table 1.2 Global energy supply in 2015 by resource type.

	Amount (HJ)	Percent (%)
Coal and coal products	160.6	28.1
Crude oil	181.1	31.7
Natural gas	123.4	21.6
Nuclear	28.0	4.9
Hydraulic	14.3	2.5
Wood, biomass and wastes	55.4	9.7
Other Renewables	8.6	1.5
Total	571.4	100

Data from [1].

globally. Petroleum products are primarily used for transportation. Electricity is used in several human activities including lighting, air-conditioning, and entertainment. Most of the other energy forms are utilized by the various industries.

The energy forms we use daily (and are depicted in Table 1.1) are not naturally occurring. They are products of the *energy resources,* often called *primary energy sources.* Some of these resources are naturally occurring minerals that are mined in a few locations (e.g. crude oil, natural gas, uranium ore, and coal), while other resources are more widely distributed and free to be used (e.g. solar energy, wind power, and hydraulic power from rivers). The mineral resources are traded daily in the international markets; they are exported and imported in all the countries of the world; and represent significant entries in the energy and financial balance sheets of nations. Table 1.2 depicts the primary energy sources that have been used globally in the year 2015 to satisfy the demand for energy [1].

It is observed in Tables 1.1 and 1.2 that 571.4 HJ of energy resources have been used to supply the human society with the needed 392.9 HJ. The difference of 178.5 HJ or 31.1% of the total resources represents the dissipation when one form of energy is converted to another. For example, approximately 60% of the chemical energy in coal is dissipated when this mineral is converted to electricity. Additional energy dissipation occurs during the consumption stage when energy forms are utilized in our engines [2]. For example, the internal combustion (IC) engines in automobiles dissipate (as heat and sound) approximately 75% of the energy in the gasoline and diesel fuels (both are petroleum products) when they convert the chemical energy of the fuel to motive power.

The subject of this book and the concept of *exergy* (the *exergy method*) comprise the energy dissipation when primary energy forms are converted to secondary and tertiary forms to fulfill the desired objectives of the human society. Chapter 1 of the book includes a summary of the fundamental thermodynamic concepts and the two principal laws of thermodynamics that govern all energy transformations. Chapter 2 introduces the *exergy* concept and derives quantitative measures for the exergy of the commonly used primary energy sources. Chapter 3 elucidates the application of exergy in several types of energy systems that produce electric and motive power and offers quantitative tools and methods for the minimization of energy resource consumption. Chapter 4 offers the quantitative tools to minimize the energy resource consumption in what is colloquially called *energy conservation*. Chapter 5 explains the exergy dissipation processes in biological systems – including humans – that produce as well as consume energy. Chapter 6 focuses on the applications of the exergy methodology in the minimization of waste, a cleaner environment, and a sustainable energy future. Finally, Chapter 7 deals with optimization methods for the minimization of exergy destruction, the economics of energy resource conservation and the uncertainty of the calculations.

1.2 Fundamental Concepts of Thermodynamics

Central to the theory of thermodynamics is the concept of the *thermodynamic system*, or simply the *system*, which is the part of the universe where our attention is focused. The system may or may not contain any material objects: A tank full of hydrogen gas is a system; a turbine where steam flows is a system; and a vacuum chamber is also a system. The system is enclosed by a *boundary*, which is permeable or impermeable to mass. Outside the boundary are the *surroundings*, which represent the part of the universe that is affected by changes in the defined thermodynamic system.

Thermodynamic systems are *closed* or *open* systems. Closed systems contain a fixed amount of molecules and mass. Open systems have one or more inlets and outlets, through which mass is allowed to flow. The quantity of mass inside open systems may be constant, but individual molecules enter and exit the systems. Work and heat cross the boundaries of both closed and open systems and are exchanged with the surroundings. In the case of closed systems, we are usually interested in the total work and heat, W and Q, or the specific work and heat, w and q, namely the work and heat per unit mass of the system. For open systems we typically perform calculations on the instantaneous rates of work and heat, $\dot{W}$ (the power) and $\dot{Q}$, as well as on the mass flow rates that enter and leave the systems. The vast majority of energy conversion machinery are open systems: pumps, boilers, turbines, compressors, nozzles, and jet engines are all open systems. At typical operating conditions these devices operate as open systems at steady state, with equal rates of masses flowing in and out.

The *properties* of the thermodynamic system are measurable variables associated with the system. Temperature, pressure, volume, enthalpy, entropy, electrical conductivity, and viscosity are examples of thermodynamic properties. When the properties of a thermodynamic system do not change with time, the system is in *thermodynamic equilibrium* or simply *at equilibrium*. At equilibrium the properties may be measured, calculated via *equations of state*, or determined from *thermodynamic tables*. For homogeneous substances (substances that have stable and uniform composition), knowledge of two independent properties is sufficient to determine all the other properties. The *equations of state* are algebraic equations that give one property in terms of other (measurable) properties, as for example the ideal gas equation of state, $Pv = RT$. When a simple equation is not adequate for the accurate determination of properties, the properties are calculated by numerical computations and become available in thermodynamic tables or via computer algorithms, for example, the *Steam Tables* for the properties of water/steam, *Refrigerant-134a Tables,* and the *REFPRO* software that calculate the properties of tens of common materials.

The *state* of the system is another fundamental concept of thermodynamics, and is simply defined as the set of all the properties of the system.[1] When the properties change, the state of the system changes too, and the system undergoes a *process.* The *thermodynamic process* is central to energy conversion and takes place within a finite amount of time. When the timescale of the change of the system's properties is much less than the characteristic time of the process, $\tau_{system} \ll \tau_{process}$, the system responds fast to the external changes and is considered to be in internal thermodynamic equilibrium during the process. In this case the process is called *reversible.* In all other situations, the system is not in internal thermodynamic equilibrium during the process and the process is considered to be *irreversible.*

Mechanical work is a primitive concept in the discipline of mechanics and is associated with the concept of force: A force performs work when its point of application moves. The amount of work performed when the force moves its point of application by a path defined by the end states 1 and 2 is:

$$W_{12} = \int_1^2 \vec{F} \cdot d\vec{x}. \tag{1.1}$$

Work is a scalar quantity and depends on the path followed by the force between the end states 1 and 2.

Let us consider a system composed of a compressible substance enclosed in a cylinder fitted with a weightless piston. The force acting on the system is equal to the external force acting on the piston. When the system is in equilibrium the external force is equal to the product of the internal pressure and the area of the piston ($F_{ext} = PA$). Under these conditions, any process that moves the piston from position 1 to position 2 is reversible and Eq. (1.1) yields an expression for the work performed in terms of the two properties of the enclosed compressible substance, the pressure, P, and the volume, V:

$$W_{12} = \int_1^2 Pd\mathrm{V}. \tag{1.2}$$

Equations such as Eq. (1.2) prove that the work, which is performed to the system or by the system, depends on the details of the process 1-2 (the path between 1-2) and does not correspond to a potential function (a function whose difference would be equal to the work regardless of the path). The subscript *12* in the symbol for work is added to denote this *path dependence.* The amount of work performed by compressible substances, such as gases and vapors, during several common processes

---

[1] Since two properties are sufficient to determine all the other properties of homogeneous substances, the state of homogeneous substances is *defined* by two properties.

may be calculated from the details of the process and is given by one of the following expressions:

isobaric (constant P) process: $W_{12} = P(V_2 - V_1)$

isothermal (constant T) process, ideal gas: $W_{12} = mRT\ln(V_2/V_1) = mRT\ln(P_1/P_2)$

polytropic process (constant PV^n): $W_{12} = \dfrac{P_1V_1 - P_2V_2}{n-1} = m\dfrac{P_1v_1 - P_2v_2}{n-1}$,

isochoric process (constant V) process: $W_{12} = 0$

(1.3)

where m is the mass of the compressible substance in the cylinder-piston system that performs the work; R is the gas constant, $R = \tilde{R}/M$ (where $\tilde{R}$ is the universal gas constant 8.314 kJ/kg K); and n is the *polytropic index*, an exponent that defines the general polytropic process (when $n = 1$, the process is isothermal and when n is equal to the ratio of the specific heats, $n = c_p/c_v$, the process is isentropic).

The rate of work is often called the *power* or the *mechanical power*:

$$\dot{W} = \frac{dW}{dt} \quad \text{with} \quad W_{12} = \int_0^t \dot{W} dt, \tag{1.4}$$

where $0 - t$ is the time of duration of the process 1-2.

Heat is transferred spontaneously from a system at higher temperature to another system at a lower temperature (a temperature difference is a prerequisite for the transfer of heat). The transfer of heat takes place via one of the following three modes:

1. *Heat Conduction* occurs when a temperature gradient exists and heat flows down this gradient. The rate of heat conducted through an area A is given by the expression:

$$\dot{Q} = -kA\frac{dT}{dx}, \tag{1.5}$$

where k is the *thermal conductivity*, a property of the materials and dT/dx is the spatial gradient of temperature. The negative sign in the r.h.s (right hand side) of Eq. (1.5) signifies that heat is transferred from the high-temperature to the lower temperature, that is, opposite to the sign of the gradient dT/dx.

2. *Heat Convection* is caused by the motion of fluids. The heat convection may be *forced convection*, (when a fluid is pumped or blown by mechanical means, as in the case of car radiators, where colder air is blown by the fan over a heat exchanger to cool the engine coolant) or *natural convection* (when the fluid moves without any mechanical forcing by density differences, as in the case of the plume generated by a fire). The rate of heat transferred by forced or natural convection from an object with an area A, which is at temperature T_H, to a fluid at temperature T_L is:

$$\dot{Q} = -hA(T_H - T_L), \tag{1.6}$$

where the coefficient h is the *convective coefficient*. This variable is a function of the flow conditions as well as of the properties of the fluid (in the boundary layer of the surface) that facilitates the heat transfer. The negative sign, again, signifies that heat is transferred *from* the system at the higher temperature T_H *to* the system at the lower temperature T_L.

3. *Heat Radiation* does not require any intermediate materials to be transferred. Radiation may pass through vacuum as well as transparent fluids and solids. All material objects radiate heat. The sun heats the earth (and the rest of the universe) by radiation to provide solar energy. Similarly, the earth transfers heat to the rest of the universe by radiation. The rate of heat absorbed by radiation by an object at temperature T_L from another at temperature T_H is:

$$\dot{Q} = \sigma A F_{HL} \alpha (T_H^4 - T_L^4), \qquad (1.7)$$

where σ is the Stefan-Boltzmann constant, $5.67 * 10^{-8}$ W/m^2 K^4; A is the surface area of the object at absolute temperature T_L; F_{HL} is a geometric factor related to the area the two objects "see" through straight radiation rays; and α is the absorptivity of the receiving object surface, an empirical factor that characterizes all types of surfaces. The absorptivity is, for most surfaces, equal to the emissivity of the surface ε.

The work and heat exchanged between a system and its surroundings during processes are not potential functions and depend on the details, such as the "path" of the processes. The mechanical engineering community has adopted the following sign convention for the quantities of work and heat exchanged by a thermodynamic system for all processes:

- A quantity of work, W, is positive when it leaves the thermodynamic system and negative when it enters the system.
- A quantity of heat, Q, is positive when it enters the system and negative when it leaves the system.

The same convention applies to the rates of work (power) and heat. Figure 1.1 illustrates schematically the sign convention for the heat and work exchanged by a thermodynamic system as it will be used in this book. The algebraic sign of the quantities of heat and work that enter or leave the system are clearly shown in Figure 1.1.

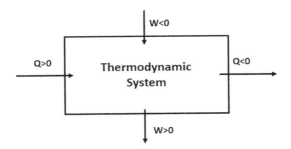

Figure 1.1 The sign convention for heat and work exchanged by a thermodynamic system.

1.3 First Law of Thermodynamics

The first law of thermodynamics defines what is commonly referred to as the *energy conservation principle*. Several formulations of the first law – some of which claim to be more mathematically rigorous than others – abound in the literature [3, 4]. All the formulations may be summarized by the general statement of energy conservation: *Energy is neither created nor destroyed. Energy may only be transformed from one form to another.* Expressions of the first law for closed systems, open systems, and systems undergoing thermodynamic cycles (all at steady state) will be given in the following subsections.

1.3.1 Closed Systems

The energy balance for a closed system is best given in terms of a thermodynamic process that takes a system from state 1 to state 2: *The heat entering a closed system minus the work produced by the system during a process 1-2 is equal to the difference of the total energy of the system between these two states.* This statement is written in symbolic form as follows:

$$Q_{12} - W_{12} = U_2^o - U_1^o, \tag{1.8}$$

where the total energy of the system U^o is a potential function defined as the sum of the internal (thermal) energy of the system, U, the potential energy, mgz, the kinetic energy, $1/2\ mV^2$, and any other form of energy the system may possess, and which may be described by potential functions as for example, electric energy, magnetic energy, surface tension energy, elastic energy, etc.[2] Thus:

$$Q_{12} - W_{12} = (U_2 - U_1) + \frac{1}{2}m(V_2^2 - V_1^2) + mg(z_2 - z_1) + \ldots. \tag{1.9}$$

In the vast majority of thermal systems operating within the terrestrial environment the internal energy difference is by far greater in magnitude than the differences of kinetic, potential, and all other forms of energy. For example, when 10 kg of water boil to produce steam, the internal energy of this mass increases by 20,880,000 J. The same mass of water would gain approximately 10,000 J if it were raised to a height of 1,000 m; and 50,000 J if it were accelerated from rest to 100 m/s (360 km/hr or 225 mph). Because for most thermal processes the changes in kinetic and potential energies are very small in comparison to the internal energy change, one may approximate Eq. (1.9) as follows:

$$Q_{12} - W_{12} \approx U_2 - U_1 = m(u_2 - u_1). \tag{1.10}$$

---

[2] The symbol E, instead of U^o, is used in several publications to denote total energy. In this book the symbol E is exclusively reserved for exergy.

1.3 First Law of Thermodynamics

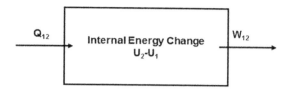

Figure 1.2 The first law of thermodynamics for closed systems.

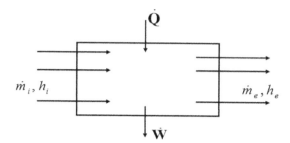

Figure 1.3 The first law of thermodynamics for open systems.

The specific internal energy, u, is a property of the system and the difference $u_2 - u_1$ may be obtained from thermodynamic tables; using a closure equation with the specific heat capacity at constant volume: $u_2 - u_1 = c_v(T_2 - T_1)$; from computer software; or from databases. Figure 1.2 depicts schematically the first law of thermodynamics for closed systems.

It must be emphasized that, unlike the internal energy, work and heat are not potential functions and depend on the process 1-2. Equation (1.10) simply states that, for a closed system at steady state undergoing a process, the energy entering the system in the form of heat, minus the energy exiting the system in the form of work, is equal to the difference of the internal energy of the system between the two end states (2 and 1) of the process.

1.3.2 Open Systems

Masses cross the boundaries of open systems through several inlets and outlets. At the same time a net rate of heat enters the open system and a net rate of work (power) is delivered by the system. Because the operations of open systems are continuous, the pertinent conservation equations are expressed in terms of rates, which are denoted by a dot (.) above the symbol of the corresponding variable.

The laws of mass and energy conservation apply to open systems. For the open system at steady state, which is depicted in Figure 1.3, the mass conservation equation is:

$$\frac{dm}{dt} = 0 = \sum_i \dot{m}_i - \sum_e \dot{m}_e, \tag{1.11}$$

where i denotes the inlets and e the exits of the system. For simple, steady-state, open systems with only one entrance and one exit, such as pumps, compressors, most turbines, nozzles, etc., the mass conservation equation simplifies as follows:

$$\dot{m}_i = \dot{m}_e = \dot{m}. \tag{1.12}$$

The first law of thermodynamics (energy conservation law) for open systems at steady state becomes:

$$\dot{Q} - \dot{W} = \sum_e \dot{m}_e h_e^o - \sum_i \dot{m}_i h_i^o, \tag{1.13}$$

where the property of specific total enthalpy, h^o, is defined as: $h^o = u^o + Pv$. The total enthalpy per unit mass incorporates the specific enthalpy, $h = u + Pv$, as well as all the other potential forms of energy that make up the total energy of the system:

$$\dot{Q} - \dot{W} = \sum_e \dot{m}_e \left(h_e + \frac{1}{2} V_e^2 + g z_e + \ldots \right) - \sum_i \dot{m}_i \left(h_i + \frac{1}{2} V_i^2 + g z_i + \ldots \right). \tag{1.14}$$

The power, $\dot{W}$, that appears in Eqs. (1.13) and (1.14) pertains to the *useful power* the system produces or consumes. The useful power produced is used for the fulfillment of engineering tasks and processes, such as the production of electric power with steam and gas turbines, the propulsion power for jet engines, and the motive power for automobiles. As in the case of internal energy and Eq. (1.10), the enthalpy difference in most open systems is by far more significant than the difference of the other forms of energy. For typical open systems – pumps, turbines, compressors, heat exchangers, etc. – the first law of thermodynamics at steady state may be simply written as:

$$\dot{Q} - \dot{W} = \sum_e \dot{m}_e h_e - \sum_i \dot{m}_i h_i. \tag{1.15}$$

For open thermodynamic systems at steady state with a single inlet and a single outlet, such as pumps, most turbines, and most compressors, Eq. (1.15) simplifies to the following form:

$$\dot{Q} - \dot{W} = \dot{m}(h_e - h_i). \tag{1.16}$$

The first law of thermodynamics is also applicable to open systems with chemical reactions, such as boilers, burners, and combustors as well as for the planet Earth. A useful property of the chemical reactions is the heat of the reaction (or heat of combustion for combustion reactions), which is expressed as the enthalpy difference between reactants and products per kg of the fuel, Δh, or as the enthalpy difference between reactants and products per kmol of the fuel, $\Delta \tilde{h}$. In both cases, the heat of reaction is calculated and tabulated when the reactants and the products are at 25°C (298 K). This quantity is sometimes called the *heating value* of the fuel and is expressed as kJ/kg of the fuel (for Δh) or as kJ/kmol of the fuel (for $\Delta \tilde{h}$).

In most industrial chemical reactions, including all combustion reactions, the useful power produced or consumed in the combustion chamber vanishes, that is $\dot{W} = 0$.

When the reactants and the products of the reaction are at 25°C the first law of thermodynamics may be used in a simplified form to calculate the rate of heat released by the open system – combustor, boiler, burner, etc. – where the combustion takes place:

$$\dot{Q} = \dot{m}\Delta h = \dot{N}\Delta \tilde{h}, \quad (1.17)$$

where $\dot{N}$ denotes the flow rate of the mols of the reactant. When $\dot{N}$ is measured in kmol/s and $\Delta \tilde{h}$ in J/kmol, the rate of heat is measured in W.

1.3.3 Systems Undergoing Cycles

A thermodynamic cycle is a series of processes with the same end states. Thermodynamic systems undergoing cycles are commonly used for the production of electric power in thermal power plants, in most of the engines used in the transportation industry, and in HVAC systems. Let us consider a system undergoing a series of n processes that constitute a cycle: 1-2, 2-3, 3-4, ... , n-1. The last process, n-1 in the cycle, ends at the initial state of the system, as shown in Figure 1.4. In this case one may write the first law of thermodynamics for each one of the n processes and add the resulting n equations to obtain the following expression:

$$Q_{12} - W_{12} + Q_{23} - W_{23} + \ldots + Q_{n1} - W_{n1} = U_2 - U_1 + U_3 - U_2 + \ldots + U_1 - U_n = 0. \quad (1.18)$$

The left hand side of Eq. (1.18) represents the difference of the net heat entering the cyclic system and the net work produced by the system when the latter performs one cycle 1-2-3- ... -n-1. The right-hand side of Eq. (1.18) is equal to zero. Therefore, Eq. (1.18) may be written succinctly as:

$$Q_{net} = W_{net}. \quad (1.19)$$

During a complete cycle, the net amount of heat that enters a cyclic engine is equal to the net amount of work performed by the engine.

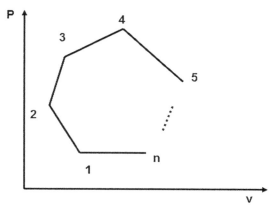

Figure 1.4 A system undergoing a cycle with n states. The $(n + 1)$th state is the same as the 1st state.

1.4 Second Law of Thermodynamics

We deduce from common experience that all the natural processes proceed only in one way and exhibit a definite *directionality*: if left unsupported, an apple always falls down; a moving billiards ball finally stops; all fluids flow from a higher pressure to a lower pressure; a drop of perfume evaporates in a room with the perfume molecules not spontaneously condensing to form the drop again; and heat always flows from hotter to colder bodies. When the natural processes are completed, it is not possible to reverse them without spending mechanical work for their reversal. For example, if we bring together two bodies, one at high temperature, T_H, and the other at lower temperature, T_L, and allow them to come to thermal equilibrium (without any heat losses or gains from other systems), they will finally come to a common temperature T_W, which is between the two original temperatures, $T_L < T_W < T_H$, as it is depicted in Figure 1.5. During this process, when the two bodies progress from state 1 to state 2, the total internal energy, which was originally contained in the two bodies, is conserved: the sum of the internal energies of the two bodies at state 1 is equal to the same sum at state 2. If we wish to reverse this process and restore the two bodies to their original temperatures, T_H and T_L, we will soon find out that this cannot be done without the use of a refrigeration devise, which consumes work. Despite the fact that the total energy of state 2 is equal to the total energy at state 1, the process 2 to 1 is impossible without the addition of work. We will draw the same conclusion when we try to reverse other naturally occurring processes. In order to restore a fallen apple from the ground (state 2) to the tree level (state 1) we must perform work by lifting it. To transport a fluid from low pressure (state 2) to a higher pressure (state 1) we must spend pumping or compression work. To reconstruct the evaporated drop of perfume, the perfume molecules in the air must be condensed (liquefied) by refrigerating the air/perfume mixture to a lower temperature, a process that only becomes possible with the consumption of work.

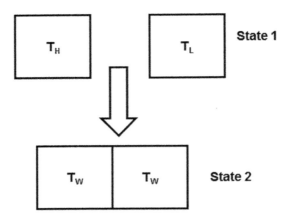

Figure 1.5 Two bodies at different temperatures, T_H and T_L, brought together adiabatically will finally attain an equilibrium common temperature, T_W, which is between the two original temperatures, $T_L < T_W < T_H$. This process cannot be reversed without work consumption.

1.4 Second Law of Thermodynamics

The second law of thermodynamics explains the directionality of natural processes, by defining the property *entropy*. Entropy increases in all the possible natural processes of adiabatic systems (systems with zero heat transfer to and from their surroundings). The second law may be expressed by the following two simple statements [3]:

1. There is a property of every system, entropy, which is defined as:

$$dS = \frac{dQ^O}{T} \Rightarrow S_2 - S_1 = \int_1^2 \frac{dQ^O}{T}. \tag{1.20}$$

2. For every natural process 1-2 that takes place in an adiabatic system:

$$S_2 - S_1 > 0. \tag{1.21}$$

The superscript "0" in the differential of heat denotes that this integral must be calculated during a *reversible* process. Although such processes are idealized processes, following standard procedures of thermodynamics [3–5], one may express the differential dQ^0 in terms of the system's properties and then carry the integration using only these properties, which are potential functions and independent of the process. For example, for a system containing a compressible substance, $dQ^0 = dU + PdV = dH - VdP$, and Eq. (1.20) may be written as:

$$S_2 - S_1 = \int_1^2 \frac{dQ^0}{T} = \int_1^2 \frac{dU + PdV}{T} = m\int_1^2 \frac{du + Pdv}{T} = \int_1^2 \frac{dH - VdP}{T} = m\int_1^2 \frac{dh - vdP}{T}. \tag{1.22}$$

The last four integrals only include material properties and may be easily calculated.

The inequality sign in Eq. (1.21) – the second part of the second law – indicates the directionality of the natural processes. Any thermodynamic system is part of a greater adiabatic system that includes the system proper and its surroundings. Processes in these greater and adiabatic thermodynamic systems always proceed in the direction where their entropy increases.[3] Actually, without any loss of generality, one may write the inequality in Eq. (1.21) as follows:

$$(S_2 - S_1)_{system} + (S_2 - S_1)_{surroundings} > 0. \tag{1.23}$$

It must be noted that there is not an *a priory* established limit on how high or low the entropy change of a system during a process may be or should be. Processes where the entropy change approaches zero ($S_2 - S_1 \rightarrow 0$) are usually idealized as *isentropic processes* with $S_2 = S_1$. In several actual processes, such as expansion in turbines or compression in pumps and compressors, the entropy differences are small and the processes are idealized as isentropic processes. In all the other processes (e.g. in

---

[3] Rudolf Clausius, who coined the word *entropy*, considered that the universe is adiabatic and expressed the second law of thermodynamics by the statement: *the entropy of the universe tends to a maximum*. However, in the 21st century we know that the universe is not an adiabatic system, and this statement may or may not be valid.

isothermal and isobaric processes) the changes of the entropy property are significant and are not neglected.

1.4.1 Implications of the Second Law on Energy Conversion

The most important implication of the second law on energy conversion processes is that *work may not be produced spontaneously by a cyclic engine, when this engine solely receives heat.*[4] Most power plants – including nuclear, gas turbines, jet engines, and car engines – are cyclic engines. As a consequence of the second law these cyclic engines must reject heat to a heat sink – typically their surroundings – a fraction of the heat received. Figure 1.6 depicts the schematic diagram of the operation of all cyclic engines. During a complete cycle, the engine is in contact with two heat reservoirs, receives heat Q_H, rejects heat Q_L, and produces net work, W_{net}, equal to the difference: $W_{net} = Q_H - Q_L$. Typically, the heat is rejected in the atmosphere (in gas turbines, jet engines, and car engines via the exhaust gases) or the hydrosphere (in larger steam power plants through their condensers and cooling systems). The rejected heat is sometimes called *waste heat*. As a consequence of the second law and the necessity for heat rejection, the thermal efficiency of all cyclic engines, η_t, cannot exceed the *Carnot efficiency*, η_C:

$$\eta_t = \frac{W_{net}}{Q_H} = \frac{\dot{W}_{net}}{\dot{Q}_H} \leq 1 - \frac{T_H}{T_L} = \eta_C. \tag{1.24}$$

Typical thermal efficiencies of fossil fuel thermal power cycles are close to 40% and typical efficiencies of nuclear power cycles are close to 33%. This implies that a coal power plant, which produces 400 MW of electric power, receives approximately 1,000 MW of heat and rejects 600 MW of heat, usually to a river or a lake. For a typical nuclear power plant, which generates 1,000 MW of electric power, the rate of heat

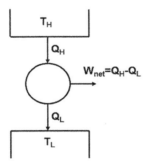

Figure 1.6 Net work produced and heat exchanged with two heat reservoirs during the operation of a cyclic engine.

---

[4] This statement is referred to as the Kelvin–Plank statement of the second law.

produced in the reactor is approximately 3,000 MW and the rate of heat rejected to the environment is close to 2,000 MW.

It is apparent that the most important consequence of the second law of thermodynamics for the energy conversion systems is that, even though heat may readily be converted to work, it is only a fraction of the heat that is actually converted into work in all thermal engines. The rest must be rejected to the environment as low temperature heat and is referred to as the *waste heat*.

Ratios, such as W_{net}/Q_H, which appears in the first part of Eq. (1.24), compare the useful work output to the heat (or energy) input of a thermal cyclic engine and are often called *first law efficiencies*. The concept of exergy generates other types of efficiencies that are usually called *exergetic efficiencies*.

1.4.2 Efficiencies of Engine Components

In addition to the overall efficiencies of cyclic engines (the power plants), engineers have defined the *component efficiencies*, to characterize the operation of the separate components that are parts of the cyclic engines. Such components are turbines, compressors, heat exchangers, fans, and pumps. The component efficiencies are defined as ratios of two variables: the actual work, W_{act}, produced or consumed by the component and the corresponding ideal work, W_{id}, which is produced or consumed in idealized, isentropic processes. All efficiencies are defined using the absolute values of work and heat (not the thermodynamic convention of Figure 1.1) in a way that their numerical values are between 0 and 1 (0–100%). The definitions of component efficiencies for commonly used equipment are:

1. For turbines:

$$\eta_T = \frac{W_{act}}{W_{id}}. \tag{1.25}$$

2. For pumps:

$$\eta_P = \frac{W_{id}}{W_{act}}. \tag{1.26}$$

3. For compressors and fans:

$$\eta_C = \frac{W_{id}}{W_{act}}. \tag{1.27}$$

4. For Solar Cells:

$$\eta_{sc} = \frac{\dot{W}_{act}}{IA}. \tag{1.28}$$

In the last expression I is the incident solar irradiance, usually expressed in W/m², and A is the area of the solar cell.

Typical values of efficiencies of industrial steam turbines are in the range 78–87% and those of gas turbines are within 80–90%. The efficiencies of large industrial compressors are in the range 75–85%, while the efficiencies of compressors used in domestic appliances (e.g. small refrigerators) may be as low as 30%. Pump efficiencies are in the range 70–82%. For commonly used, commercial solar cells, typical efficiencies are in the range 14–25%, with some prototype, experimental cells having efficiencies as high as 40%. For larger industrial units, component efficiency charts are supplied by the manufacturers for the entire range of the operation of the components. For the calculations, an engineer first calculates the ideal isentropic work, W_{id}, using the first and second laws of thermodynamics and then calculates the actual work, W_{act}, using the pertinent expression from Eqs. (1.25)–(1.28). It must be noted that the concept of exergy, which is central for this book, gives rise to several other efficiencies and figures of merit.

1.5 Practical Cycles for Power Production and Refrigeration

The vast majority of thermal electric power plants utilize one of two generic types of cycles: vapor cycles and gas cycles. This section provides a succinct description of the essential components and processes of the two types of cycles that will assist with the applications of the exergy method. More detailed descriptions, improvements of the basic cycles, and details of practical cycles may be found in textbooks of *Engineering Thermodynamics*, such as [5].

1.5.1 Vapor Power Cycles – The Rankine Cycle

The *Rankine cycle* is the most commonly used vapor power cycle. In most practical power systems, water is the working fluid of the cycle and the produced vapor is steam. A few systems in operation utilize an organic fluid, typically a hydrocarbon and produce the vapor of the organic fluid. These cycles are referred to as *Organic Rankine Cycles* (*ORCs*). Water-steam Rankine cycles and their variations are the principally used cycles of all coal and nuclear power plants, which currently produce more than one-third of the global electricity [1].

The schematic diagram of this thermodynamic cycle with its four basic components is shown in Figure 1.7. Liquid water at state 1 is pressurized in the pump from where it exits at state 2, which is at much higher pressure. The pressurized water enters the boiler, where it absorbs heat, and exits as superheated steam at high temperature, state 3. The high-temperature and high-pressure steam enters the vapor turbine, where it expands to a much lower pressure (usually subatmospheric) and its temperature drops to almost the ambient temperature. From the turbine the steam exhausts at state 4 into the condenser where the spent steam exits as liquid water and is fed back to the pump at state 1 to repeat the cycle. The condenser is cooled by the cooling system of the power plant, typically a water-cooling system.

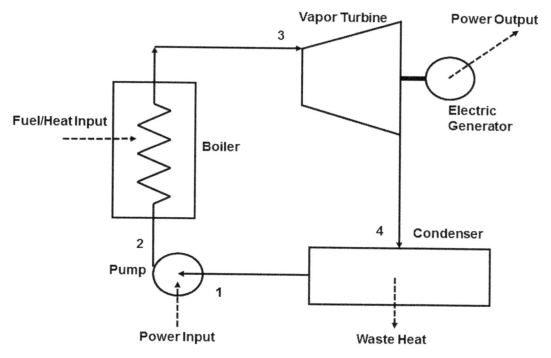

Figure 1.7 Schematic diagram of the basic Rankine cycle with its four essential processes.

The P,v and T,s diagrams that represent the four processes of the basic Rankine cycle and the four states of the working fluid are shown in Figure 1.8, which also includes the saturation curve (liquid-vapor dome) for water. The basic processes in this cycle are:

1. Process 1-2 is the pressurization of the liquid water effluent from the condenser in the pump and is almost isentropic. In large steam units, typical inlet conditions at state 1 are 6–10 kPa (6–10% of atmospheric pressure) and the outlet pressures at state 2 vary from a few MPa to 30 MPa for supercritical cycles. The pump is driven by a motor that consumes a small fraction (on the order of 1%) of the power produced by the steam turbine.
2. The boiler (sometimes called steam generator, burner, or combustion chamber) is the component where fossil fuel combustion (coal, petroleum, natural gas) or nuclear reactions produce a large amount of heat (Q_H in Figure 1.6), which is transferred to the pressurized water to produce steam. The process, 2-3 in the diagrams, produces steam at high temperature and pressure. The boiler is essentially the high temperature reservoir for the cycle. The heating process 2-3 occurs with very low pressure loss and is considered to be isobaric. Therefore, $P_2 \approx P_3$.
3. After the boiler, the steam enters a single turbine in smaller units, or several steam turbines in larger power plants. The expansion of the steam in the turbine(s) provides the motive power that drives an electricity generator via the *prime shaft*. The pressure and temperature of the steam are significantly reduced in the turbine

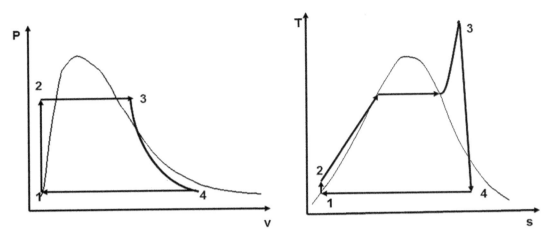

Figure 1.8 Thermodynamic states of the basic Rankine cycle in the P,v and T,s coordinates.

and the steam exhausts in the condenser at very low pressure, in the range 6–10 kPa, and low temperature, in the range 32–50°C. The steam expansion process, 3-4, is almost isentropic. The net work produced by the ideal cycle, W_{net}, is equal to the difference of the work produced by the turbine and the work consumed by the pump. In actual cycles a small amount of work is also consumed for the circulation of the cooling water in the condenser.

4. The condenser is a heat exchanger that receives steam from the turbine and typically cooling water from the cooling system of the power plant. The steam transfers its latent heat to the cooling water and condenses to become liquid. Heat is rejected in this process from the cycle to the cooling water and, finally, to the environment. This is the waste heat of the power plant, Q_L, in Figure 1.6. The condensation process, 4-1, occurs at constant pressure, which implies that $P_4 \approx P_1$. If cooling water is not available, the condenser is cooled by air.

If the mass flow rate of the circulating water in the cycle is denoted by $\dot{m}$, the power consumed by the pump is: $\dot{W}_P = \dot{m}(h_2 - h_1)$;
the rate of heat transferred to the water/steam in the boiler is: $\dot{Q}_H = \dot{m}(h_3 - h_2)$;
the power produced by the turbine is: $\dot{W}_T = \dot{m}(h_3 - h_4)$;
and the rate of heat rejected by the condenser via the cooling water to the environment is: $\dot{Q}_L = \dot{m}(h_4 - h_1)$.

The net power produced by the basic Rankine cycle is equal to the difference of the power produced by the turbine and the power consumed by the pump: $\dot{W}_{net} = \dot{m}[(h_3 - h_4) - (h_2 - h_1)]$. Hence, the thermal efficiency (first-law efficiency) of the cycle is:

$$\eta_t = \frac{(h_3 - h_4) - (h_2 - h_1)}{h_3 - h_2}. \tag{1.29}$$

The enthalpy and the other properties of the water and steam at the states 1, 2, 3, and 4 may be obtained from *steam tables,* which are standard appendices in books of Thermodynamics [3, 5].

1.5.2 Gas Cycles – The Brayton Cycle

All the gas turbine cycles are variations of the basic Brayton cycle and typically use air as the working fluid. A few advanced gas cycles for nuclear reactors have used carbon dioxide, argon, and helium. The arrangement of the basic components of the Brayton cycle is shown in Figure 1.9, while the pressure-volume and temperature-entropy diagrams of this cycle are depicted in Figure 1.10. The four processes of the Brayton cycle are:

1. Air at ambient temperature and pressure at state 1 enters the compressor where its pressure and temperature increase to state 2. Typical pressures at the exit of the compression process are 8–30 atm. Oftentimes intercoolers are used to reduce the work input for the compression. The compressor consumes a significant fraction (between 30 and 45%) of the power produced by the turbine. In order to avoid

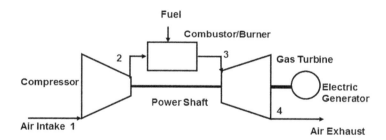

Figure 1.9 The basic components of a Brayton cycle.

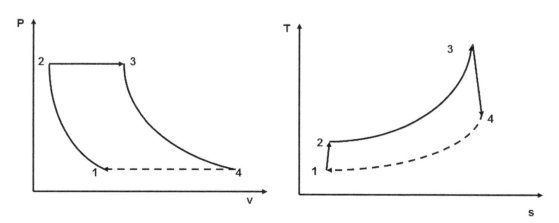

Figure 1.10 Thermodynamic states of the basic Brayton cycle in the *P,v* and *T,s* coordinates.

additional losses, the compressor is coupled mechanically to the turbine by the *prime shaft* and is driven directly by the turbine.

2. The compressed air enters the combustor (burner) where a fossil fuel (pulverized coal, natural gas, or liquid hydrocarbons) is injected and burns by combining with the oxygen in the air. The combustion product gases are at very high temperatures in the range 1,000–1,800°C.

3. The high-temperature combustion products are fed to the gas turbine where they expand to atmospheric pressure and are discharged to the environment at state 4. The temperature at this state is significantly higher than the ambient temperature T_1.

The electric generator – coupled to the turbine and the compressor via the prime shaft – produces electric power equal to the difference of the power produced by the turbine and the power consumed by the compressor.

The gas cycle is not a cycle *per se* because the turbine exhaust is at a different state and has different composition (it contains the combustion products) than the ambient air at the compressor input. This series of processes is treated as a cycle, because the input to the compressor is always air at ambient temperature. In this case, the atmosphere plays the role of a heat and mass reservoir, which receives the hotter output of the turbine, cools it, purifies it to the ambient temperature and composition and, finally, allows it to be fed back to the compressor at the prevailing, ambient pressure, temperature, and composition. The fictitious cooling process in the atmosphere that completes the cycle is denoted by the broken lines 4-1 in the diagrams of Figure 1.10. The thermal efficiency of the Brayton cycle is defined in the same way as in the vapor cycles and may be expressed as follows in terms of the states in Figure 1.10:

$$\eta_t = \frac{(h_3 - h_4) - (h_2 - h_1)}{h_3 - h_2}. \tag{1.30}$$

As with the vapor cycles, the enthalpy and other properties of the air at the states 1-2-3-4 may be obtained from *air tables* that are standard appendices in Engineering Thermodynamics books [3, 5]. Oftentimes the air and the combustion products are modeled as ideal gases and the ideal gas relations are used for approximate calculations.

1.5.3 Refrigeration, Heat Pump, and Air-Conditioning Cycles

If the operation of the power generation cycle in Figure 1.6 is reversed, the cycle absorbs heat, Q_L, from a colder heat reservoir and transfers a higher amount of heat, Q_H, to a hotter reservoir, while simultaneously consuming work, W_{net}. Such reversed cycles are used in refrigerators, heat pumps, and air-conditioning systems. The working fluid in the refrigeration cycles is one of the common refrigerants, which are chemical compounds of carbon, hydrogen, chlorine, and fluorine. The four basic components of the refrigeration cycle are shown in Figure 1.11 and the thermodynamic diagram – in the T, s coordinates – is shown in Figure 1.12. A comparison of Figures 1.12 and 1.8 proves

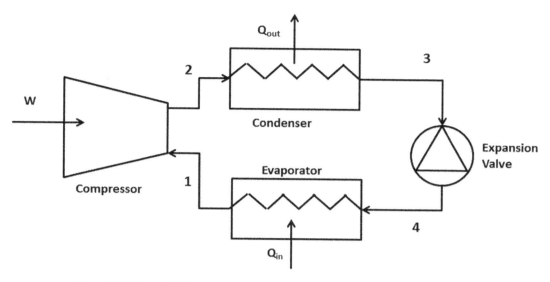

Figure 1.11 Schematic diagram of the components and states in a refrigeration cycle.

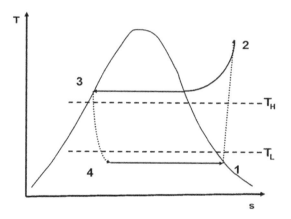

Figure 1.12 Thermodynamic (T,s) diagram of the refrigeration cycle.

that the refrigeration cycle is a modified reversed Rankine cycle that comprises the following processes:

1. The compressor admits the vapor refrigerant at state 1 and raises its pressure and temperature to those of state 2. The compressor is driven by an electric motor, which consumes power, $\dot{W}$. The refrigerant exits the compressor as superheated vapor at higher temperature.
2. The superheated refrigerant is directed to the condenser, where it undergoes the condensation process 2-3. During this process the refrigerant dissipates heat to a heat sink at the higher temperature T_H. In a heat pump, the heat sink is the interior of the

building. In a refrigerator, the heat sink is the room (or kitchen) where the refrigerator is located.[5] In an air-conditioner, the heat sink is the outside air or the ground.

3. Process 3-4 is caused by the expansion valve. During this process, the pressure of the refrigerant is reduced abruptly to the original pressure, P_1, and a mixture of liquid and vapor at significantly lower temperature, $T_4 = T_1$ is produced. The expansion valve is small and well insulated for this process to be considered isenthalpic, $h_3 = h_4$. Because this process entails significant irreversibilities, it is usually denoted with a broken line.

4. The refrigerant passes through the evaporator and undergoes the evaporation process 4-1. Because the temperature of the refrigerant at this stage is very low, it evaporates by absorbing heat from the heat source at the lower temperature, T_L, and leaves the evaporator at the original state 1, typically as saturated vapor or slightly superheated vapor. In an air-conditioning cycle, the lower temperature heat source is the air in the interior of the building. In a refrigeration cycle it is the interior of the refrigerator, which is kept at lower temperature. In a heat pump cycle the lower temperature heat source is the outside ambient air or the ground.

The main difference between a reversed Rankine cycle and the refrigeration cycle is that the process 3-4 of the refrigeration cycle is isenthalpic and does not produce work. The work that could be produced from the expansion of the liquid refrigerant at state 3 is too low to justify the additional cost of an expander/turbine. The simple expansion valve adopted for the process 3-4 is by far cheaper and fulfills the function to significantly lower the temperature of the refrigerant.

The figure of merit most often used in refrigeration cycles is commonly referred to as the *coefficient of performance* (COP) and is defined as the ratio of the rate of heat removal (the benefit of the cycle) and the power input (the cost) to the cycle:

$$COP = \frac{\dot{Q}_L}{\dot{W}} = \frac{q_{41}}{w_{12}} = \frac{h_1 - h_4}{h_2 - h_1}. \qquad (1.31)$$

As with the heat engines, the second law of thermodynamics also imposes a limit on the COP of refrigerator cycles:

$$COP \leq \frac{T_L}{T_H - T_L}. \qquad (1.32)$$

1.6 A Note on the Heat Reservoirs

The notion of heat reservoirs is central to the theory of thermodynamics and essential in the application of the second law. A heat reservoir is a natural or artificial thermodynamic system that may supply or absorb any quantity of heat without appreciable change of its temperature. Natural heat reservoirs (e.g. the atmosphere and the

---

[5] This transfer of heat occurs through the coils at the back of the refrigerator, which always feel warmer to the touch.

hydrosphere) are massive systems that receive and supply heat to the thermal engines. As a result of the heat exchange, the extensive properties of the heat reservoir change according to whether heat is received or supplied, but their temperature remains approximately the same. For example, when during a process 1-2 heat Q_{12} is supplied to a heat reservoir by a thermodynamic system, the properties of the reservoir change according to the two laws of thermodynamics:

$$\begin{aligned}(U_2 - U_1)_R &= Q_{12} \\ (H_2 - H_1)_R &= Q_{12} \\ (S_2 - S_1)_R &\geq \frac{Q_{12}}{T_R}\end{aligned} \quad (1.33)$$

If the heat exchange is accomplished without phase change and chemical reactions (e.g. only sensible heat is exchanged between the system and the reservoir), the first law of thermodynamics yields the following relationship between the temperature changes of the system and of the reservoir:

$$|Q_{12}| = |m_S c_{PS}(T_2 - T_1)_S| = |m_R c_{PR}(T_2 - T_1)_R|, \quad (1.34)$$

where the subscripts S and R denote the system and the reservoir, respectively. For most thermodynamic systems and reservoirs, the two specific heats, c_{PS} and c_{PR}, are of the same order of magnitude. If the temperature of the heat reservoir is to remain approximately constant, $(T_2 - T_1)_R \approx 0$, Eq. (1.34) leads to the following result:

$$\frac{m_S}{m_R} \approx 0 \Rightarrow m_R \to \infty, \quad (1.35)$$

which implies that the mass of the heat reservoir is much greater than the mass of the system.

Natural heat reservoirs – the atmosphere, large lakes, rivers, and oceans – are massive and satisfy this condition. The sun and other stars, which are also massive, may be considered as heat reservoirs at high temperatures but, at present, we do not utilize these extraterrestrial, high-temperature heat reservoirs for the production of work and mechanical power. Instead, we have developed artificial systems, within the biosphere, that are often modeled as heat reservoirs: boilers, burners, superheaters, and combustion chambers, which are components of power plants, jet engines, and automobiles. They supply the engines with heat at high temperature and are often modeled as heat reservoirs. Energy is continuously supplied to these "reservoirs" by the combustion of fuels that maintain their "reservoir's" temperature at high levels. Effectively, it is the continuous supply of the fuels (the natural energy resources) that creates and maintains the terrestrial high-temperature heat "reservoirs."

Problems

1. A waterfall has 50 m height. Assuming that there is no other change and zero evaporation during the fall, what is the increase of the water temperature at the bottom?

2. A nuclear power plant generates 1,200 MW of electric power and has 32.5% overall thermal efficiency. What are the rate of heat input from the nuclear fuel and the rate of waste heat from this power plant? How much heat, in J, does the power plant reject every day?

3. A gas turbine produces 80 MW of power and has 52% thermal efficiency. What is the thermal power input to the turbine? The fuel of the gas turbine is methane (CH_4) with heating value 802 MJ/kmol. Determine the rate of fuel supply in kmol/s and kg/s.

4. The interior of a refrigerator is kept at 4°C, while the ambient temperature is 32°C. What is the maximum COP for the refrigerator?

5. A pump is used to elevate 15 kg/s of water by 52 m. Frictional and other losses are 18% of the ideal power required to lift the water. If the combined efficiency of the pump and its motor is 75%, what is the rate of electric power to be supplied?

6. Convert to J the following: 3.4 MeV; 15 Btu; 35 kWh; 98 Quads; 450 TJ; 115 bbl of crude oil.

References

[1] International Energy Agency, *Key World Energy Statistics-2019* (IEA: Paris, 2019).
[2] E. E. Michaelides, *Energy, the Environment, and Sustainability* (Boca Raton, FL: CRC Press, 2018).
[3] J. Kestin, *A Course in Thermodynamics*, vol. 1. (New York: Hemisphere, 1979).
[4] A. Bejan, *Advanced Thermodynamics*, 3rd ed. (New York, Wiley: 2006).
[5] M. J. Moran and H. N. Shapiro *Fundamentals of Engineering Thermodynamics*, 6th ed. (New York, Wiley: 2008).

2 Exergy

Summary

Of all the energy forms, mechanical energy (e.g. as motive energy in vehicles, and for electricity generation) is the form most useful to humans and of high demand in our society. This fact bears the question: Given the amount of energy in a resource – fossil fuel, nuclear fuel, wind, solar, geothermal – what is the maximum mechanical energy one may extract? A similar question is: What is the quantitative difference between the low-temperature waste heat of a nuclear power plant and the high-temperature heat in the reactor of this plant? The combination of the first and second laws of thermodynamics, in conjunction with the characteristics of the environment where energy conversion processes occur, offer a definitive answer to these and similar questions: exergy is the thermodynamic variable that describes the maximum mechanical work, which may be extracted from energy resources, the concept that quantifies the *quality of energy*. This chapter elucidates the concept of exergy and its relationship with the properties of the energy resources and the environment. It also derives useful expressions for the exergy of several primary energy sources including fossil fuels, geothermal, solar irradiance, wind, hydraulic, tidal, wave, and nuclear. The effects of the environment on the exergy of the several energy sources, the energy conversion processes, and the exergetic efficiencies of the processes are also elucidated.

2.1 General Observations on the Capacity of Engines to Perform Work

The laws of thermodynamics require that all cyclic thermal engines, which produce and consume work, are in contact with at least two heat reservoirs, as depicted in Figure 1.6. At least one of the heat reservoirs is at a high temperature, T_H, and supplies heat to the engine. At least one other heat reservoir is at a low temperature, T_L, and receives the waste heat from the cyclic engine. All the heat reservoirs must be large enough so that their temperatures remain constant regardless of the quantities of heat exchanged with the cyclic engine. It is apparent that, for this to happen, the mass of the heat reservoirs must be much higher than the mass of the working fluid in the engine, as indicated in Eqs. (1.34) and (1.35).

Regarding the high-temperature heat reservoir, a moment's reflection proves that there is not a single natural system that has the characteristics of a high-temperature heat reservoir that is available for engines operating in the terrestrial environment. The sun and other stars have the characteristics of high-temperature heat reservoirs and may qualify as such, but they are not accessible to be used with our thermal engines.[1] The earth's interior, which is also at high temperature, may be considered as a high-temperature heat reservoir. However, the earth's interior – at least at present and in the near future – is not accessible to be used with thermal engines. In most practical thermal engines, it is the consumption of a fuel (coal, gasoline, diesel, natural gas, nuclear, geothermal, etc.) derived from a primary energy resource that constantly provides the needed heat to the engines. Effectively, our thermal engines convert the energy resources to mechanical and electrical power.

Typically, the fuel enters the cyclic engines at the ambient temperature and undergoes combustion. The products of the fossil fuel combustion are created at high temperature and do not remain at that high temperature, but cool as they go through the several heat exchange processes in the engines. Figure 2.1 shows the production of steam in a power plant. The fuel – coal or natural gas at ambient conditions – is mixed with the ambient air and subsequently burns, creating the combustion products, typically CO_2, water vapor, nitrogen, and excess oxygen. The combustion products, which are initially at very high temperature, transfer their enthalpy to the working fluid of the power plant,

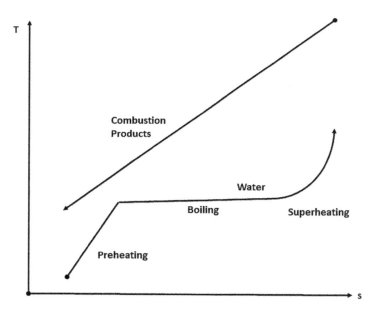

Figure 2.1 Superheated steam is produced in a power plant by heat exchange with the fuel combustion products.

---

[1] In thermal solar power plants, such as *Solar Two* and *Gemasolar,* it is the central receiver of solar energy that is considered to be the high temperature heat reservoir.

2.1 General Observations on the Capacity of Engines to Perform Work

typically water. During this process, the water preheats to its saturation temperature, boils and evaporates, and finally becomes superheated steam to be subsequently fed to the turbine, where power is produced. As Figure 2.1 shows, the combustion products are cooled during this heat transfer process and exit the combustion chamber at a significantly lower temperature. In well-designed power plants the exit temperature of the combustion products is in the range 60–100°C. In general, the lower the exit temperature of the combustion products – that is the closer the combustion products are to the ambient temperature – the more electricity the power plant produces per unit mass of fuel consumed.

It is apparent in Figure 2.1 that steam is not produced by a high-temperature heat reservoir operating autonomously, but by the consumption of the fuel, which enters the engine at ambient temperature. Also, there is no constant temperature, T_H, from which the water in the cycle receives the needed heat, but a range of temperatures with the exit temperature being close to the ambient temperature.

The same observation will be made when one considers all the other cyclic engines we have invented and use for the production of energy and power. The engines do not receive heat from a high-temperature reservoir. Instead it is the consumption of an energy resource, which enters the engine (usually at ambient temperature), that produces the high temperature of the working fluid in the cyclic engine: The consumption of nuclear fuels generates steam in nuclear power plants; the consumption of gasoline or diesel raises the temperature of the gases in car engines to be followed by a power-producing expansion; and the combustion of kerosene is the cause of the higher gas temperatures in jet engines that generate the aircraft propulsion.[2]

While the high-temperature heat reservoir is at best an idealization, the low-temperature reservoir is realistic. Low-temperature heat reservoirs are parts of the natural environment. The atmosphere and hydrosphere, which are parts of the natural environment, are massive and their temperatures fluctuate daily and seasonally within relatively narrow limits. Cyclic engines, regardless of how large they are, may reject to the natural environment any quantity of heat without significant effects on its temperature. Thus, the spent steam in the coal and nuclear power plants condenses and all its latent heat is transferred to the cooling water of the power plants. The cooling water is discharged and subsequently cooled in the hydrosphere (rivers, lakes, and sea) or in cooling towers and the heat is rejected in the atmosphere. Similarly, automobile internal combustion engines and jet engines exhaust hot combustion products in the atmosphere. The hydrosphere and the atmosphere have all the characteristics of natural heat reservoirs with (almost) constant temperatures.[3]

---

[2] The very idea that an airplane engine at 10,000 m above ground may be thermally connected to a high-temperature heat reservoir with an enormous mass is, by itself, absurd.

[3] The rate of heat rejected in the natural environment by all the anthropogenic activities is several orders of magnitude lower than the solar irradiance and has an imperceptible effect on the temperature of the environment. In 2015 all the thermal power plants globally rejected approximately $127 * 10^{15}$ kJ of heat to the environment. During the same year, the energy received from the sun was $5.46 * 10^{24}$ kJ and the annual amount of heat radiated by the earth to the outer space was of a comparable magnitude. Clearly, the waste heat of the electric power plants is eight orders of magnitude lower than the irradiance and does not

It must be emphasized that thermal engines operate because they consume fuels (the natural resources), which in their natural condition are at the environmental temperature and pressure, but they are not in chemical equilibrium with the environment. For example, H_2O and CO_2 may be in chemical equilibrium with the environment (when their concentration is the same as that of the environment), but methane fuel (CH_4) invariably is not. When methane is used as a fuel in an engine, it is chemically transformed to different components (H_2O and CO_2) that are constituents of the natural environment and are closer in thermodynamic equilibrium with the environment. The chemical transformations in all combustion processes alter the state of the fuels to a thermodynamic state that is closer to the state of the environment. Similarly, the nuclear fuel in reactors is transformed to other nuclei that are closer to the isotopes in the natural environment. The products of all nuclear and combustion reactions are in closer equilibrium with the environment than the original nuclear and chemical fuels.

One may also consider the geothermal power plants, where water or steam at high pressure and temperature from a deep aquifer, the geothermal fluid, is the "fuel." The energy conversion processes in all geothermal power plants have two effects: The reduction of both the pressure and temperature of the geothermal fluid, which approaches the temperature and pressure of the environment. At the end of all the processes, this "fuel" exits the power plant at a thermodynamic state that is closer to the state of the environment.

From the above observations that pertain to all terrestrial thermal power plants and thermal energy conversion processes one may draw the following conclusions [1, 2].

1. There are no naturally available and accessible high-temperature heat reservoirs on earth.
2. Heat is supplied to energy conversion processes because of the consumption of naturally occurring substances, the energy resources. The energy resources are not in thermodynamic equilibrium with the terrestrial environment.
3. The only naturally occurring heat reservoir is the terrestrial environment, which is also a reservoir of mass, and work. The average pressure and temperature of the environment vary within narrow limits. Energy conversion engines may absorb and reject heat to the environment without any perceptible change on the intensive properties of the environment. Thermal engines intake and exhaust substantial quantities of volumes and masses to the natural environment (primarily the atmosphere and the hydrosphere) without significant changes in its chemical composition.[4]

---

have an effect on the environmental temperature. When we consider the effect of the waste heat of all the thermal engines, the total amount of heat rejected to the environment is on the order of the total energy supply, $571.4 * 10^{15}$ kJ, 7 orders of magnitude less than the annual irradiance. The observed global temperature increase, the signal of GCC, is not due to waste heat or other energy dissipation, but to the emissions of GHGs.

[4] The increase in the atmospheric composition of CO_2 from 0.028% at the onset of the industrial revolution to 0.041% in 2018 occurred during a timescale of more than 250 years. This is long enough for the change to be considered as extremely slow.

4. When natural resources (fuels) are used by thermal engines for the production of work or heat, the thermodynamic states of the resources are altered to states that are closer in thermodynamic equilibrium with the terrestrial environment.
5. When the state of the energy resources is reduced to the state of the environment, the energy resources are unable to produce additional work or heat.
6. A corollary of the last observation is that the most we can achieve with the energy resources is to reduce their thermodynamic state to a state in equilibrium with the environment. At the state of the environment, the resources have been "spent" or "exhausted" and it is not feasible to extract additional work or heat from them.
7. The kind of interactions (mass exchanges, chemical reactions, heat transfers, etc.) that occur between thermal engines and the terrestrial environment, and the thermodynamic irreversibilities in the engines determine the final thermodynamic state of the energy resources and, by extent, the work or heat that may be extracted from them.

As a consequence of these observations, when we consider the useful work that may be extracted from a naturally occurring energy resource we may a priori stipulate that the work will depend on the following variables:

1. The thermodynamic state of the environment.
2. The type and thermodynamic state of resources considered.
3. The types of interaction(s) between the resource and the environment.
4. The final state of the resource and its proximity to the state of the environment.
5. The irreversibilities of the energy conversion processes.

In the light of these observations, one realizes that the optimum work to be extracted from energy resources cannot be given by a simple formula. The optimum work depends on the thermodynamic states of the resource and the environment as well as on the feasible (or thermodynamically allowable) interactions between the energy resource and the environment. As a consequence, the derived expressions for the optimum work are different for the several types of energy resources.

2.2 The Model Environment

Since its inception the discipline of thermodynamics has considered idealized systems and processes, which are commonly called *models*. The use of idealized models – a few of which appear to be counterintuitive – has long been verified and almost sanctified because models have allowed scientists to perform calculations on practical systems and derive experimentally verifiable results. The thermodynamic models have contributed to the design and optimization of thermal engines – locomotives, diesel engines, jet engines, and nuclear power plants – that have satisfactorily operated for centuries. In this section we will introduce the idealized model of the terrestrial environment, where all the anthropogenic energy conversion processes take place.

The model environment of the energy conversion processes and systems is characterized by the pressure, P_0, the temperature, T_0, and the chemical potentials of its several constitutive components, μ_{i0}. The model of the environment is a combination of four separate reservoirs:

1. *Heat reservoir.* The environment will absorb any quantity of heat from engines and will transfer heat to artificially cold systems (e.g. refrigerators) without any perceivable effect on its local temperature, T_0. When a system rejects heat Q to the environment, the overall internal energy of the environment increases, $(U_2 - U_1 = Q)$ as in Eq. (1.33), but its temperature remains unaffected because the mass of the environment is very large.
2. *Work reservoir.* All the work produced by the engines is transferred to the environment (e.g. electricity) and is transmitted to the locations of consumption, which are also parts of the environment. The transfer and transmission of vast quantities of work do not alter the properties of the environment.
3. *Volume reservoir.* Systems and their components may expand and contract within the environment. The environment will accommodate any change in their volume without any perceivable effect on the local environmental pressure, P_0.
4. *Mass reservoir.* All engines receive materials from their immediate environment and exhaust combustion products, cooling water, and other less common materials. The environment supplies and receives all this matter without any perceivable effects on the chemical potential of its components, μ_{i0}, and its overall composition. When the environment receives and supplies matter to the engineering systems it should be considered as an open system.

During the processes of heat, work, volume, and mass exchanges, the entropy of the environment and the system do not remain constant. All these processes are inherently irreversible. As a result, the sum of the entropies of the system proper and of the environment increases. In addition, when combustion products are exhausted from the system and attain equilibrium with the environment, the mixing processes, which occur isothermally and without the performance of additional work, are irreversible and increase the entropy of the environment.

2.3 Maximum Work – Exergy of Closed Systems

Let us consider a closed system (a system of a fixed quantity of matter) which is not in equilibrium with the environment. The system is at temperature T ($T \neq T_0$) and pressure P ($P \neq P_0$). The closed system may expand or contract and may transfer heat to the environment or receive heat from the environment. There are no chemical interactions in this system, which does not exchange matter with the environment. An example of such a system is a mass of hot rock in the lithosphere. Despite the fact that such "hot rock" systems are massive, they are not heat reservoirs because the extraction of heat (e.g. by circulating a fluid in their interiors) brings about significant temperature reduction at the locations of heat extraction.

2.3 Maximum Work – Exergy of Closed Systems

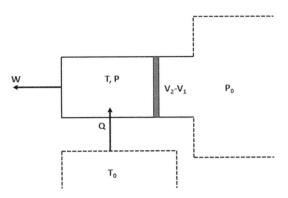

Figure 2.2 Interactions of a closed thermodynamic system with its environment, which in this case is a combination of heat, work, and volume reservoirs.

Figure 2.2 depicts the interactions of this type of closed systems with the environment, which in this case is a combination of heat, work, and volume reservoirs. The system at temperature T and pressure P undergoes a process during which it exchanges heat, Q, with the environment that has constant temperature, T_0, and pressure, P_0. The volume of the system changes from V_1 to V_2 during this process. As a result of the heat transfer and volume change, the system performs useful work, W. The arrows for the exchanges of heat and work in Figure 2.2 point to the direction of positive Q and W according to thermodynamic convention.

We are, naturally, interested to calculate the maximum amount of work that may be extracted from this system, during this process 1-2. The process is irreversible, the internal energy of the system changes from U_1 to U_2 and its volume from V_1 to V_2. The system delivers work W and, in addition, the volume change includes work against the environmental pressure, $P_0(V_2 - V_1)$. The first law of thermodynamics for this process yields:

$$Q - [W + P_0(V_2 - V_1)] = U_2^0 - U_1^0. \tag{2.1}$$

As in Eq. (1.8) the superscript 0 denotes the presence and summation of all the types of energy, including potential and kinetic energies. During this process the entropy of the system changes from S_1 to S_2. At the same time the entropy of the environment, which is at constant temperature T_0, changes by $-Q/T_0$ (minus sign because of the convention for heat). The second law of thermodynamics for this system and its surrounding environment may be written as follows:

$$S_2 - S_1 - \frac{Q}{T_0} \geq 0 \quad or \quad S_2 - S_1 - \frac{Q}{T_0} = \Theta \geq 0, \tag{2.2}$$

where Θ denotes the entropy production during the process 1-2 and is nonnegative. We may now eliminate the heat Q in Eqs. (2.1) and (2.2) to obtain the following equation for the work, W [1–3]:

$$W = U_1^0 - U_2^0 + P_0(V_1 - V_2) - T_0(S_1 - S_2) - T_0\Theta. \tag{2.3}$$

Since $T_0 > 0$ and $\Theta \geq 0$, the maximum work during the process 1-2 is obtained when the process occurs reversibly and $\Theta = 0$. In this case, we may write the reversible work as follows:

$$W_{rev} = U_1^0 - U_2^0 + P_0(V_1 - V_2) - T_0(S_1 - S_2). \tag{2.4}$$

The maximum work performed by the system during the process 1-2 may be written succinctly in terms of the *exergy*, a function of the properties of the system and of the environment:

$$W_{rev} = E_1 - E_2. \tag{2.5}$$

An inspection of Eqs. (2.4) and (2.5) leads to the following definition for the *exergy of closed systems*:

$$E = U^0 + P_0 V - T_0 S. \tag{2.6}$$

It must be noted that the sum $U + P_0 V$ is different from the enthalpy of the closed system, $H = U + PV$. It is also apparent that the exergy is not a property of the system, which is given simply by the state of the system, but a combination of the state of the system and the characteristics (model) of the environment. Equation (2.5) shows that the maximum work, which may be delivered by a closed system during the process 1-2 that has specified end points, can be simply written as the difference of the exergy function between the two end points.

In our analysis so far, we have assumed that states 1 and 2 are prescribed. With natural energy resources, such as the mass of hot rock envisioned here, it is only the initial state, 1, which is prescribed by nature. State 2 is determined by the details of the energy conversion process. We now ask the question: Of all the available processes for the extraction of work from this energy resource and all the allowable states 2, which process is optimal? A moment's reflection proves that as long as state 2 of the resource is not in equilibrium with the environment (that is as long as $T_2 \neq T_0$ and $P_2 \neq P_0$) then we may devise another process, 2-3, bringing the system into equilibrium with the environment and augmenting the total work W [3–5]. The option to devise such a process 2-3 that produces additional work is nullified when the system at state 2 attains the environmental temperature T_0, the pressure P_0, zero elevation (to nullify its potential energy), and zero velocity (to nullify its kinetic energy). At this state, the closed system is in thermodynamic equilibrium with the environment and no further work or heat may be extracted from it. This equilibrium state of the system is referred to as the *dead state* [6], a name that has become common in the engineering literature and is usually denoted by the subscript 0, as in E_0. Therefore, the maximum work we may obtain from a resource utilized in a closed system is given when the system is brought into thermodynamic equilibrium with the environment and the processes occur reversibly. Hence, the maximum work for the closed system is:

$$\begin{aligned} W_{max} &= U_1^0 - U_0^0 + P_0(V_1 - V_0) - T_0(S_1 - S_0) = E_1 - E_0 \\ \text{or} \quad W_{max} &= m\big[u_1^0 - u_0^0 + P_0(v_1 - v_0) - T_0(s_1 - s_0)\big] = m[e_1 - e_0] \end{aligned}, \tag{2.7}$$

where the total energy, volume, and entropy of the system at the end state, 0, are defined at T_0, P_0, zero elevation, and zero velocity, respectively. Following the suggestion by Keenan [6] we may define the exergy of the system at this state to be equal to zero ($E_0 \equiv 0$) and obtain the final expression for the maximum work that may be obtained from a resource utilized as a closed system:

$$W_{max} = E_1 = U_1^0 + P_0 V_1 - T_0 S_1. \tag{2.8}$$

Therefore, the maximum work obtained from a closed system is equal to the initial exergy of the system. Following thermodynamic notation, we may write the extensive variables of Eq. (2.8) in terms of the mass of the system and the specific variables to obtain the expression:

$$W_{max} = m e_1 = m \left(u_1^0 + P_0 v_1 - T_0 s_1 \right). \tag{2.9}$$

The function e_1 is the *specific exergy of the closed system*.

A note on numerical calculations: Even though by convention we have agreed that the exergy of the environment is zero ($E_0 \equiv 0$), the numerical values of the properties we use are obtained from thermodynamic tables and equations that do not have the same reference points as "our" model environment. For example, the reference state of the internal energy and entropy in most steam tables is the triple point of water ($P_{tr} = 0.611$ kPa and $T_{tr} = 0.01°C$) and **not** at the temperature and pressure of "our" environment. For this reason, when values from thermodynamic tables are used, one must use differences in the calculations for all the extensive and specific properties, such as Eq. (2.7).

2.3.1 The Exergy of Hot Rock

Let us consider as an example the maximum work to be obtained from a mass of hot rock at a known depth from the surface of the earth, where the pressure is P and the temperature is T. Oftentimes hot rock systems are called *dry rock geothermal* systems. One may extract heat from the mass of the rock using an arrangement of heat exchangers with a working fluid (water, organic fluid, refrigerant, etc.), evaporate the fluid, and expand it in a turbine. Two such energy conversion systems have been developed near the towns of Soultz-sous-Forêts, Alsace, France; and Landau, Pfalz, Germany, both close to the French–German border. The mass of the rock is finite. Therefore, the temperature of the rock will be continuously decreasing as the working fluid extracts energy (and exergy from it). For a hot rock system of mass m, which does not have any potential or kinetic energy, Eq. (2.7) yields:

$$W_{max} = m[u_1 - u_0 + P_0(v_1 - v_0) - T_0(s_1 - s_0)]. \tag{2.10}$$

Because the solid rock is incompressible and its density is constant, the product $P_0(v_1 - v_0)$ vanishes. The specific internal energy difference and the entropy difference of the solid rock may be expressed in terms of the specific heat capacity, $c = c_v = c_p$, using the thermodynamic relationship $cdT = Tds$:

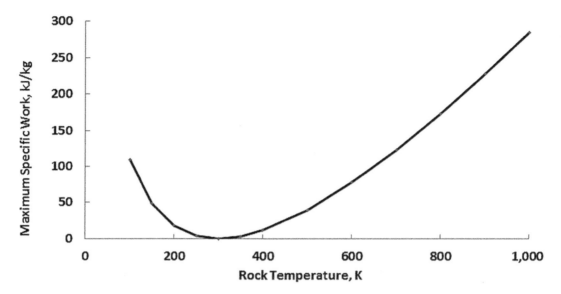

Figure 2.3 Maximum specific work from a basaltic rock in the range $100\ K < T_1 < 1{,}000\ K$.

$$u_1 - u_0 = c(T_1 - T_0) \quad \text{and} \quad s_1 - s_0 = c \ln \frac{T_1}{T_0}. \qquad (2.11)$$

The substitution of the last equation into Eq. (2.10) yields the following expression for the maximum work obtained per unit mass of the rock:

$$w_{max} = \frac{W_{max}}{m} = c\left(T_1 - T_0 - T_0 \ln \frac{T_1}{T_0}\right) = cT_0\left(\frac{T_1}{T_0} - 1 - \ln \frac{T_1}{T_0}\right). \qquad (2.12)$$

Figure 2.3 depicts the maximum specific work from a solid as a function of its temperature, T_1. The specific heat capacity of the solid is 0.84 kJ/kgK (approximately equal to the specific heat of basaltic rock) and the environmental temperature is $T_0 = 300$ K. For the generalization of the results, the temperature T_1 is in the range $100\ K < T_1 < 1{,}000\ K$, which includes applications with hot, geothermal rocks as well as solids in very cold and cryogenic conditions. It is observed in Figure 2.3 that the maximum specific work has a minimum, equal to 0, when $T_1 = T_0$, and that it attains positive values when the solid is colder than the environment, $T_1 < T_0$. In this case, work may be extracted with a thermal engine that receives heat from the environment and rejects heat to the colder solid. This is an indication that the exergy function is positive for all the values of the resource temperature T_1 and only becomes zero when $T_1 = T_0$. In the latter case, there is neither heat transferred to nor from the system, there is no thermal interaction and the work is null.

Example 2.1: Geologists have discovered a solid "pocket" of impermeable rock at 3 km beneath the surface, with dimensions $2.5 * 3 * 6$ km and almost uniform temperature 235°C. The average density of the rock is 2,700 kg/m³ and its specific heat capacity

0.84 kJ/kg K. It is proposed that several wells be drilled to circulate water and transfer the heat of the rock to thermal engines that would produce power. Calculate the maximum amount of work that may be extracted from this project.

Solution: The maximum work per unit mass of the rock corresponds to the specific exergy of the rock, which is given by Eq. (2.12). Using the temperature $T_1 = 235°C = 508$ K and an environmental temperature $T_1 = 300$ K, we obtain $w_{max} = 42.0$ kJ/kg.

Since 1 km^3 is 10^9 m^3 the mass of the rock is $2.5 * 3 * 6 * 10^9 * 2{,}700 = 121.5 * 10^{12}$ kg. Therefore, the maximum work that may be generated from this hot pocket in the interior of the earth is: $5.1 * 10^{15}$ kJ or 1,417 TWh.

2.4 Maximum Power – Exergy of Open Systems

Let us consider the operation of an open system that utilizes an energy resource and produces mechanical power (work per unit time). For simplicity, we will first consider an open system where no chemical reactions occur, with only physical changes and thermal interactions taking place. The system is supplied with materials from a group of pipelines with different mass flow rates, pressures, and temperatures as shown in Figure 2.4. In the most general case, such an open system receives mass flow rates $\dot{m}_{i1}, \dot{m}_{i2}, \ldots, \dot{m}_{ik}$ from k inlets and rejects to the environment n streams/outlets with mass flow rates $\dot{m}_{e1}, \dot{m}_{e2}, \ldots, \dot{m}_{en}$. Heat rate $\dot{Q}$ enters the system from the environment and the system produces useful mechanical power $\dot{W}$. As with the closed system, the arrow for the rate of heat points to the positive direction of heat transfer. In this case, the environment acts as a heat reservoir, a work reservoir, a volume reservoir, and a mass reservoir.

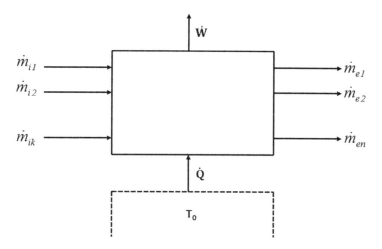

Figure 2.4 An open system with several inlets and outlets. The directions of the arrows follow standard thermodynamic convention for positive rates of work and heat.

The first law of thermodynamics for the steady state operation of this open system is:

$$\dot{Q} - \dot{W} = \sum_{j=1}^{n} \dot{m}_{ej} h_{ej}^o - \sum_{l=1}^{k} \dot{m}_{il} h_{il}^o, \quad (2.13)$$

where h_{ej}^o represents the total specific enthalpy of the exiting stream, j; and h_{il}^o represents the specific enthalpy of the inlet stream, l. The total enthalpy includes all potential and kinetic energy, of the entering and exiting streams.[5]

Similarly, the second law of thermodynamics may be written as:

$$-\frac{\dot{Q}}{T_0} + \sum_{j=1}^{n} \dot{m}_{ej} s_{ej} - \sum_{l=1}^{k} \dot{m}_{il} s_{il} = \dot{\Theta} \geq 0. \quad (2.14)$$

In Eq. (2.14), s represents the specific entropy of the corresponding inlets and outlets and $\dot{\Theta}$ is the rate of entropy production (irreversibilities) associated with the operation of the system. One may eliminate the rate of heat, $\dot{Q}$, in Eqs. (2.13) and (2.14) to obtain the following expression for the power produced by this system, in terms of the inlet and outlet states and the rate of irreversibilities:

$$\dot{W} = \sum_{l=1}^{k} \dot{m}_{il}\left(h_{il}^o - T_0 s_{il}\right) - \sum_{j=1}^{n} \dot{m}_{ej}\left(h_{ej}^o - T_0 s_{ej}\right) - T_0 \dot{\Theta}. \quad (2.15)$$

The quantities inside the parentheses of the last equation ($h^o - T_0 s$) are defined as the specific exergy, e, of the corresponding inlets and outlets. With the states of the outlets specified, the open system produces maximum power when its operation is reversible, that is when the rate of entropy production, $\dot{\Theta}$, is zero. In this case the maximum power produced is:

$$\dot{W}_{rev} = \sum_{l=1}^{k} \dot{m}_{il} e_{il}^o - \sum_{j=1}^{n} \dot{m}_{ej} e_{ej}^o. \quad (2.16)$$

The reversible power may be considered as a local maximum because it is only the operation of the system that has been optimized by making the processes reversible and the rate of entropy production zero. As in the case of the closed system, one may consider under what conditions (states) of the outlets the reversible power becomes a global maximum. Because additional power may be extracted from any exhaust stream, which is not in thermal and mechanical equilibrium with the environment, maximum power is achieved when all the outlets are at the dead state that is at the environmental pressure P_0, temperature T_0, with zero velocity and zero potential energy.

$$\dot{W}_{max} = \sum_{l=1}^{k} \dot{m}_{il} e_{il}^o - \sum_{j=1}^{n} \dot{m}_{ej} e_{0j}^o = \sum_{l=1}^{k} \dot{m}_{il}\left(h_{il}^o - T_0 s_{il}\right) - \sum_{j=1}^{n} \dot{m}_{ej}\left(h_{0j}^o - T_0 s_{0j}\right), \quad (2.17)$$

where the subscript 0 at the outlets signifies that each outlet is in equilibrium with the local environment. As with the closed systems we may use Keenan's convention [6] and

---

[5] In Section 2.4 we do not include any additional work that may be obtained by the possible expansion of the outlet streams through semi-permeable membranes to their partial pressures in the environment. Such expansions are discussed in Sections 2.5 and 2.6.

define the exergy of all effluents/outlets at the environmental pressure P_0, and temperature T_0 to be zero. Under these conditions, the last sum in Eq. (2.17) vanishes and the expression for the maximum power produced by the open system becomes:

$$\dot{W}_{max} = \sum_{l=1}^{k} \dot{m}_{il} e_{il}^o. \tag{2.18}$$

The product $\dot{m}e$ is the rate of exergy for each one of the streams. Therefore, the maximum work per unit time (power) that may be produced by an open system is equal to the sum of the rates of the exergy supplied by its inlet streams. It must be noted that the specific exergy that corresponds to the open systems, $h^o - T_0 s = u^o + Pv - T_0 s$, is similar but not exactly the same as the specific exergy that corresponds to the closed systems, $u^o + P_o v - T_0 s$, as it becomes apparent from an inspection of Eqs. (2.9) and (2.15).

2.4.1 A Note on Numerical Calculations

As with the properties of the closed systems the numerical values of the properties we use for open systems are oftentimes obtained from thermodynamic tables and equations that do not have the same reference points as the "environment" of the system under consideration. For this reason, we must use algebraic differences in the calculations of all the extensive and specific properties, for example use expressions with property differences, such as Eq. (2.17), rather than Eq. (2.18).

Of specific interest in numerical calculations is the steady operation of commonly used engines with one inlet and one outlet, such as pumps, compressors, and turbines. In this case, the mass flow rates at the inlet and the outlet are the same, $\dot{m}_i = \dot{m}_o = \dot{m}$, and Eq. (2.17) is reduced to:

$$\dot{W}_{max} = \dot{m}(e_1 - e_0) = \dot{m}[h_1 - h_0 - T_0(s_1 - s_0)], \tag{2.19}$$

The exergy difference term in the square brackets of Eq. (2.19) is usually obtained from property tables and property software. The term may be also analytically calculated for particular fluids, such as ideal gases with constant specific heat, c_P:

$$e_1 - e_0 = h_1 - h_0 - T_0(s_1 - s_0) = c_P(T_1 - T_0) - T_0\left(c_P \ln \frac{T_1}{T_0} - R \ln \frac{P_1}{P_0}\right), \tag{2.20}$$

and for incompressible fluids with constant specific heat, c:

$$e_1 - e_0 = h_1 - h_0 - T_0(s_1 - s_0) = c(T_1 - T_0) + v(P_1 - P_0) - T_0 c \ln \frac{T_1}{T_0}. \tag{2.21}$$

Because the exergy associated with the expansion of the chemical species of the system to their concentrations in the environment, is not taken into account, the exergy differences in the last two sections are oftentimes referred to as the *physical exergy*.

One notices that the difference in the specific exergy expressions of Eqs. (2.12) and (2.21) is the term $v(P_1 - P_0)$. For most incompressible fluids and solids this term is typically very small in comparison to the other terms.

Example 2.2: A geothermal field has 12 production wells. Each geothermal well produces 58 kg/s of a steam-water mixture at 190°C, and dryness fraction 22%. Because water recirculates in the down-hole aquifer, the temperature and mass flow rates of the wells are constant. Calculate the maximum electric power the field may produce. The environment is at $T_0 = 300$ K and $P_0 = 1$ atm.

Solution: This system has one inlet, the geothermal fluid from all the wells; and one outlet, the water effluent after the production of power. The total amount of water/steam at the inlet is $12 * 58 = 696$ kg/s. By the mass conservation principle, the mass flow rates at the outlets is the same. The geothermal fluid at the inlet is a mixture of water and steam with $x_1 = 0.22$ and $T_1 = 190°C$. Using thermodynamic tables, we calculate $h_1 = 1,243$ kJ/kg and $s_1 = 3.1757$ kJ/kg K. Maximum work is produced when the state of the effluents of the power plant is in equilibrium with the local environment. At $T_0 = 300$ K and $P_0 = 1$ atm the water effluent is a compressed liquid with $h_0 = 113$ kJ/kg and $s_0 = 0.3954$ kJ/kg K. From Eq. (2.19) we obtain:

$$\dot{W}_{max} = \dot{m}[h_1 - h_0 - T_0(s_1 - s_0)] = 696 * [1,243 - 113 - 300 \\ * (3.1757 - 0.3954)] = 205,953 \text{ kW} \approx 206 \text{ MW}.$$

2.4.2 The Specific Exergy of Common Fluids – Water, Air, and Refrigerant-134a

Most of the energy conversion systems – turbines, compressors, burners, and heat exchangers – are open thermodynamic systems that operate with a working fluid: water/steam; hydrocarbons; refrigerants; air and combustion products, etc. The first three fluids may be in the liquid or the vapor/gas phase, while air and the products of combustion are typically in the vapor/gas phase. The thermodynamic properties of the working fluids are different and their specific exergies are also expected to differ. Figure 2.5a–c depicts the specific exergy associated with the operation of open systems of water/steam, refrigerant 134a, and standard air. The environmental temperature in all the figures is 300 K and the pressure 101.3 kPa (1 atmosphere). For the fluids that may exist in the liquid and vapor/gas phases, the exergies of the saturated liquid and the saturated vapor are depicted, as well as an isobar, which is approximately at a pressure close to 20% of the critical pressure. The saturated liquid curves are denoted by the broken lines for both water and refrigerant 134a; and the saturated vapor curves by the solid lines that complete the loops. The critical point for both is the point where the broken and the solid lines meet. For standard air, which is depicted in Figure 2.5c, several isobars are shown.

The following observations are made for the numerical values of the exergy of these fluids:

1. At states where the fluid exists as saturated vapor and saturated liquid, the specific exergy of the liquid is significantly lower than the specific exergy of the vapor.
2. The saturated liquid exhibits maximum exergy at the critical point. The maximum exergy of the saturated vapor is at a lower temperature. This primarily happens

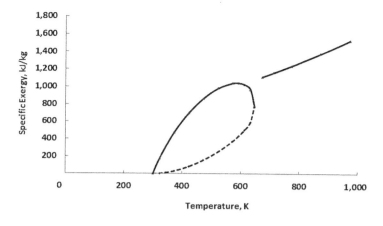
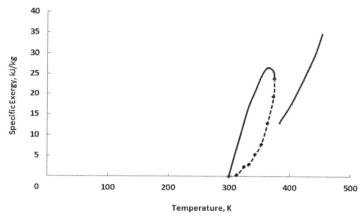
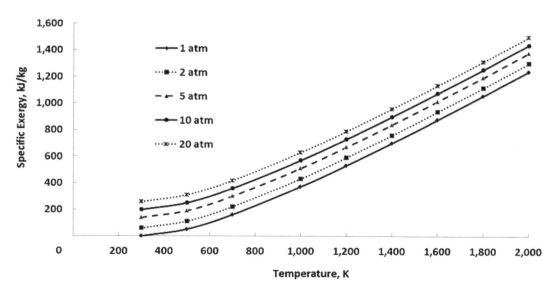

Figure 2.5 Specific exergy of: (a) saturated liquid water and steam (loop) and of superheated steam at 20 bar (at the right). $P_0 = 1$ atm, $T_0 = 300$ K; (b) Specific exergy of saturated Refrigerant 134a (loop) and of superheated Refrigerant 134a vapor at 8 bar (at the right). $P_0 = 1$ atm, $T_0 = 300$ K; (c) Specific exergy of standard air at several pressures. $P_0 = 1$ atm, $T_0 = 300$ K.

because the enthalpy of the saturated vapor also exhibits a maximum before the critical point.
3. The specific exergy of superheated vapor increases with the temperature and pressure.
4. The exergy of air (and of all gases) increases monotonically with temperature and pressure.
5. In the range of temperatures of Figures 2.5a–c, the values of the specific exergy for refrigerant 134a (and also for all other refrigerants) are almost two orders of magnitude lower than the specific exergy of steam and air.

The last observation explains why steam (water vapor) and air are primarily used as the working fluids in power production engines. A glance at Eqs. (2.17) and (2.18) proves that, for the generation of a desired quantity of power, the high specific exergy of the working fluid indicates that low mass flow rate is needed to flow through the engine. If the working fluid also has high density relative to other fluids, the volumetric flow rate of the working fluid is low, the size of the engine is smaller, and the pumping requirements are lower. Both steam and compressed air have these thermodynamic characteristics in addition to being chemically very stable (to undergo several thermodynamic cycles per minute) and practically free of cost. These three characteristics are the main reason why water/steam and air are almost exclusively used as the working fluids in the electric power production industry. Refrigerant 134a, the other refrigerants, and some hydrocarbons are selectively used for refrigeration cycles, which require relatively low work input. The primary reason for this selection is the desirable saturation pressure-temperature properties that allow subzero temperatures to be readily obtained.

2.5 Exergy of Chemical Resources – Fossil Fuels

In Sections 2.3 and 2.4 we have considered closed and open systems without any chemical reactions. Work and power were generated because the systems were at different temperatures and pressures than the environment. Let us now consider a system where chemical reactions may take place, such as the combustion of fossil fuels. The fuels are typically introduced into the combustor/boiler at the ambient temperature and pressure. The combustion process generates the combustion products, typically carbon dioxide and water; and the chemical energy of the fuel is released to increase the temperature of the products in the combustor. The hot combustion products are directed to a gas turbine (in the gas cycles) or they heat up water (in the steam cycles) to produce steam. Any chemical reaction, including all combustion reactions, may be written in a general form as follows:

$$\sum_r \nu_r R_r = \sum_p \nu_{pr} P_p, \qquad (2.22)$$

where ν_r and ν_p are the stoichiometric coefficients of the reactants and products respectively, R_r denotes the reactant compounds, and P_p the product chemical compounds. Figure 2.6 is a schematic diagram of a chemical reaction vessel. In this case the

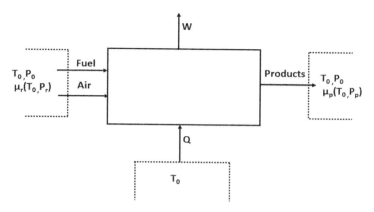

Figure 2.6 A chemically reactive system as an open thermodynamic system. The environment is a reservoir of heat, work, volume, and matter.

environment acts as a reservoir of heat, work, volume, and matter. The reactants, typically the fuel and air from the atmosphere, enter the combustor and the products exit. As in Sections 2.3 and 2.4, the system may receive or reject heat, Q, (the direction of the arrow in Figure 2.6 corresponds to positive heat as per the convention of thermodynamics)[6] and produce work, W, (the direction of the arrow corresponds to positive work). The environment compensates for any change of volume as a result of the conversion of reactants into products. The reactants enter the thermodynamic system at temperature T_0 and total pressure P_0. The chemical potentials of the reactants, $\mu_r(T_0, P_r)$ are calculated at the prevailing temperature, T_0, and at their partial pressures in the environment, P_r. Similarly, the chemical potentials of the products, $\mu_p(T_0, P_p)$, are calculated at the prevailing environmental temperature, T_0, and at their partial pressures in the environment, P_p.

It must be recalled that, for incompressible solids and liquids, their chemical potentials are only functions of temperature. Therefore, for solid and liquid reactants and products we have the relations: $\mu_r(T_0, P_r) = \mu_r(T_0, P_0) = \mu_r(T_0)$ and $\mu_p(T_0, P_p) = \mu_p(T_0, P_0) = \mu_p(T_0)$. The chemical potential is equal to the molar Gibbs free energy, $\tilde{g}$:

$$\mu_i(T_0, P_i) = \mu_i(T_0) = \tilde{g}(T_0) \quad for \quad solids \quad and \quad liquids. \tag{2.23}$$

If the chemical species in the reactions are gaseous, the dependence of their chemical potentials on the partial pressure may be approximated using the Gibbs' free energy function and the ideal gas model [5, 7, 8]:

$$\mu_i(T_0, P_i) = \tilde{g}(T_0, P_0) + \tilde{R}T_0 \ln(x_i) \quad for \quad gases, \tag{2.24}$$

where x_i is the mole fraction of the gas i in the environment; $\tilde{g}$ is the molar Gibbs free energy of the gas i, evaluated at the temperature and pressure of the environment; and $\tilde{R}$ is the universal gas constant (8.314 kJ/kmol/K). Table 2.1 shows the mole fraction

---

[6] Typical energy conversion systems reject heat to the environment, in which case $Q < 0$.

Table 2.1 Mole fractions, x_h of the constituents of the environment. $T_0 = 25°C$, $P_0 = 101.13$ kPa.

Gaseous Component	Mole Fraction
Nitrogen	0.7562
Oxygen	0.2031
Water vapor	0.0313
Carbon dioxide	0.0004
Argon and other gases	0.0090

values of the most common gases in the atmospheric environment. The value for the water vapor is based on the saturation pressure of water at $T_0 = 25°C$ and the carbon dioxide concentration on atmospheric measurements of 2018.

Since we already have calculated the exergy of substances at temperatures and pressures other than those of the environment (often referred to as *physical exergy*), we now calculate the exergy associated only with the chemical potentials of the reactants and products, which is often referred to as *chemical exergy*. For this reason, we stipulate that the reactants enter the system of Figure 2.6 at T_0 and total pressure P_0 and that the products exit also at temperature T_0 and total pressure P_0. The molar internal energy of the reactants and products as they enter and leave the system of Figure 2.6 includes their chemical energy, which is equal to the chemical potential of the material, $\mu_i(T_0, P_i)$, and is measured in J/kmol. Therefore, the total internal energy of each one of reactants and products may be given by the expression:

$$U_i^0(T_0, P_i) = U_i(T_0, P_i) + \frac{1}{2}m_i V^2 + m_i g z + N_i \mu_i(T_0, P_i), \qquad (2.25)$$

where N_i is the number of kmols of the material i and the variables in the parentheses denote the arguments of the corresponding functions. A similar expression applies to the enthalpy of the reactants and products. Using the method of Sections 2.3 and 2.4, after writing the first and second laws of thermodynamics as equations, eliminating the heat, Q, among the two equations, and stipulating that the entropy production vanishes, we obtain the following expression for the maximum work obtained from one kmol of the fuel ($N_{fuel} = 1$):

$$W_{max} = \sum_r N_r \mu_r(T_0, P_i) - \sum_p N_p \mu_p(T_0, P_i) \equiv E_{chem}(T_0, P_0). \qquad (2.26)$$

Equation (2.26) implies that the maximum work to be obtained from a chemical, the fuel, is equal to the chemical exergy of the fuel. It is observed that the expression for the chemical exergy of the fuel is similar (but not the same) to the expression of the negative of the Gibbs free energy of the reaction:

$$-\Delta G^o = \sum_r N_r \tilde{g}_r(T_0, P_0) - \sum_p N_p \tilde{g}_p(T_0, P_0) = \sum_r N_r \mu_r(T_0, P_0) - \sum_p N_p \mu_p(T_0, P_0).$$

$$(2.27)$$

The difference between the maximum work (exergy) and the Gibbs' free energy of the reaction – a property of all chemical reactions, with numerical values printed in most chemical reference publications – may be calculated using Eqs. (2.24), (2.26), and (2.27).

Equation (2.26) is applicable to all chemical reactions and may be used for the calculation of the chemical exergy of all substances. As an example, let us consider the general combustion reaction of a hydrocarbon fuel with oxygen as the oxidant:

$$C_aH_b + \left(a + \frac{b}{4}\right)O_2 = aCO_2 + \frac{b}{2}H_2O. \tag{2.28}$$

The chemical exergy per kmol of this fuel is:

$$E_{chem}(T_0, P_0) = \mu_{C_aH_b}(T_0, P_{C_aH_b}) + \left(a + \frac{b}{4}\right)\mu_{O_2}(T_0, P_{O_2})$$
$$- a\mu_{CO_2}(T_0, P_{CO_2}) - \frac{b}{2}\mu_{H_2O}(T_0, P_{H_2O}). \tag{2.29}$$

Several studies [1, 8, 9] have proceeded with the assumption that the fuel is solid or liquid (that is, its chemical potential is only a function of the temperature) and that it is oxidized in pure oxygen. Following the set of Eqs. (2.24), (2.26), and (2.27) one may calculate the molar exergy (exergy per kmol) of the fuel using the fractions of oxygen, carbon dioxide, and water in the atmosphere as follows:

$$W_{max} = E_{chem}(T_0, P_0) = -\Delta G^o + \tilde{R}T_0 \ln\left(\frac{x_{O_2}^{(a+b/4)}}{x_{CO_2}^a x_{H_2O}^{b/2}}\right). \tag{2.30}$$

This is the maximum work that may be extracted from the fuel when pure oxygen is used in the combustion process. The first term in the r.h.s., ΔG^o, denotes the Gibbs free energy of the reaction – a property of every reaction, which is reported in chemical tables. It represents the work that may be obtained if all the reactants and products entered and exited the reaction vessel at T_0 and P_0. The second term represents the additional work that could be recovered from the expansion of the products to their partial pressures, P_i, in the atmosphere. This work is recovered with the use of semi-permeable membranes, which in principle allow the molecules of certain materials to pass through, but do not allow other molecules to pass and block them. The preferential blocking allows this type of membranes to facilitate the expansion of certain gases at their partial pressures in the environment.

A moment's reflection on the combustion of the fuels and Eq. (2.28) proves that pure oxygen is not a natural resource. Pure oxygen may not be obtained to be used without the expense of a significant amount of work. Fuels are invariably oxidized with atmospheric air, which includes nitrogen and traces of other gases. Also, water at the environmental temperature and pressure is liquid, and this implies that the value of ΔG^o pertains to liquid water. With this oxidation model, and given that there are approximately 3.76 kmol of nitrogen in the air for every kmol of oxygen, the actual combustion reaction becomes:

$$C_aH_b + \left(a+\frac{b}{4}\right)O_2 + 3.76\left(a+\frac{b}{4}\right)N_2 = aCO_2 + \frac{b}{2}H_2O + 3.76\left(a+\frac{b}{4}\right)N_2. \tag{2.31}$$

Nitrogen is both a reactant and a product in this reaction. Following the expressions for the chemical potentials, a more realistic expression for the chemical exergy of the hydrocarbon fuel that burns in an atmosphere of nitrogen and oxygen would be:

$$E_{chem}(T_0, P_0) = -\Delta G^o + \tilde{R}T_0 \ln\left(\frac{x_{O_2}^{(a+b/4)}}{x_{CO_2}^a} * \left(\frac{(4.76a+0.94b)x_{N_2}}{3.76a+0.94b}\right)^{3.76a+0.94b}\right). \tag{2.32}$$

The Gibbs' free energy in the last expression, ΔG^o, pertains to the water product being in the liquid state. Again, the work denoted by the last term of Eq. (2.32) may only be recovered with the use of semi-permeable membranes, when the gaseous components expand to their partial pressures in the atmosphere, the environment of the combustion process.

The magnitude of the last term in both Eqs. (2.30) and (2.32) is very small in comparison to the magnitude of ΔG^o. Table 2.2 shows the heat of combustion, ΔH^o, and the Gibbs' free energy, ΔG^o, of several fuels as well as their chemical exergy, calculated with Eqs. (2.30) and (2.32). The fractional differences of the two calculated exergies and the standard Gibbs' free energy, ΔG^o, are also shown in Table 2.2.

Table 2.2 Heat of combustion, Gibbs' free energy, and chemical exergy for several commonly used fuels (T_0 = 298 K, P_0 = 1 atm).

$-\Delta H^o$ (MJ/ kmol)	$-\Delta G^o$ (MJ/ kmol)	E_{chem}, Eq. (2.32) (MJ/kmol)	Difference (%)	E_{chem}, Eq. (2.30) (MJ/kmol)	Difference (%)
Carbon	393.5	394.4	410	3.9	409.8	3.8
Hydrogen	286	237	236	−0.7	236.2	−0.4
Methane	890	818	827	1.1	829.5	1.4
Ethane	1,559	1,468	1,489	1.4	1,493.0	1.7
Propane	2,220	2,108	2,141	1.6	2,146.4	1.8
Butane	2,879	2,748	2,793	1.6	2,799.9	1.9
Pentane	3,510	3,386	3,443	1.7	3,451.3	1.9
Hexane	4,163	4,023	4,092	1.7	4,101.8	2.0
Octane	5,471	5,297	5,390	1.8	5,402.7	2.0
Ethylene	1,411	1,331	1,355	1.8	1,357.9	2.0
Propylene	2,059	1,957	1,993	1.8	1,997.4	2.1
Butene	2,719	2,598	2,646	1.8	2,651.9	2.1
Pentene	3,376	3,237	3,297	1.9	3,304.3	2.1
Benzene	3,302	3,208	3,289	2.5	3,294.7	2.7
Methanol[1]	733	711	720	1.2	7,22.2	1.6
Ethanol[1]	1,365	1,342	1,363	1.6	1,367.0	1.9

1. The molecules of ethanol and methanol contain one oxygen atom. This is taken into account in the calculations

It is observed in Table 2.2 that the fractional difference between the Gibbs' free energy, ΔG^o, and the chemical exergy of the fuel is very small, typically less than 2%. Therefore, the maximum mechanical work to be obtained from a fuel is approximately equal to the Gibbs' free energy of the fuel, ΔG^o. This is a property of the fuel, its value does not depend on the composition of the environment, and may be easily found or calculated from information commonly available in most textbooks of thermodynamics [4, 5, 7, 8, 10]. Another observation in Table 2.2 is that the heat of combustion of the fuels, $-\Delta H^o$, is also approximately equal to the chemical exergy and to the Gibbs free energy of the fuels. This implies that the maximum work, which may be obtained from these fuels is also very close to the heat of combustion, $-\Delta H^o$. An immediate conclusion of this is that burning the fuels in gas and steam cycles that have thermal efficiencies on the order of 40% – and only deliver approximately 40% of the maximum work they may produce – is not the best way to utilize these energy resources. Exergy considerations imply that there are better methods than combustion to extract the available work from these common fuels, for example with direct energy conversion devices, such as fuel cells.

Example 2.3: 1 kmol of methane gas at 298 K is burned with excess air in an adiabatic chamber. Calculate the exergy of the combustion products and the exergy destruction of the combustion process when the excess air varies from 0% to 200%.

Solution: The combustion equation of methane with excess air a ($2 \geq a \geq 0$) is:

$$CH_4 + 2(1+a)O_2 + 2*3.76*(1+a)N_2 = CO_2 + 2H_2O + 2aO_2 + 2*3.76*(1+a)N_2.$$

At the temperatures expected inside the burner, the water product is vapor and the combustion heat of methane is the low heating value of the gas, 800,320 kJ/kmol [5]. The temperature of the combustion products is obtained by iteration using the energy balance of the above chemical reaction:

$$LHV = \int_{298}^{T} c_{PCO_2} dT + 2\int_{298}^{T} c_{PH_2O} dT + 2a \int_{298}^{T} c_{PO_2} dT + 7.52*(1+a)\int_{298}^{T} c_{N_2} dT.$$

For $a = 0$ (adiabatic flame temperature), this expression yields $T \approx 2,189$ K and the exergy of the products is approximately 579,962 kJ/kmol. Since the original exergy of methane gas is 818,000 kJ/kmol, 238,038 kJ/kmol of exergy (approximately 29.1%) is lost during the combustion process. When $a = 0.5$ (50% excess air) $T \approx 1,786$ K, the exergy of the products is 541,516 kJ/kmol (33.8% exergy destruction) and when $a = 1$ (100% excess air) $T \approx 1,480$ K the exergy of the products is 498,600 kJ/kmol (39% exergy destruction). Figure 2.7 shows the temperature of the combustion products and the percent destruction of the exergy of methane during the combustion process as a function of the excess air. It is observed that the temperature of the products is a monotonically decreasing function of the excess air, a, and that a significant amount of the exergy of methane (29–46%) is dissipated during the combustion process.

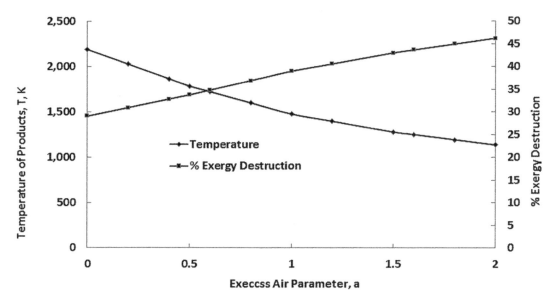

Figure 2.7 Temperature of combustion products and fraction of exergy destroyed during the combustion of methane with excess air. $P_0 = 1$ atm, $T_0 = 300K$.

2.6 A Note on Semipermeable Membranes

Semipermeable membranes allow the molecules of a number of substances to pass through their pores and restrict other molecules. Several biological and polymeric membranes are partly semi-permeable to gases and ions. For example, water molecules pass through a category of membranes while sugar molecules do not, a phenomenon known as *osmosis*. The biological membranes of the living cells are semipermeable – they have selective permeability – and allow the preferential diffusion of only certain chemical species into the cells.

The concept of semipermeable membranes (or semipermeable partitions) has been used in thermodynamics for more than a century to model separation processes and the production of work from gases expanding to their partial pressures. It is, in principle, possible to produce a membrane that only allows the molecules of a single chemical species (e.g. carbon dioxide) to pass through its pores to the exclusion of all the other species. Two of these membranes may be used with a piston to produce work as shown in the model mechanism depicted in Figure 2.8: The membrane on the left side allows only one of the reaction products, the gas i, to pass through its pores and enter the cylinder. Then the pressure at the left part of the cylinder is equal to the partial pressure of this gas in the reaction products, P_i. Similarly, the membrane on the right side of the cylinder allows only the same gas i to pass through. The pressure on the right part of the cylinder is $x_i^0 P_0$, the partial pressure of this gas in the environment. The difference of the two pressures in the cylinder "pushes" the piston to the side of the lower pressure and isothermal work is produced [5, 8, 10]. The reversible expansion of all the reactants

2.6 A Note on Semipermeable Membranes

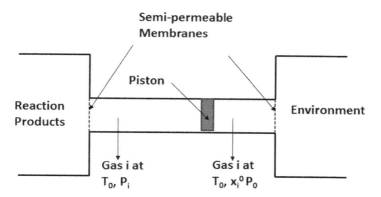

Figure 2.8 Semipermeable membranes may be used, in principle, to produce work when there is a partial pressure differential.

and products of the chemical combustion of Eq. (2.28) generates work, equal to the last terms of Eq. (2.30) or (2.32), depending on whether the reaction is accomplished with pure oxygen or with atmospheric air that contains nitrogen.

While semipermeable and selectively permeable membranes are useful theoretical tools and have been used as models in the theory of classical thermodynamics, such membranes are very difficult to construct and operate for the production of power: At first, the operation of the semipermeable membranes is based on the molecular diffusion process, which is very slow and – even in theory – would produce very small amounts of power. Second, the technology to produce selectively permeable membranes that would allow only a single (desired) species to pass, to the exclusion of all the other species is still in its embryonic form and applies only to very small molecules. While we have membranes that would allow the small molecules of hydrogen to diffuse through their pores, we do not have workable membranes that would allow carbon dioxide to pass through, while excluding water vapor and nitrogen. Third, the operation of selectively permeable membranes is currently based on the exclusion of molecules because the size of the pores of the membranes is very small or because there are electric charges on the sides of the pores. Even if selectively permeable membranes were to be developed, their application in the power industry would be problematic because impurities would quickly clog the pores and chemical reaction products would modify or eliminate the electric charges that control the permeability (unless substantial amounts of work are spent for cleaning the pores of the membranes).

The only known engineering system that was designed to produce energy using a kind of semi-permeable membranes was a pilot plant that attempted to produce power using the salinity gradient between fresh water and salt water. The plant was constructed in the island of Tofte, Norway by *Statkraft*, the Norwegian state energy corporation. One side of the membranes was in contact with seawater and the other with freshwater. The diffusion of freshwater through the semipermeable membranes created a height differential across the membranes, effectively an osmotic pressure, which was used for the production of low quantities of power in a small turbine. The system periodically produced 2–4 kW of electric power, a very low production for the size of the system and the investment. For this reason, this power plant was decommissioned four years after its construction [11]. Based on the results of this prototype, a few months after the decommissioning, Statkraft announced that it discontinued any investment in this energy source that utilizes semipermeable membranes.

Simply put, the technology of the semipermeable membranes has not been developed for systems, such as the one depicted in Figure 2.8, to be technically feasible, and to reliably produce power. Our experience shows that, while very small amounts of power are possible to be produced using semipermeable membranes with water systems, the actual work produced is very low and does not justify the investment. The additional production of work from gaseous systems, such as the combustion products of fuels, via semi-permeable membranes has not been demonstrated and does not appear to be feasible in the foreseeable future.

In summary, the work denoted by the last terms of Eqs. (2.30) and (2.32) is very small – on the order of 2% of $-\Delta H^o$ and $-\Delta G^o$ – as it is also shown in Table 2.2. An addition, engineering systems for its harnessing do not appear to be feasible at least in the near future. Because of this, it is reasonable to stipulate that the chemical exergy of all substances may be well approximated by the Gibbs' free energy of the reaction, and thus the following expression can be used as the maximum work to be obtained from a fuel:

$$W_{max} = E_{chem}(T_0, P_0) = -\Delta G^o. \tag{2.33}$$

The Gibbs' free energy, $-\Delta G^o$, is a well-defined property of all chemical reactions, its numerical values are available in most chemistry and thermodynamics textbooks or may be easily calculated from available information in these textbooks. Unlike Eqs. (2.30) and (2.32), ΔG^o is unambiguously defined for all the fuels and is used in a plethora of other computations without any controversy.

2.7 Exergy of Black Body Radiation

Among the first studies on the exergy of black-body radiation is that of Petela [12] who used the analogy that the constituent elements of radiation, the photons, may be modeled as material particles. A quantity of radiation – the irradiance – is composed of a number of photons/particles that behave as the molecules of an ideal gas. Hence, the quantity of radiation is characterized by its temperature, T, pressure, P, volume, V, internal energy, U, and entropy, S. The internal energy of the quantity of black-body radiation, which is enclosed in a volume V and has temperature T, is given by the expression:

$$U = \frac{8\pi^5 k_B^4}{15 h^3 c^3} V T^4, \tag{2.34}$$

where k_B is the Boltzmann constant, $1.38 * 10^{-23}$ J/K; h is the Planck constant, $6.626 * 10^{-34}$ Js; and c is the velocity of light in vacuum, $2.998 * 10^8$ m/s. The term in the fraction of Eq. (2.34) is a physical constant equal to $7.565 * 10^{-16}$ J/(m^3 K^4).

2.7 Exergy of Black Body Radiation

For the quantity of black-body radiation/photons, which is "enclosed" in a volume V with internal energy U, the maximum work that may be obtained from the expansion of the photon gas to the environmental pressure and temperature is calculated as [12]:

$$W_{\max} = E_{rad} = U\left[1 - \frac{4T_0}{3T} + \frac{1}{3}\left(\frac{T_0}{T}\right)^4\right]. \tag{2.35}$$

This is often expressed as an efficiency ratio:

$$\eta_P = \frac{E_{rad}}{U} = 1 - \frac{4T_0}{3T} + \frac{1}{3}\left(\frac{T_0}{T}\right)^4. \tag{2.36}$$

The main reason one is interested in the exergy of (black-body or grey-body) radiation is that the energy of the sun, which may be approximated as a radiating black-body, is transmitted to the earth and is widely available as an energy resource, the solar energy. This resource is available in the terrestrial environment as the energy flux from the sun, usually called irradiance and measured in W/m². Solar irradiance is the source of energy for all the plants; it generates food for all the living organisms; and is also harnessed by photovoltaics (PV) and solar thermal power plants. While solar energy has been harnessed as a source of heat and electricity for several centuries, in the second decade of the 21st century, solar energy is the fastest growing source of energy: In the period 2010–15, the electricity produced by PV cells has grown by an average annual rate of 42% and the equivalent number of solar thermal collectors is 35% [13]. Because irradiance is abundant and free, it is considered as the energy source for the future, especially in developing countries that have few other primary energy sources.

When the objective is the production of maximum energy or power for the solar installations in our planet, rather than asking the question: "What is the maximum work a quantity of radiation enclosed in a volume V may produce?" one should ask the different question: "What is the maximum work we may obtain from solar energy in the terrestrial environment?" Given that the sun may be modeled as a heat reservoir at temperature T, in principle, one may construct a large thermal engine that receives heat from the sun/reservoir, produces work/power, and rejects waste heat to the terrestrial environment at temperature T_0. The maximum efficiency of this cyclic engine would be equal to the Carnot efficiency:

$$\eta_C = 1 - \frac{T_0}{T}. \tag{2.37}$$

Jeter [14] arrived at this expression for the efficiency of solar energy using different reasoning.

An inspection of Eqs. (2.36) and (2.37) proves that $\eta_C > \eta_P$ for all temperatures $T > T_0$. The difference between the two expressions of the efficiency is a few percentage points at the temperature of the sun, ($T = 5{,}800$ K, or $T = 5{,}760$ K when the dilution of irradiance by the terrestrial atmosphere is considered), but becomes significant when the irradiance emanates from a lower temperature source, as it may be seen in Figure 2.9. The difference in the two expressions for solar energy was explained by Bejan [15] who

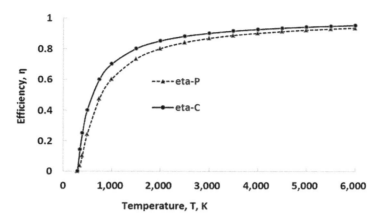

Figure 2.9 Maximum efficiency of radiation energy utilization, according to Eqs. (2.36) and (2.37).

demonstrated analytically that, when the radiation is "enclosed" and communicates with two temperature reservoirs, at T and T_0, a reversible cycle may be executed that produces additional work. When this work is added to the maximum work of Eq. (2.35), the result is consistent with the efficiency of Eq. (2.37).

When the energy of solar radiation is written in terms of the solar energy flux, σT^4, one may calculate the rate of exergy associated with solar irradiance as:

$$\dot{E}_{sol} = A\left(1 - \frac{T_0}{T}\right)\left(\frac{2\pi^5 k_B^4}{15 h^3 c^2}\right) T^4 = \left(1 - \frac{T_0}{T}\right) A\sigma T^4, \qquad (2.38)$$

where σ is the *Stefan–Boltzmann constant*, $5.670 * 10^{-8}$ W/(m² K⁴) and A is the area of the solar panels.

2.8 Exergy of the Water and the Wind

Mechanical energy is a "pure" form of energy, unencumbered by entropy. Consequently, the mechanical energy of any system is equal to the corresponding exergy. Wind energy has been used for ship propulsion for more than 3 millennia and hydraulic (water) energy has been used in water mills for more than 1,000 years. In the 21st Century both of these forms of renewable energy are widely harnessed to produce electric power. It is, therefore, important to determine what the maximum amount of energy is that may be extracted from the wind and the water.

Water energy is available in three forms: the hydraulic energy of the rivers, which is primarily potential energy; the energy of the tides, which is both kinetic and potential energy; and the energy of the sea waves, also kinetic and potential energy. All these forms of energy are mechanical energy and, in principle, may be converted in their entirety to mechanical work in reversible engines. However, the types of the engines used for the harnessing of water and wind power have operational limitations that reduce the maximum power generated.

2.8.1 Hydraulic Exergy

The schematic diagram of a typical hydroelectric power plant that utilizes the potential energy of stagnant water in a large reservoir is shown in Figure 2.10. An artificial dam creates a water reservoir upstream. Water from the reservoir enters a large pipe, the *penstock*, and is directed to one or more large hydraulic turbines that are connected to electric generators and produce electric power. The water effluent of the turbines is discharged downstream at a lower elevation. The maximum power that may be produced from this open system, when there is no entropy generation is:

$$\dot{W}_{max} = \dot{E}_{hyd} = \dot{m}g\Delta z, \qquad (2.39)$$

where $\dot{m}$ is the mass flow rate of water through the turbine; g is the gravitational constant, 9.81 m/s^2; and Δz is the water height difference between the reservoir and the turbine. In hydroelectric power plants the frictional losses, the irreversibilities in the penstock, and in the turbine/generator systems are small. Typically, 70–80% of the water exergy is recovered and converted to electricity in such power plants.

Example 2.4: A small river passes through the outskirts of a town carrying on average 45 m^3/s of water. The river has a small waterfall downstream and the town is considering building a dam and a hydroelectric power plant. The dam will increase the height of the fall to 9.5 m. Determine the maximum power of this small hydroelectric project. Also, calculated what the generated power will be, if all the irreversibilities amount to 22% of the total exergy of the water.

Solution: The 45 m^3/s volumetric flow rate of water corresponds to a mass flow rate of 45,000 kg/s. From Eq. (2.39) the maximum average power produced by this plant will be: $45{,}000 * 9.81 * 9.5 = 4.19 * 10^6$ W (4.19 MW). If the total irreversibilities are 22% of the total, the actual power produced is 3.27 MW.

Note that this is sufficient power to satisfy the demand of about 3,270 homes in most OECD (Organization of Economic Cooperation and Development) countries.

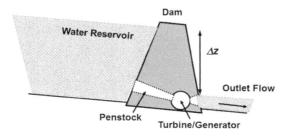

Figure 2.10 Schematic diagram of a hydroelectric power plant that utilizes the potential energy of water in a reservoir.

2.8.2 Tidal Exergy

The relative positions of the earth, the moon, and the sun, in combination with the rotation of the earth are the causes of the tides. The masses of the moon and the sun "pull" the masses of the ocean water and create a "bulge" on the surface of the oceans. The rotation of the earth causes this "bulge" to propagate on the surface of the oceans as a moving wave. Depending on the position of the three celestial bodies we have the stronger *spring tides* and the weaker *neap tides* [16]. Since the tides primarily depend on the relative position of the celestial earth-moon-sun system as well as on the rotation of the earth, they occur periodically with a longer period of approximately 29.5 days (the lunar cycle) and a shorter period of 12.5 hours (45,000 s, approximately ½ the period of the earth's rotation).

The globally-averaged amplitude – referred to as the *range* of the tidal wave – is approximately 0.6 m, which is very short for the production of significant power [16]. However, the existence of the continental shelves, the variable ocean depth, and coastline irregularities combine to produce distortions and resonances of the tidal wave on several locations on the surface of the earth. The local terrain characteristics create tidal waves with high ranges, close to 10 m in a few prime locations.

The range of the tide corresponds to potential energy, which may be stored in a *basin* of surface A and converted to mechanical work. The maximum work recovered from a basin with surface area A and range Δz is [16, 17]:

$$W_{max} = E_{tidal} = \frac{1}{2} A \rho g (\Delta z)^2, \qquad (2.40)$$

where ρ is the seawater density, approximately 1,025 kg/m^3. The tidal power plants utilize the potential energy of the water mass within the enclosed basin of surface area A and their power generation system is similar to that of hydroelectric plants, with the only difference that the height, Δz, in the tidal plants varies when water enters the basin and when it is extracted from the basin to produce power [16]. Optimized tidal installations may recover 70–80% of the exergy of the water in the enclosed basin.

2.8.3 Wave Exergy

Ocean waves are caused by wind shear. Because the wind is the result of thermal instabilities generated from uneven heating of the earth by the sun, the ocean waves are, indirectly, caused by the solar irradiance. Sea waves are characterized by their amplitude, α, their wavelength, λ, their phase velocity, c, and their frequency, f. The total power of the waves, which may be harnessed by a reversible engine, is [16]:

$$\dot{W}_{max} = \dot{E}_{wave} = \frac{\rho L \alpha^2 g^2}{4\pi f}, \qquad (2.41)$$

where ρ is the density of seawater, approximately 1,025 kg/m^3; L is the length of the area where the waves are; α is the amplitude (height) of the waves; and f is the wave frequency.

It must be noted that, although in principle all of this power may be utilized to produce mechanical work, the current state of technology of wave engines is not advanced and current designs of wave engines only convert a small fraction of the maximum wave power (on the order of 10%) to electricity.

2.8.4 Wind Exergy

The energy of the wind is kinetic energy. Since this is a form of mechanical energy, in principle, it may be converted in its entirety in a reversible engine to produce maximum power:

$$\dot{W}_{max} = \dot{E} = \frac{1}{2}\dot{m}V_i^2 = \frac{\pi}{8}\rho D^2 V_i^3, \qquad (2.42)$$

where V_i is the incoming velocity of the wind; ρ is the density of the air, approximately 1.2 kg/m^3; D is the diameter of the engine; and $\dot{m}$ is the mass flow rate, which is intercepted by the power producing engine – the wind turbine. This power would be produced when the air masses deliver power and come to a complete stop.

A moment's reflection, however, proves that, if the engine/turbine operated continuously and the air masses came to a complete stop and remained stagnant downstream, there would be accumulation of air and an increase of the air pressure at the back of the engine that would stop the flow of air. A necessary condition for the wind engines to operate and continuously produce power is that the mass flow rate $\dot{m}$ (not necessarily the velocity V) be maintained downstream the engine. This implies that the wind will have finite velocity downstream the turbine and will carry a fraction of its kinetic energy and exergy. The continuous (not necessarily constant) flow through the wind engine is a constraint imposed by the atmospheric environment, because far from the engine, in all directions, the local environmental pressure is P_0. It is easy to prove that for the horizontal axis wind turbines, which are most commonly used for the production of electric power from the wind, maximum power is obtained when the air velocity far downstream the turbine is equal to $V_i/3$. The maximum power that may be continuously derived from the wind with velocity V_i at the hub of the turbine and with blade diameter D (area A) is [16–18]:

$$\dot{W}_{max}^{tur} = \frac{8}{27}\rho A V_i^3 = \frac{2\pi}{27}\rho D^2 V_i^3, \qquad (2.43)$$

where ρ is the density of the air, which may be calculated from the ideal gas equation at the local pressure P_0 and temperature T_0. At typical wind turbine operations, $\rho \approx 1.2$ kg/m^3. A comparison of Eqs. (2.42) and (2.43) reveals that only 16/27 or 59.3% of the available energy of the wind may be continuously converted to power. This is called *the Betz's limit* or *Betz's law* [19]. It must be noted that the conversion of the entire amount of the wind exergy, as in Eq. (2.42), to electric power may be accomplished with an infinite system of horizontal axis turbines in series – not a practical scheme to be used for the commercial production of power.

One distinguishes in this case the maximum mechanical power (exergy) that is available in the wind, which is given by Eq. (2.42), and the maximum power that may be recovered from a wind turbine, which is given by Eq. (2.43). When assessing the potential of a wind site for the production of electric power, Eq. (2.43) is preferable to be used for the following reasons:

1. The extraction of the entire exergy of the wind as in Eq. (2.42) involves an infinite number of turbines in tandem, which is unfeasible. Using a large number of turbines in tandem with the output of the first turbine being the input of the second, the output of the second being the input to the third, etc. is uneconomical.
2. The expression in Eq. (2.43) includes the constraint of the necessity of downstream flow, which is imposed by the terrestrial environment.
3. All the practical studies for the reduction of irreversibilities and optimization of the wind turbine performance are based on Eq. (2.43).
4. There is no other type of practical device that may produce (even in theory) more power from the wind than that shown in Eq. (2.43).

It must be noted that the same two expressions, Eqs. (2.42) and (2.43) may be used for the harnessing of energy from ocean currents, such as the Gulf Stream. The ocean currents are masses of moving seawater and their energy characteristics are similar to those of the wind [16]. For ocean currents, the density, ρ, is that of the seawater, approximately 1,025 kg/m^3, and is almost constant from the sea surface to the depths of the ocean currents.

2.9 Exergy of Nuclear Fuel

Conventional nuclear reactors use fissile materials (uranium-235, uranium-233, and plutonium-239) and fertile materials (primarily thorium-232 and uranium-238) as their fuel. The nuclear fuel is split by thermal (slow) neutrons in the thermal reactors, which comprise the vast majority of nuclear reactors currently in operation, and by fast neutrons in the few fast reactors. The splitting (fission) of the atoms of the nuclear fuel creates a myriad of nuclei and subatomic particles at very high velocities [16, 17, 20]. These particles slow-down in the bulk of the reactor by inelastic collisions with other atoms and their mechanical energy is dissipated to heat. Thus, nuclear reactors produce heat, which is transferred to the working fluids of thermodynamic cycles that are either water to produce steam (in the PWR, BWR, CANDU, LMFBR, and RBMK types of reactors) or a gas (in the AGR, HTGR, and PBR types of reactors).

The heat produced per unit mass of the fissile nuclear materials is six orders of magnitude higher than that of fossil fuels. One kg of the fissile uranium-235 produces $81.8 * 10^9$ kJ of heat, and the heat produced per unit mass by the other pure fissile materials (uranium-233 and plutonium-239) is approximately the same. For comparison, the same quantity of heat is produced by the combustion of 2,500,000 kg of carbon or by 1,640,000 kg of methane. A relatively small amount of fissile material generates a large quantity of heat that is transferred to the reactor coolant and the working fluid of the thermodynamic cycle for the continuous production of power.

2.9 Exergy of Nuclear Fuel

The exergy of the other primary energy sources was calculated in Sections 2.3–2.7 using the transformation of the state of the fuel (the energy producing process) and applying the first and second laws of thermodynamics. When it comes to nuclear fuels, however, the thermodynamic states of the fuel and its products are not well defined. A complication for any deterministic calculations is that the radionuclide production in a thermal reactor (the products of the fission process) is stochastic and the actual number of the radionuclide products exceeds 1,600 [20]. This makes any exergetic calculations, based on the nature and the thermodynamic state of the nuclei products, complex and stochastic.

A practical way to define and calculate the exergy of the nuclear fuels is to use *their exergetic effect* on the nuclear reactor, which is to increase the temperature of the reactor and supply heat to the coolant at a high temperature, T. With this definition, the exergy of the nuclear fuels is:

$$E_{nuc} = me = mq\left(1 - \frac{T_0}{T}\right), \qquad (2.44)$$

where q is the heat supplied by 1 kg of the nuclear fuel for example $81.8 * 10^9$ kJ/kg for uranium-235 and plutonium-239; and $0.70 * 10^9$ kJ/kg for natural uranium, which contains 0.715% of the fissile uranium-235.

While T_0 is the known environmental temperature of approximately 300 K, the choice of the high temperature in the interior of the reactor is not as simple. Candidate values for the temperature T are:

1. The calculated temperature of the fission products before any collisions, which is calculated on the basis of their kinetic energy to be on the order of $600 * 10^9$ K. This implies $T_0/T \approx 0$ and that the exergy of the nuclear material is equal to the heat supplied. However, there are no practical energy conversion devices that may harness the kinetic energy of these fast radionuclides.
2. The melting point of uranium oxide (3,138 K), which is commonly used as a reactor fuel.
3. The melting point of the zirconium-steel alloy (2,400 K), which is commonly used to contain the reactor fuel in long tubes – the cladding of the reactor.
4. The high temperature achieved by the gas – the coolant in the HTAGR types of reactors – which is approximately 1,200 K.
5. The high temperature achieved by the gas in the commonly used AGRs, which is approximately 1,100 K.
6. The upper temperature of the water or steam in the water-cooled reactors (PWR, BWR, CANDU, RBMK, etc.), which is approximately 600 K.

It is apparent that the choice of the upper temperature, T, significantly effects the specific exergy of nuclear fuels. For the pure fissile materials (uranium-235, uranium-233, and plutonium-239) their specific exergy is in the range $81.8 * 10^9$ kJ/kg to $40.9 * 10^9$ kJ/kg. For natural uranium the specific exergy is in the range $0.70 * 10^9$ kJ/kg to $0.35 * 10^9$ kJ/kg. Figure 2.11 depicts the specific exergy of natural uranium and uranium-235 in the range of

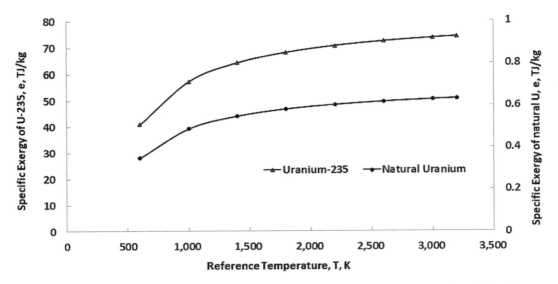

Figure 2.11 Specific exergy of natural uranium and uranium-235 in the range of possible reference temperatures 600–3,200 K. Safety considerations restrict the temperatures of nuclear power plants to the lower part of this range. (1 TJ = 10^{12} J).

temperatures 600–3,200 K, which includes the last 5 potential values of the reference temperature, T, (for the conversion of units: 1 TJ = 10^{12} J). It is observed that the choice of this temperature affects significantly the specific exergy of both nuclear fuels. Consequently, reactors with high coolant temperatures would produce higher electric energy per unit mass of fuel. However, safety considerations and practical limitations constrain the temperatures of the water-cooled reactors to approximately 600 K and of gas-cooled reactors to approximately 1,100 K.

The proposed definition for the exergy of nuclear fuels does not include any work that may be recovered from the products of nuclear reactions, when they come in chemical equilibrium with the environment – the chemical exergy of the radionuclides. Bringing the nuclear products in equilibrium with the environment, necessarily involves their release and eventual dispersion in the environment. Most of the radionuclide products are highly radioactive; they constitute a severe health hazard for humans and animals; and are isolated to be stored in secure facilities. Release of radionuclides in the environment is an environmental calamity – as the several nuclear accidents have proven it to be. The production of additional work by releasing and dispersing the products of the nuclear fuel in the environment is simply absurd. This demonstrates that using the chemical part of the exergy of certain fuels in the calculations for the maximum work potential may lead to absurd conclusions. One has to make sensible choices about the form of the exergy function to be used – especially when this involves the chemical exergy of pollutants. The choice of the exergy function should lead to the improvement of the energy conversion processes without any hazardous environmental consequences.

2.10 Lost Work and Power – Exergy Destruction

Exergy analyses of energy conversion systems indicate the maximum work or maximum power the systems may produce when the conversion processes occur with zero entropy production, $\Theta = 0$ or $\dot{\Theta} = 0$. However, real engines do not operate reversibly. The entropy production associated with the operation of real engines implies that the actual work or power performed by the engine is always less than what the exergy calculations indicate. The difference between the theoretical maximum and the actual work or power is called the *lost work* and sometimes *exergy destruction*. From Eqs. (2.3) and (2.15) the following expressions are derived for the lost work and lost power in closed and open thermodynamic systems:

$$W_{lost} = E_1 - E_2 - W = T_0 \Theta$$
$$\text{and} \quad \dot{W}_{lost} = \sum_{l=1}^{k} \dot{m}_{il} e_{il} - \sum_{j=1}^{n} \dot{m}_{ej} e_{ej} - \dot{W} = T_0 \dot{\Theta}. \quad (2.45)$$

Engineers always strive to minimize the lost work or power by minimizing the entropy production in all the processes and making energy conversion engines more efficient.

Example 2.5: A small electric power plant produces 150 MW of power and operates with a Rankine cycle with superheat. Steam enters the turbine at 8 MPa and 520°C, and exits at 40°C. Cooling water enters the condenser at 27°C and exits at 37°C. The fuel is coal, for which carbon is the only combustible material. The isentropic efficiencies of the pump and the turbine are 80% and 82%, respectively. Determine the lost power (rate of exergy destruction) in each of the components of this power plant.

Solution: The schematic diagram and the thermodynamic (T, s) diagram for the operation of this power plant are given in Figures 1.7 and 1.8. Using standard methods of thermodynamics and the steam tables, we obtain the following numerical values for the properties of the working fluid and the specific exergy of steam/water at each state ($T_0 = 27°C = 300$ K and $P_0 = 1$ atm):

$T_1 = 40°C, x_1 = 0 \rightarrow P_1 = 7.384 \text{ kPa}, h_1 = 168 \text{ kJ/kg} \rightarrow e_1 = 1.2 \text{ kJ/kg}$,

$P_2 = 8 \text{ MPa}, \eta_P = 0.8 \rightarrow h_2 = 178 \text{ kJ/kg} \rightarrow e_2 = 9.2 \text{ kJ/kg}$,

$T_3 = 520°C, P_3 = 8 \text{ MPa} \rightarrow h_3 = 3,448 \text{ kJ/kg}, s_3 = 6.7871 \text{ kJ/kg K} \rightarrow e_3 = 1,417.2 \text{ kJ/kg}$,

$T_4 = 40°C, \eta_T = 0.8, S_{4S} = 6.7871 \text{ kJ/kg K} \rightarrow h_{4s} = 2,114 \text{ kJ/kg}, h_4 = 2.354 \text{ kJ/kg}$,

$x_4 = 0.9082 \rightarrow e_4 = 93.8 \text{ kJ/kg}$.

The net specific work of the cycle is $w_{net} = 1,094$ kJ/kg and the specific heat input $q_{in} = 3,280$ kJ/kg. The overall thermal efficiency of this cycle is 33.4%.

For the production of 150 MW of electric power, the mass flow rate of the water/steam in the cycle is $150,000/1,094 = 137.1$ kg/s and the rate of heat input is 449.7 MW. Since the heat of combustion of carbon is $-\Delta H^o = 393.5$ MJ/kmol, the rate of heat input is provided to the burner from the combustion of $449.7/393.5 = 1.1428$ kmol/s of carbon or 13.71 kg/s, with an exergy rate input to the burner ($-\Delta G^o = 394.4$ MJ/kmol) 450.7 MW.

With the 10°C difference of the cooling water temperature, the condenser handles a mass flow rate of 7,163 kg/s of cooling water, which enters with 0 kJ/kg of exergy (the temperature and pressure of the cooling water at the inlet are those of the environment) and leaves with specific exergy 0.7 kJ/kg.

The lost power (rate of exergy destruction) in the four components of the cycle is:

1. **The pump:** It receives power $137.1 * (178 - 168) = 1,371$ kW. The difference of the exergy of the water between outlet and inlet is $137.1 * (9.2 - 1.2) = 1,096.8$ kW. Hence, the lost power in the pump is 274.2 kW.
2. **The burner/boiler:** It receives exergy from the coal equal to 450,700 kW; the difference of the exergy of the water/steam between the outlet and the inlet is: $137.1 * (1,417.2 - 9.2) = 193,037$ kW. Hence, the lost power in the burner is 253,663 kW.
3. **The turbine:** It produces 150,000 kW of power; its inlet is steam with exergy $1,417.2 * 137.1$ kW and its outlet is steam with exergy $93.8 * 137.1$ kW. Hence, the lost power in the turbine is $137.1 * (1,417.2 - 93.8) - 150,000 = 31,438$ kW.
4. **The condenser:** The inlets are the spent steam from the turbine with $137.1 * 93.8$ kW exergy and the cold water with $0 * 7,163$ kW $= 0$ rate of exergy. The outlets are the water for the cycle with exergy $137.1 * 1.2$ and the warmer cooling water with exergy $0.7 * 7,163$ kW. Hence, the lost power in the condenser is 7,681 kW.

The following table portrays the summary of the rates of exergy inputs and outputs of the four components of this small power plant.

	Pump	Boiler	Turbine	Condenser
Exergy rate inputs (MW)	1.37	450.7	194.3	12.9
Exergy rate outputs (MW)	1.10	193.0	162.9	5.2
Exergy rate destruction (MW)	0.27	257.7	31.4	7.7

It is observed in this example that most of the lost power occurs in the boiler, where a high fraction of the exergy of the fuel is destroyed in the combustion process. This is an indication that the combustion of chemicals (fuels) to produce heat and then mechanical and electrical power is an inefficient way to utilize these energy resources.

2.10.1 Lost Work in Nonequilibrium Processes

The subject of *Nonequilibrium Thermodynamics* (NET) or *Irreversible Thermodynamics* (IT) provides a methodology for the calculation of the rate of entropy production and the lost work in processes. The subject has its origins in the work of Onsager and Kasimir [21–23] and offers a powerful method for the development of the phenomenological relationships for irreversible processes including coupled processes [10, 24].

2.10 Lost Work and Power – Exergy Destruction

The rate of entropy production in all irreversible processes may be written as the product of several *generalized fluxes*, J_i, and the corresponding *generalized forces (or conjugate forces)*, F_i. For example, for the heat transfer between two systems at temperatures T_1 and T_2 ($T_1 > T_2$), the instantaneous rate of entropy production and the associated entropy flux may be written as:

$$\dot{\theta} = \frac{\dot{\Theta}}{A} = \frac{\dot{Q}}{A}\left(\frac{1}{T_2} - \frac{1}{T_1}\right) = \frac{\dot{Q}}{A}\left(\frac{T_2 - T_1}{T_1 T_2}\right) = J_Q F_{1/T}, \qquad (2.46)$$

where A is the area through which heat is transferred. The rate of heat transfer is the generalized flux for the heat transfer process and the function in the parenthesis is the generalized force.

When this result is generalized to n coupled processes that occur simultaneously (e.g. several coupled chemical reactions, mass diffusion combined with heat transfer, etc.) the entropy flux is the sum of the products of all the generalized fluxes (or conjugate fluxes) and the corresponding generalized forces (or conjugate forces) [21]:

$$\dot{\theta} = \sum_{i=1}^{n} J_i F_i \geq 0. \qquad (2.47)$$

The entropy flux is not simply the sum of arbitrary products, $J_i F_i$. Every conjugate force is derived from the corresponding conjugate flux via the equation known as the Gibbs' equation, which emanates from the combination of the first and second laws of thermodynamics. For a reacting mixture that also exchanges heat and work, the Gibbs' equation is:

$$dS = \frac{1}{T}dU + \frac{P}{T}dV + \frac{\left(\sum_r v_r \mu_r - \sum_p v_p \mu_p\right)}{T}dN. \qquad (2.48)$$

When written in terms of the entropy flux, $\dot{\theta}$, this equation yields [10, 24]:

$$\dot{\theta} = \frac{dS}{A dt} = \frac{1}{T}\frac{dU}{A dt} + \frac{P}{T}\frac{dV}{A dt} + \frac{\left(\sum_r v_r \mu_r - \sum_p v_p \mu_p\right)}{T}\frac{dN_i}{A dt}, \qquad (2.49)$$

where N_i in the equation represents both reactants and products. The three terms of the entropy flux correspond to the thermal energy (heat) transfer, the work transfer, and the chemical reactions that take place in the system under consideration. Equation (2.49) may be written in the flux-force notation of Eq. (2.47) as follows:

$$\dot{\theta} = F_{1/T} J_Q + F_{P/T} J_V + \sum_{r,p} F_\mu J_N \geq 0. \qquad (2.50)$$

Because the last relationship emanates from the Gibbs' equation, the two laws of thermodynamics are implicitly satisfied with the choices of the pairs of generalized fluxes and the corresponding generalized forces.

The generalized forces and the corresponding generalized fluxes vanish simultaneously when the thermodynamic system attains equilibrium, and the rate of entropy production vanishes. For nonequilibrium thermodynamic systems that participate in physical processes, the conjugate forces drive the corresponding fluxes and there is positive entropy flux. Experimental evidence proves that the generalized fluxes are driven not only by their corresponding conjugate forces, but by all the other forces as well [21–23]. Therefore, one may write the following expression for the functions of all the fluxes:

$$J_i = J_i(F_1, F_2, F_3, \ldots, F_n) \tag{2.51}$$

Onsager [21, 22] stipulated that any flux may be written as the sum of the products of all the conjugate forces and the derivatives of that flux with respect to the forces. The latter are called *phenomenological coefficients* and are denoted by the symbols L_{ij}. Hence, the expression for the generalized flux J_i, may be written succinctly as:

$$J_i = \sum_{j=1}^{n} L_{ij} F_j, \tag{2.52}$$

and the expression for the entropy flux of Eq. (2.47) becomes:

$$\dot{\theta} = \sum_{i=1}^{n} \sum_{j=1}^{n} F_i L_{ij} F_j \geq 0. \tag{2.53}$$

Accordingly, the lost power (exergy destruction) in the coupled processes is:

$$\dot{W}_{lost} = T_0 \dot{\Theta} = A T_0 \dot{\theta} = A T_0 \sum_{i=1}^{n} \sum_{j=1}^{n} F_i L_{ij} F_j \geq 0. \tag{2.54}$$

Based on experimental evidence from systems of several conjugate forces and statistical mechanical arguments, Onsager postulated that the tensor L_{ij} is symmetric, that is $L_{ij} = L_{ji}$ [21, 22]. Later, Kasimir [23] discovered that there is a class of force-flux relationships – primarily associated with fluid dynamics processes – where the tensor of the phenomenological coefficients is anti-symmetric, that is: $L_{ij} = -L_{ji}$ [23, 24]. The conjugate forces and fluxes in the processes where the tensor L_{ij} is symmetric are even functions of the molecular velocities. In these processes the conjugate forces remain unchanged when all the molecular velocities reverse (e.g. when the conjugate forces are temperature gradients, concentration gradients, and concentration gradients of electric charges that generate an electric current). The conjugate forces in the class of processes where the tensor L_{ij} is anti-symmetric are odd functions of the molecular velocities. Such conjugate forces reverse sign when all molecular velocities are reversed (e.g. the shear stresses and the viscous pressure of fluids). A general method to write the symmetry relationship of the phenomenological coefficients, L_{ij}, for all the processes is [24]:

$$L_{ij} = e(F_i) e(F_j) L_{ji}, \tag{2.55}$$

where the functions $e(F_k)$ are defined as [24]:

$$e(F_k) = \begin{cases} +1 \text{ if } F_k \text{ is an even function of molecular velocities} \\ -1 \text{ if } F_k \text{ is an odd function of molecular velocities} \end{cases}. \quad (2.56)$$

Equations (2.55) and (2.56) are sometimes referred to as the *Onsager–Kasimir theorem*. The theorem has been supported by all the available experimental evidence [25, 26] and has become the foundation of the theory of *Nonequilibrium Thermodynamics*.

2.10.2 Lost Power and Exergy Destruction in Continua

Let us consider a continuum in a control volume V, which is enclosed by the area A, as shown in Figure 2.12. The local instantaneous velocity of the continuum is denoted by the vector $\vec{V}$, with components V_x, V_y, and V_z, and the local instantaneous temperature by T. Both T and $\vec{V}$ are functions of time and space, that is $\vec{V} = \vec{V}(x,y,z,t)$ and $T = T(x,y,z,t)$. The continuum governing equations – continuity, linear momentum conservation, angular momentum conservation, and energy conservation – apply to this control volume. For an incompressible fluid, the rate of entropy production in the control volume is [27, 28]:

$$\dot{\Theta} = \int_V \frac{k}{T}\left[\left(\frac{\partial T}{\partial x}\right)^2 + \left(\frac{\partial T}{\partial y}\right)^2 + \left(\frac{\partial T}{\partial z}\right)^2\right]dV$$
$$+ \int_V \frac{2\mu}{T}\left\{\left[\left(\frac{\partial V_x}{\partial x}\right)^2 + \left(\frac{\partial V_y}{\partial y}\right)^2 + \left(\frac{\partial V_z}{\partial z}\right)^2\right] + \left[\frac{\partial V_x}{\partial y} + \frac{\partial V_y}{\partial x}\right]^2 + \left[\frac{\partial V_y}{\partial z} + \frac{\partial V_z}{\partial y}\right]^2 \right.$$
$$\left. + \left[\frac{\partial V_z}{\partial x} + \frac{\partial V_x}{\partial z}\right]^2\right\}dV, \quad (2.57)$$

where k is the conductivity of the continuum and μ is its dynamic viscosity. The exergy destruction is always equal to the product $T_0\dot{\Theta}$. It is apparent that in all continua in motion the rate of entropy production is always positive definite. The first integral in the last equation is associated to the heat transfer and the second to friction. The rate of

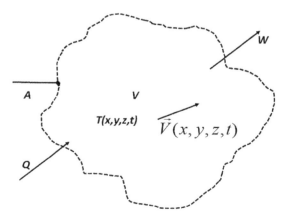

Figure 2.12 A continuum, enclosed in the control volume V and surrounded by the area A.

entropy production in multiphase continua is determined by a similar expression with the variables representing the corresponding properties and velocities of the distinguishable phases (e.g. liquid/gas, different immiscible liquids, etc.) [29].

2.11 Exergetic Efficiency – Second Law Efficiency

Unlike the expressions for internal energy, enthalpy, and exergy, which have been defined according to natural laws – the first and second laws of thermodynamics – the efficiencies of processes, engine components, engines, and systems are figures of merit that have been conveniently defined by scientists and engineers. They are ratios of variables associated with energy or power and have a specific meaning for those who defined and use them. For example, the thermal efficiency of the cyclic engines, defined in Eq. (1.24), represents a benefit-cost ratio for thermal power plants: the numerator, W, is a measure of the revenue derived by the sale of electricity and the denominator, Q_H, is a measure of the cost of the fuel. Improving the efficiency of the cycle always increases the benefit-cost ratio for the thermal power plant. The efficiency (coefficient of performance) of refrigeration systems, defined in Eq. (1.31), is also a benefit (the coolness effect) to cost (electricity input) ratio. All the efficiency expressions in Eqs. (1.24)–(1.32) represent benefit/cost ratios for the operators of the engines. A characteristic of these expressions is the usage of heat and work, quantities related to the first law of thermodynamics. These efficiencies are referred to as the *first law efficiencies*.

Exergy represents the maximum work and power and one may use these maxima as the benchmark for processes and systems and define the ratios:

$$\frac{W_{act}}{E} \quad \text{and} \quad \frac{W_{act}}{\dot{E}},$$

as figures of merit for the operation of engines. These figures of merit signify how closely the operation of the system/engine is to producing maximum work or power. The figures of merit, which are based on exergy – and, therefore, make use of the first and second laws of thermodynamics – are called *exergetic efficiency*. The term *utilization factor* has also been used in the past [1, 3], but the use of this term has faded. In general, the exergetic efficiency of any system is defined as:[7]

$$\eta_{II} \equiv \frac{\text{Sum of all work and exergies leaving}}{\text{Sum of all work and exergies entering}}. \quad (2.58)$$

For simple closed and open systems that only produce work or power, Eq. (2.58) may be expressed as follows:

---

[7] A few authors have used the symbol ε for exergetic efficiency. This symbol has been historically used for the effectiveness of heat exchangers as well as for small quantities in calculus. The symbol η_{II} will be used in this book to avoid confusion.

$$\eta_{II} \equiv \frac{W_{act}}{E_{in}} = 1 - \frac{T_0 \Theta}{E_{in}} \quad \text{and} \quad \eta_{II} \equiv \frac{\dot{W}_{act}}{\dot{E}_{in}} = 1 - \frac{T_0 \dot{\Theta}}{E_{in}}. \tag{2.59}$$

The exergetic efficiencies offer more insight on the operation of systems and engines, because they offer information on the actual entropy production (the irreversibilities) and denote how close to the ideal operation the systems/engines are operating. Using the exergetic efficiencies the engineers may pinpoint where high irreversibilities occur within systems and identify the most critical components of the systems and processes for improvement.

The *second law efficiency,* defined in terms of reversible processes, is also occasionally used with energy conversion systems [30]:

$$\eta_s = \frac{W_{act}}{W_s} \quad \text{or} \quad \eta_s = \frac{\dot{W}_{act}}{\dot{W}_s}, \tag{2.60}$$

where the subscript s denotes isentropic processes. The second law efficiencies are numerically equal to the exergetic efficiencies when the low temperature reservoir of the reversible process is equal to the temperature of the environment, T_0. For this reason, oftentimes, the term second law efficiency is considered to be synonymous with the exergetic efficiency and the two names are used interchangeably.

As with the first law efficiencies that have been conveniently defined to characterize a myriad of systems and processes [e.g. Eqs. (1.24)–(1.31)] several figures of merit, often called second law efficiencies, have been defined to characterize systems and processes. Among these figures of merit, which are based on exergy and the two laws of thermodynamics, one finds in the literature other figures of merit emanating from exergy considerations, such as:

1. The degree of perfection (for the production of chemicals).
2. The efficiency of mixing.
3. The efficiency of separation.
4. The ratio of the first law efficiency to the Carnot efficiency.
5. The environmental exergy efficiency.
6. The sustainability efficiency.

The numerical values of the above figures of merit are used to characterize the performance of engines and systems and oftentimes to improve this performance. While using any figure of merit, one must be cognizant of its underlining definition to avoid confusion and possible calculation errors.

It must be noted that, while the first law, the exergetic, and the second law efficiencies may be associated with a benefit/cost ratio, any comparison of thermodynamic efficiencies for economic purposes is only meaningful for the same type of power production units. Commercial PV cells and nuclear power plants typically have lower efficiencies than fossil fuel plants. However, the cost of the nuclear fuel is much lower than that of fossil fuels and the cost of irradiance for photovoltaics is zero. The comparison of the thermal efficiency of a gas turbine (approximately 50%) to that of a solar thermal power plant (approximately 18%) does not lead to any meaningful economic comparison because the primary energy resource for the latter is free. A more meaningful figure of merit in this case – and one that is commonly used by electric utilities and consumers – is the actual cost per kWh (e.g. in units of $/kWh or ¢/kWh) after considering all the costs associated with the production of energy [16].

2.12 Characteristics of the Exergy Function

A glance at all the final expressions for the exergy of power production systems – from hot rock to nuclear fuels – proves that the exergy function is different for the systems that have been considered and depends on: (a) the state of the system; (b) the state of the environment; and (c) the ways of interaction of the system with its environment. Because of this, exergy is not a property of the system and cannot be defined solely by the other properties of the system. Even though the exergy function is not a system property, it corresponds to a potential function when the state of the environment does not change, because for all thermodynamic systems undergoing a cycle:

$$\oint dE = m \oint de \equiv 0. \qquad (2.61)$$

Since it is known that part of the exergy is destroyed in the processes of a thermodynamic cycle, the last equation implies that the lost exergy in the cyclic system is replenished by exergy input from another system. This system is the energy (and exergy) resource of the cycle, typically a fuel, solar energy, geothermal fluid, etc. With the injection of exergy from the exergy resource, the exergy of the cyclic engine is restored to its original value at the end of every cycle.

By its definition exergy is nonnegative. Since exergy is defined as the maximum work that may be obtained from a given thermodynamic system, systems not in equilibrium with their environment produce positive work, while systems in equilibrium with their environment do not produce any work (zero work). This precludes the exergy function from attaining negative values. For example, a system with $T \neq T_0$ will produce work regardless of whether $T > T_0$ or $T < T_0$, as the two parts of Figure 2.13 show: If $T > T_0$, a heat engine may be constructed that receives heat Q from the system, rejects heat Q_0 to the environment and produces work $W > 0$, as in the left part of Figure 2.13. If $T < T_0$, another heat engine may be constructed to receive heat Q_0 from the environment and reject heat Q to the (colder) system, thus again producing positive work, $W > 0$, as in the right part of Figure 2.13. In both cases work will be produced until the system's temperature becomes equal to that of the environment, in which case no further work may be produced. Actually, and because the mass of the system is finite, the removal of heat (in case a) and addition of heat (in case b) will continuously bring the temperature of the system closer to that of the environment, until thermal equilibrium is reached. In both cases, $T > T_0$ and $T < T_0$, the final state of the system is $T = T_0$, and the system stops producing work, because it is in thermal equilibrium with the environment.

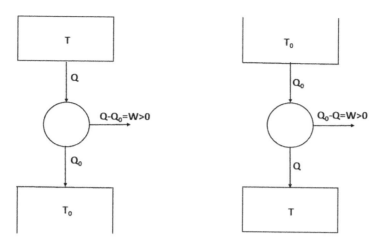

Figure 2.13 A system not in equilibrium with the environment ($T \neq T_0$) will produce work regardless of whether $T > T_0$ or $T < T_0$.

Unlike the other energy-related properties of systems (internal energy, enthalpy, the free energies) that do not have a natural zero-point and their datum needs to be defined by convention (e.g. by convention the datum for the enthalpy of homogeneous substances is the enthalpy of the liquid phase at their triple point), the exergy function has a natural zero, the state of the system in equilibrium with its environment.

2.13 Models for the Reference Environment

It was postulated in Section 2.2 that the model for the terrestrial environment – where all the energy conversion processes take place – is a reservoir of heat, work, volume, and mass. The four-reservoir model is a good approximation to the terrestrial environment (our natural environment). The terrestrial environment is very large, massive, and exchanges daily enormous quantities of mass, volume, heat, and work (in the form of electricity) with a myriad of engineering systems and without any noticeable change of its state.

The annual anthropogenic rejection of waste heat from electricity generation to the atmosphere and the hydrosphere – the *thermal pollution* – is approximately $127 * 10^{15}$ kJ annually [16]. The annual quantity of dissipated energy from the other anthropogenic activities is on the order of 10^{16} kJ. This is seven orders of magnitude less than the annual amount of radiation energy received from the sun and of the annual amount of energy emitted by the earth to the outer space as infrared radiation. It is apparent – and there is unanimous agreement by the scientific community on this – that the anthropogenic rejection of heat/energy in the natural environment does not have any perceivable effect on the average environmental temperature. One may conclude that the model of a heat reservoir describes very well the terrestrial environment.

The terrestrial environment has also been the sole depository of the exhaust products of fossil fuel combustion since the beginning of the industrial revolution, in particular the sole depository of CO_2. The atmospheric composition of this gas has increased from approximately 280 parts per million (ppm) in the year 1750, to 410 ppm in 2018. Because the composition of the atmosphere has changed with the CO_2 emissions, strictly speaking, the atmosphere is not a mass reservoir. A moment's reflection, however, will prove that the change of the CO_2 composition (from 0.0280% to 0.0410%) in almost 250 years (0.000052%/yr) is extremely slow. Very large quantities of CO_2 are continuously absorbed by the plants and the oceans [31]. The latter are converted to salts, such as $CaCO_3$, and precipitate on the ocean bottom, without appreciably changing the composition of the hydrosphere.[8] Since there is no noticeable change in the globally-averaged CO_2 composition in the hydrosphere and very slow change of the CO_2 composition in the atmosphere, the modeling of the terrestrial environment as a mass reservoir is reasonable. More important – as with other thermodynamic idealizations that are now considered sanctified because of the accuracy of their predictions – the calculations, conclusions, and predictions based on a model of the environment as a mass reservoir, have led to accurate and verifiable outcomes.

2.13.1 Local and Temporal Nonequilibrium

An apparent objection to modeling the terrestrial environment as a combination of thermodynamic reservoirs is that the environment itself is not in equilibrium. Everyday observations with thermometers and barometers prove that the local temperature and pressure of the atmosphere varies spatially as well as daily, and seasonally, that is, $T_0 = T_0(x,y,z,t)$ and $P_0 = P_0(x,y,z,t)$. The state of the hydrosphere exhibits similar spatial and temporal variability. The temperature of the upper part of the lithosphere – down to an approximate depth of 20 m – also varies, following the local air temperature [32]. The rest of the lithosphere exhibits a high temperature gradient, from 5,000 K at the center of the earth to approximately 300 K close to the surface. This gradient is the driving force of the $44 * 10^{12}$ W rate of heat transferred from the core to the surface of the earth, which created the geothermal resources.

When large segments of the terrestrial environment are considered, it becomes apparent that the environment is not in thermal and mechanical equilibrium: on a given hour the temperature and pressure in New York City are different than the temperature and pressure in Buenos Aires and in Beijing. The coastal waters in Greenland have always lower temperature than the waters in the Canary Islands. The atmospheric temperature 10 km above a given location is always lower than that at the surface, while the temperature 10 km deeper in the lithosphere is always higher. The terrestrial environment is spatially inhomogeneous and it is not in thermodynamic equilibrium.

---

[8] The artificial formation of calcium and magnesium carbonates has been also proposed as a method for carbon sequestration [52].

This presents the conundrum as to how a thermodynamic system (e.g. a thermal engine) may attain "thermodynamic equilibrium" with the environment, when the terrestrial environment is not in equilibrium in itself.

A moment's reflection on the environment of energy conversion engines will resolve the conundrum. The engines interact locally with a very small part of the environment, the surroundings of the engines: A gas turbine in Dallas, Texas is affected only by the local atmospheric temperature and pressure, which at a given time are almost constant. The different temperatures and pressures in Mexico City, in Beijing or at a depth of 10 km, are irrelevant to the operation of the engine. Faraway locations should not be considered as the "environment" of the gas turbine in Dallas. Similarly, a Boeing-777 aircraft at the cruising altitude of 10.3 km is only affected by the temperature and pressure at that altitude, and not by the conditions on the surface of the earth. The larger parts of the terrestrial environment (where appreciable differences of temperature and pressure are manifested) are of no significance to the thermal engines of the aircraft.

Such considerations offer a working definition for the "environment" of an energy conversion system – the engine – as *the part of the Universe by which the defined thermodynamic system's operation is affected*. This definition is analogous to the model of the surroundings of a thermodynamic system as *the part of the Universe, which is affected by changes in the defined thermodynamic system*. This model has been used since the inception of thermodynamics and has led scientists to accurate calculations and proven conclusions. The local definition of the environment of a thermal engine satisfies the characteristics of the environment that come into the derivation of the exergy function: The part of the Universe that affects a thermal engine – the environment of the engine – is small enough in scale to be considered homogeneous, with constant and well-defined properties at all instants during short periods of time.

It is also well known that the local environment – homogeneous as it may be locally – exhibits temporal variability, with two timescales diurnal (daily) and seasonal: The night temperature of the environment of a gas turbine is usually lower than the daily temperature; the river water temperature that removes the waste heat from the condensers of a large nuclear power plant is colder in the winter months and warmer in the summer. The local atmosphere near the gas turbine and the hydrosphere in the vicinity of the nuclear power plant are the environments of the two thermal engines and, apparently, exhibit significant temporal variability. Figure 2.14 depicts the hourly temperature in Fort Worth, Texas during the year 2010 (8,760 hours) [33]. The temperature variability in this city is typical of places on the planet with continental climate. The diurnal and seasonal variability of the local temperature are apparent in Figure 2.14. The broken line, at 292.2 K, is the average temperature during the year and the standard deviation of the data is 9.8 K, 3.4% of the mean. It is observed in Figure 2.14 that the temporal variability of the local temperature is very low and that the local temperature during most of the hours of the year is close to its mean value of 292.2 K. The very low standard deviation, 3.4% of the mean, supports this observation. In addition, the two timescales of the temporal temperature variability – 24 hours and

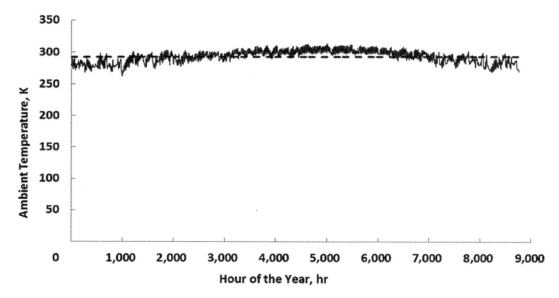

Figure 2.14 Hourly temperature in Fort Worth, Texas during the 8,760 hours of the year 2010. The standard deviation of the data is 9.8 K or 3.4% of the average temperature.

12 months – are by far longer than the timescales of all thermal engines, which are on the order of seconds. Since the timescales of the engines are much less than those of the environment, $\tau_{eng} \ll \tau_{env}$, one concludes that, from the point of view of the thermal engines, their respective environments are in equilibrium (see Section 1.2 for the definition of thermodynamic equilibrium and reversible processes).

In conclusion, because energy conversion systems (engines) are small in comparison to the terrestrial length-scales and because their operation is fast in comparison to the local terrestrial environmental timescales, the defined models for their environment are locally homogeneous and in equilibrium. The stipulation that a thermodynamic system will come to equilibrium with such an environment is rational and has generated thermodynamic models that produce verifiably accurate results and conclusions about the operation of engines and the conversion of energy resources.

2.13.2 Models for the Environment

It is apparent from the consideration of all the expressions for the exergy of resources and energy systems that the concept of exergy is meaningless without the definition of the state of the environment, often referred to as *the reference environment*. While it is easy to agree on a model for the reference environment for the physical part of the exergy (e.g. the local pressure and temperature P_0 and T_0 respectively, which have very low variability and are *quasi-static*), there is no general agreement on a suitable environmental model for the calculation of chemical exergy that appears in Eqs. (2.29), (2.30), and (2.32). The main issue here is the chemical composition of the

2.13 Models for the Reference Environment

reference environment and more specifically, which chemical components and at what compositions constitute this environment.

Ahrends [34] postulated an environment that includes the atmosphere, the hydrosphere, and 15 of the most abundant elements of the lithosphere, which make up 99% of the crust of the earth. He also postulated that the environment should be in internal thermodynamic equilibrium; it should be defined in a way that the resultant exergy of all chemical products should not be as high as to obscure irreversibilities in processes; and that the resultant exergy of chemicals should represent their value as thermodynamic and economic resources. By arbitrarily inventing pathways for the production of chemicals, Arhends [34] was able to calculate the exergy of all common chemicals. However, the arbitrary chemical pathways inflated the exergy of a few substances (e.g. of oxygen) and underestimated the exergy of the fossil fuels. Because the exergies of many chemicals calculated by Ahrends' model disagree with empirical observations of chemistry, this model has been highly criticized. Another criticism of the model is the intention to calculate the thermodynamic and economic value of scarce resources using a single value, the exergy value. The two concepts – one is based on the laws of Physics, the second on market demand and supply considerations – are entirely different and cannot possibly be measured by a single number, the exergy.

A second model for the chemical environment has been formulated by Szargut and his coworkers [34, 36]. They postulated "reference substances" for all the elements in the environment. From these reference states of the elements are derived and the exergy of all the chemical compounds is calculated by a series of chemical reactions. While this model for the chemical environment is conceptually simple, it has proven to be vague in its application and suffers from several drawbacks among which are:

1. The exergies of several common substances are calculated to be negative, a contradiction with the very definition of exergy.
2. The exergy of several reference substances is arbitrarily chosen to fit empirical data. For example, the model fails to calculate the exergies of common fossil fuels and determines the fuel exergy using the heating values, ΔH^o, of the fuels and empirically defined constants, a method that is inconsistent with the model's premises.

A model that relies solely on thermodynamic theory is the one proposed by van Gool [37], who disregarded any connection between the (economic) prices and the exergy of the resources. Van Gool calculated the exergies of chemicals based on a model for the atmosphere, the chemical properties of pure compounds, and linear optimization techniques to define the environment. While this model is based on sound physical principles and data, it has often been criticized for its use of linear optimization, which does not guarantee that the calculated exergy actually represents the maximum work of the resource.

Several other models for the environment have been proposed in the literature. The practical significance of some of these models on power generation calculations has been studied in [38]. Most of the efforts to define a model environment for the materials of the planet stem from attempts to quantify the "value"[9] of the natural resources and of wastes (effluents of industrial processes) in terms of exergy. In several of the proposed models, the degradation of the resource and the deterioration of the environment are associated with exergy destruction and lost work. One quickly realizes that these are three different issues with the models for the environment: (1) the maximum work that may be obtained from a material; (2) the "value" of the material (in any way "value" is defined); and (3) and the environmental degradation triggered when the material is released in the environment. These are three entirely different concepts that cannot be quantitatively described by a single number, the material's exergy. The relationship of exergy to ecological and environmental impacts is extensively discussed in Chapter 6.

Any kind of stipulated environment that would result from the quantification of maximum work, "value," and environmental degradation will have very few similarities with the terrestrial environment, where most energy conversion activities occur. Actually, such a *universal reference environment* that quantifies all three issues – the maximum obtainable work, the "value" of the resources, and the environmental degradation – is not possible to be formulated, even as a mathematical tool.

2.13.3 Effects on Calculations

In addition to the three models of the environment that were formulated to calculate the chemical exergy of substances, several other models for a reference environment have been formulated. Because the exergy of all substances depends on the reference states, each one of the models generates different values (numbers) for the chemical exergy. It is axiomatic that arbitrary reference states (dead states) would yield arbitrary values for the exergies. This presents a significant drawback for the application of the exergy concept in energy engineering systems.

One quickly recognizes that the arbitrary values of exergy and the uncertainty associated with exergy calculations only pertain to the chemical part of the exergy. The vast majority of the environmental models do not affect the numerical values of the physical exergy of the resources, $E(P,T)$, which is of importance in thermal energy conversion processes. The differences in the several proposed models of the environment specifically pertain to the second terms of Eqs. (2.30) and (2.32). When the environment is defined as having the composition of the atmosphere, these terms represent very small fractions of the total exergy of chemical substances, as it is apparent in Table 2.2. The Gibbs' free energy of a combustion reaction, ΔG^o, which is a well-defined thermodynamic property of the reaction and readily available in thermodynamic tables, accounts for approximately 98% of the chemical exergy of the fuels and represents the dominant part in the numerical calculation of exergy and of the maximum work or power.

Munoz and Michaelides [38] conducted a comparative study to determine the effect of 11 environmental models on the engineering analysis of three energy systems:

---

[9] The "value" of the resource appears to be different than the price and not rigorously defined in the publications that use the concept.

1. A 30-MW methane-fired cogeneration gas turbine.
2. A 60-MW coal-fired electric power plant.
3. A 30-MW dual-flashing geothermal power plant.

The conclusions of the study are that the physical exergy of all the states of the working fluid in the three electricity production units is not affected at all by the choice of the model for the environment. Small differences were observed with the chemical exergy of the states of the working fluids. The differences largely cancelled out and did not significantly affect the calculated quantities of heat and work exchanged in the several components of the three units, thus making the choice of the chemical description of the reference environment irrelevant to the calculations of the energy conversion systems [38]. In particular, the choice of the chemical composition of the environmental model made very small differences in the calculations of the exergetic efficiency of the several components of the three systems that were examined. Table 2.3 shows the calculated exergetic efficiencies for the 5 components of the 30 MW cogeneration gas turbine as well as the overall efficiency of the entire power plant. The 11 columns, labeled A to K, represent the calculated efficiencies according to each one of the 11 models for the environment. It is observed in Table 2.3 that the 11 models generated almost identical values for the efficiency of every component as well as for the overall efficiency of the cogeneration plant. The corresponding results for the coal-fired plant and the geothermal unit are very similar: the exergy efficiencies of the components of the energy conversion systems are almost unaffected by the chosen model for the environment.

The agreement of the numerical values in Table 2.3 is very significant in the exergy analysis of energy conversion systems: Exergy analyses are used in practice to deduce which are the inefficient components of systems and then strive to improve them. If the values for the exergy efficiencies of the components significantly depended on the environmental model, it would be a very challenging task to discern the inefficient components, and to invest in their improvements. It becomes clear from the values of Table 2.3 that the steam generator, the combustor, and the preheater are the most inefficient components and, therefore, engineers should strive to improve a more reversible operation of these components. This fortuitous agreement happens because it is the physical exergy and the Gibbs free energy that play the most important roles in energy conversion processes. These two quantities are unambiguously defined and well-known to be used in all exergy analyses.

2.13.4 Energy Conversion and the Terrestrial Environment

When we consider the energy conversion processes, it quickly becomes apparent that most of them occur in a small part of the terrestrial environment, the environment of the biosphere. This is a narrow layer, surrounding the earth's surface, where most human activities occur. This layer of the biosphere is essentially composed of three parts:

1. The lower atmosphere, which has a homogeneous gaseous composition and is at almost constant pressure, one atmospheric pressure. The temperature of the lower

Table 2.3 Exergetic efficiency for the 5 components of the cogeneration gas turbine and overall efficiency as generated by 11 (A–K) models for the environment [38].

Component	Exergetic Efficiency (%) Model										
	A	B	C	D	E	F	G	H	I	J	K
Turbine	95.20	95.20	95.20	95.20	95.20	95.20	95.20	95.20	95.20	95.20	95.20
Compressor	92.80	92.80	92.80	92.80	92.80	92.80	92.80	92.80	92.80	92.80	92.80
Combustor	80.09	79.77	80.59	80.09	79.17	79.96	80.05	80.01	80.12	80.10	80.50
Preheater	84.55	84.55	84.55	84.55	84.55	84.55	84.55	84.55	84.55	84.55	84.55
Steam Generator	67.26	67.26	67.26	67.26	67.26	67.26	67.26	67.26	67.26	67.26	67.26
Overall Power Plant	52.04	52.03	51.99	52.04	52.03	52.03	52.00	51.99	52.04	53.97	52.01

atmosphere exhibits low enough variability and high enough timescales to be considered constant from the vantage point of energy conversion systems.
2. The near-surface hydrosphere, which is almost homogeneous, and where the temperature exhibits lower variability than that of the atmosphere. While the pressure of the hydrosphere varies significantly with depth, most of the human energy conversion activities occur in a shallow layer where the pressure is approximately 1 atm. The atmosphere and the near-surface hydrosphere are in chemical equilibrium.
3. The near-surface lithosphere, which is highly inhomogeneous. Chemical substances in this part of the environment are not in chemical equilibrium with each other and they are not in equilibrium with the hydrosphere and the atmosphere. For example, sulfur in the lithosphere is not in chemical equilibrium with the oxygen of the atmosphere and limestone is not in equilibrium with water in the hydrosphere.

The vast majority of the energy conversion processes occur within this thin layer of the biosphere and most of the energy conversion engines are affected by the states of the near-surface atmosphere and hydrosphere – because they admit air for combustion and, oftentimes water for cooling – but not of the lithosphere. The lithosphere supplies the fuels used in the energy conversion processes, but it very seldom influences the processes themselves, which primarily involve the local air and water sources. As long as energy conversion processes and energy conversion engines are considered, their environment is local and composed of (very) small parts of the atmosphere and the hydrosphere. This environment is chemically homogeneous and is at (almost) constant temperature and pressure. This part of the terrestrial environment is a very good approximation for the model environment of Section 2.2, which is composed of the four reservoirs of work, heat, volume, and mass.

2.14 An Operational Definition of Chemical Exergy

In all processes where fuels are used for energy conversion (e.g. coal power plants, gas turbines, jet engines) the definition of the fuel's exergy must satisfy the following conditions:

1. It must be uniquely and unambiguously defined.
2. It must model realistically the maximum work that may be produced by the fuel.
3. It must be calculated from first principles using thermodynamic properties that are well defined and easily obtainable from commonly available chemical tables.

The Gibbs' free energy, ΔG^o, clearly satisfies the first and third condition. It also reasonably satisfies the second condition because:

1. The recoverable work associated with the second terms of Eqs. (2.30) and (2.32) pertains to the use of selectively permeable membranes. While such membranes are currently used for the separation of light gases (e.g. hydrogen and helium), the

use of membranes for power production from the exhaust gases of combustion has not been demonstrated even in the controlled conditions of a laboratory.

2. For the commonly used fuels, as Table 2.2 shows, the second terms of Eqs. (2.30) and (2.32) represent only a small fraction (close to 2%) of the magnitude of ΔG^o. The magnitude of the Gibbs' free energy, ΔG^o, accounts for almost all the chemical exergy of commonly used fuels.
3. Experimental evidence with batteries and fuel cells indicates that ΔG^o is a very good approximation of the maximum voltage and the maximum work that may be recovered from a chemical compound.

The empirically justified stipulation that the chemical exergy of energy resources is equal to the Gibbs free energy of the combustion reaction was stipulated in Eq. (2.33): $W_{\max} = E_{chem}(T_0, P_0) = -\Delta G^o$.

This choice assigns a unique and unambiguous value to the fuel exergy, which is equal to the maximum work that is realistically recovered from a fuel and is readily available in textbooks and chemical tables. The choice eliminates the ambiguity, arbitrariness, and uncertainty in the calculations pertaining to chemical exergy and allows a realistic application of the exergy method in energy conversion systems and processes. For this reason, the property, $-\Delta G^o$, will be used for the calculation of the exergy and the maximum work associated with chemical species in the remainder of this book. An exception is reactions in biological systems (e.g. human cells) that do not occur at the atmospheric temperature, T_0, and pressure, P_0.

2.15 A Brief Historical Background

Since the beginning of the industrial revolution scientists and engineers strived to understand the operation of heat engines that utilize the various energy resources (primarily wood and coal at the time) and endeavored to improve their performance. The quantitative knowledge of the maximum mechanical work that may be performed by the utilization of a given energy resource – the upper limit of the energy conversion engines – is of paramount importance in all attempts to optimize the performance of engines. The work of Biot [39] who stated the importance of the high temperature of the vapor in the performance of steam engines may be considered as the earliest analytical work on the subject. Sadi Carnot's [40] treatise on the motive power of heat engines is also considered one of the pioneering works on the optimization of thermal engines. J. W. Gibbs is credited for drawing our attention to what is now called the Gibbs' free energy, $G = TS - PV + \Sigma(\mu_i N_i)$, and to the role of this property in energy conversion and the stability of materials [41, 42]. Gibbs coined the term *free energy* to represent the work that may be extracted from an energy resource (fuel). A few years before him, Tait expressed the desirability " ... to have a word to express the availability for work of the heat ... " and, thus, coined the term *availability,* which was widely used until the middle of the 20th century [43]. The French physicist Gouy expanded on the ideas of energy conversion at the end of the 19th century and postulated that the amount of work recovered from an energy resource is always less than the free energy [44], a postulate that also appears in the seminal book of Stodola on turbines [45]. This postulate, which introduces the concept of *lost work,* is sometimes referred to as the *Gouy–Stodola theorem.*[10] The concept of the maximum mechanical work and the quest to achieve it in thermal engines became part of the curriculum of thermodynamics in the 1930's with the term *availability* following the works of Darrieus [46] and Keenan [6] and was introduced in monographs and treatises of thermodynamics [4, 5, 7, 8, 10, 47].

The term *exergy (exergie* in German*)* was introduced by Rant [48] in the 1950s as a new word in the terminology of thermodynamics – roughly translated from Greek as "extracted work." This term was adopted by several authors in Europe. Among those, Baehr differentiated between energy and exergy, and also introduced the word *anergy (anergie)* for the work dissipated in irreversibilities, the lost work [49]. Evans introduced the term *essergy* (from the essence of energy) but this term is not currently in use [50].

Following the "energy crisis" of the 1970's and the worldwide realization that the human society must engage in conservation measures for the preservation of the global energy resources and a sustainable future, the concept of maximum work – derived from either the *availability* or the *exergy* concept – was widely adopted by the scientific community; it became part of the undergraduate curricula and was introduced in textbooks; and it became the tool in thousands of optimization studies for thermal and chemical systems. In the early part of the 21st century, the concept of maximum mechanical work is widely used globally under the term *exergy,* which has persevered in the scientific and engineering literatures. The numerous applications of the exergy concept in the optimization of energy conversion engines; the efforts for energy conservation; the conservation of natural resources; and environmental protection have made the *exergy method* one of the key engineering tools for the optimization of energy conversion systems and processes. An indication of the wide acceptance of the term exergy and the adoption of exergy analyses is the unanimous opinion paper issued by the Physical, Chemical and Mathematical Sciences Committee of *Science Europe* that exergy and the *exergy destruction footprint* (the lost work) be used exclusively in the policy making arena [51] with the slogan *"Forget Energy, Think Exergy."*

Problems

1. A covered solar pond with dimensions $20*50*3$ m^3 is filled with water. During the summer days the temperature of the water reaches 58°C and the environment is at 30°C. Determine the maximum work that may be produced by the water in this pond.

2. A geothermal well produces 62 kg/s of water and steam at 220°C with dryness fraction 15%. Determine the maximum power this geothermal resource may produce. The environment is at 30°C and 0.1 MPa.

---

[10] There is no indication that Gouy and Stodola collaborated to derive this theorem. The two worked independently and arrived to similar conclusions.

3. Calculate the Gibbs' free energy, ΔG^o, and the chemical exergy of methane and pentane according to Eqs. (2.30) and (2.32) and verify that $-\Delta G^o \approx E_{chem}$.

4. What is the maximum power that may be derived from 4 m² of solar cells at two locations: (a) the North Pole where the temperature is $-35°C$ and the irradiance 250 W/m²; and (b) at the Equator where the temperature is 40°C and the irradiance 1,050 W/m²?

5. A manufacturing facility requires 8.6 TJ of work approximately every 12 hours. Since the facility is in a tidal estuary with 5.2 m range, it is proposed that a tidal power plant be built to provide the needed energy. Calculate the minimum area of the basin of this tidal plant.

6. A coal power plant produces 250 MW of power and operates with a Rankine cycle with superheat. Steam enters the turbine at 10 MPa and 480°C, and exits at 40°C. Cooling water enters the condenser at 25°C and exits at 35°C. The fuel is coal, for which carbon is the only combustible material. The isentropic efficiencies of the pump and the turbine-generator system are 80% and 82% respectively. Determine the lost power (rate of exergy destruction) in each of the components of this power plant. The environment is at 25°C and 0.1 MPa.

7. A nuclear power plant produces 2,500 MW of power with thermal (first law) efficiency 31.8%. Determine the rate of heat input to this power plant and its exergetic efficiency, based on the six suggested temperatures of Section 2.9. The environment is at 25°C and 0.1 MPa.

8. Determine the specific exergy of steam at 400°C and 6 MPa, with three different environmental temperatures: 0°C, 25°C, and 40°C. $P_0 = 0.1$ MPa. What do you observe?

9. A turbine receives steam at 560°C, 10 MPa and exhausts at 45°C with $x = 95\%$. Determine the exergetic efficiency and the second law efficiency of this turbine. The environment is at 25°C and 0.1 MPa.

10. A gas turbine receives air at 2,000 K, 5 MPa and exhausts at 860 K, 0.1 MPa. What are the exergetic and second law efficiencies of this turbine? The environment is at 27°C and 0.1 MPa.

11. The gears of a wind turbine operate steadily at $T = 65°C$. The turbine produces 1,500 kW mechanical work and the gears transmit 1,430 kW to the generator. The environment is at 27°C. Determine the rate of exergy destruction in the gear box.

12. Prove that it is impossible to develop an engine with exergetic efficiency higher than 100%.

References

[1] J. Kestin, Availability – The Concept and Associated Terminology. *Energy*, **5** (1980), 679–92.

[2] E. E. Michaelides, Exergy and the Conversion of Energy. *International Journal of Mechanical Engineering Education*, **12** (1984), 65–9.

[3] E. E. Michaelides, The Concept of Available Work as Applied to the Conservation of Fuel Resources, *Proc. 14th Intersociety Energy Conversion Engineering Conference*, paper 799376, ACS, 1162–6, Boston, August 1979.

[4] J. H. Keenan, *Thermodynamics*, reprint 1970 (New York: Willey, 1941).
[5] J. Kestin, *A Course in Thermodynamics*, vol. 1 (Waltham, MA: Blaisdell, 1966).
[6] J. H. Keenan, A Steam Chart for Second Law Analysis. A Study of Thermodynamics Availability in a Steam Power Plant. *Mechanical Engineering*, **54** (1932), 195–204.
[7] M. Modell and R. C. Reid, *Thermodynamics and Its Applications*, 2nd ed. (Englewood Cliffs, NJ: Prentice Hall, 1983).
[8] M. J. Moran, *Availability Analysis* (New York: ASME Press, 1989).
[9] I. Dincer and M. A. Rosen, *Exergy – Energy, Environment and Sustainable Development* (Amsterdam: Elsevier, 2007).
[10] A. Bejan, *Advanced Engineering Thermodynamics* (Hoboken, NJ: Wiley, 2006).
[11] S. Patel, Statkraft Shelves Osmotic Power Project, *Power Magazine,* March 1, 2014.
[12] R. Petela, Exergy of Heat Radiation. *Journal of Heat Transfer*, **86** (1964),187–92.
[13] REN21, *Renewables 2016 Global Status Report* (Paris: REN21 Secretariat, 2016).
[14] S. J. Jeter, Maximum Conversion Efficiency for the Utilization of Direct Solar Radiation. *Journal of Solar Energy Engineering*, **26** (1981), 231–6.
[15] A. Bejan, Unification of Three Different Theories Concerning the Ideal Conversion of Enclosed Radiation. *Journal of Solar Energy Engineering*, **109** (1987), 46–51.
[16] E. E. Michaelides, *Energy, the Environment, and Sustainability* (Boca Raton, FL: CRC Press, 2018).
[17] E. E. Michaelides, *Alternative Energy Sources* (New York: Springer, 2012).
[18] M. M. El-Wakil, *Power Plant Technology* (New York: McGraw Hill, 1984).
[19] A. Betz. Schraubenpropeller mit Geringstem Energieverlust, (mit einem Zusatz von L. Prandtl) *Nachrichten von der Gesellschaft der Wissenschaften zu Göttingen, Mathematisch-Physikalische Klasse,* (1919) 193–217.
[20] D. J. Bennet, *The Elements of Nuclear Power* (London: Longmans, 1972).
[21] L. Onsager, Reciprocal Relations in Irreversible Processes. *I. Physical Review Journal*, **37** (1931), 405–26.
[22] L. Onsager, Reciprocal Relations in Irreversible Processes. *II. Physical Review Journal*, **37** (1931), 2265–79.
[23] H. B. G. Kasimir, On Onsager's Principle of Microscopic Reversibility. *Reviews of Modern Physics*, **17**(1945), 343–50.
[24] J. Kestin, *A Course in Thermodynamics*, vol. 2, (Washington DC: Hemisphere, 1979).
[25] D. G. Miller, Thermodynamics of Irreversible Processes: The Experimental Verification of the Onsager Reciprocal Relations. *Chemical Reviews*, **60** (1960), 15–37.
[26] Y. Demirel and S. I. Sandler, Nonequilibrium Thermodynamics in Engineering and Science. *Journal of Chemical Physics*, **33** (1960), 28–31.
[27] H. Schlichting, *Boundary Layer Theory*, 8th ed. (New York: McGraw-Hill, 1978).
[28] R. B. Bird, W. E. Stewart, and E. N. Lightfoot, *Transport Phenomena* (New York: Wiley, 1960).
[29] E. E. Michaelides, *Particles, Bubbles, and Drops – Their Motion, Heat and Mass Transfer* (New Jersey: World Scientific, 2006).
[30] N. Lior and N. Zhang, Energy, Exergy, and Second Law Performance Criteria. *Energy*, **32** (2007), 281–96.
[31] S. L. Sarmiento and S. C. Wofsy, *A US Carbon Cycle Plan* (Washington DC: US Global Change Research Program, 1999).
[32] K. Ochsner, *Geothermal Heat Pumps – A Guide for Planning and Installing* (London: Earthscan, 2008).
[33] S. Wilcox, *National Solar Radiation Database 1991–2010 Update: User's Manual*, Technical Report NREL/TP-5500-54824 (August 2012).

[34] J. Ahrendts, Reference States. *Energy-The International Journal*, **5** (1980), 667–77.
[35] J. Szargut, D. R. Morris, and F. Steward, *Exergy Analysis of Thermal, Chemical, and Metallurgical Processes* (New York: Hemisphere, 1988).
[36] J. Szargut, *Exergy Method – Technical and Ecological Applications* (Boston, MA: MIT Press, 2005).
[37] W. Van Gool, Thermodynamics and Chemical References for Exergy Analysis, Florence World Energy Research Symposium, Florence, Italy, SGE Editoriali (1997), 949–57.
[38] J. R. Munoz and E. E. Michaelides, The Impact of the Model of the Environment in Exergy Analyses. *Journal of Energy Resources Technology*, **121** (1999), 268–76.
[39] J. B. Biot, *Traité de physique expérimentale et mathématique*, 4th vol. (Paris: Detervill, 1816).
[40] S. Carnot, *Reflexions sur la Puissance Motrice du Feu, et sur les Machines Propres à developer ceite Puissance* (Paris: Bachelier, 1824).
[41] J. W. Gibbs, A Method of Geometrical Representation of the Thermodynamic Properties of Substances by Means of Surfaces. *Transactions of the Connecticut Academy II*, (1873), 382–404.
[42] J. W. Gibbs, *The Collected Works of J. W. Gibbs* (New York: Green and Co., 1928).
[43] P. G. Tait, *Sketch of Thermodynamics* (Edinburgh: Edinburgh University Press, 1868).
[44] M. Gouy, Sur l' Energie Utilisable. *Journal de Physique Archives*, **8** (1889), 501–18.
[45] A. Stodola, *Steam and Gas Turbines* (transl. by L.K. Lowenstein), vol. II (New York: McGraw-Hill, 1927).
[46] G. Darrieus, The Rational Definition of Steam Turbine Efficiencies. *Engineering*, **130** (1930), 283–85.
[47] G. N. Hatsopoulos and J. H. Keenan, *Principles of General Thermodynamics* (Malabar, FL: Krieger, 1965).
[48] Z. Rant, Exergie, Ein Neues Wort für "Technische Arbeitsfähigkeit." *Forschung im Ingenieurwesen*, **22** (1956), 36–42.
[49] H. D. Baehr, Definition und Berechnung von Exergie und Anergie. *Brennstoff-Wärme-Kraft*, **17** (1965), 1–6.
[50] R. B. Evans, A Proof that Essergy Is the Only Consistent Measure of Potential Work (for Chemical Systems), Ph. D. Thesis, Dartmouth College, Hanover, NH (1969). (University Microfilms 70–188, Ann Arbor, MI).
[51] Physical, Chemical and Mathematical Sciences Committee, A Common Scale for Our Common Future: Exergy, a Thermodynamic Metric for Energy, Physical, Chemical and Mathematical Sciences Committee Opinion Paper, *Science Europe*, Brussels (Sept. 2015).
[52] R. Zevenhoven, J. Fagerlund, and J. K. Songok, CO_2 Mineral Sequestration: Developments toward Large-Scale Application. *GHG: Science and Technology*, **1**, (2011), 48–57.

3 Energy Conversion Systems and Processes

Summary

The application of the exergy concept in energy conversion systems and processes reveals the components where high exergy dissipation occurs and improvements may be accomplished to conserve energy resources. The exergy method is applied to several energy conversion systems: heat exchangers, including boilers and condensers; vapor power cycles; gas power cycles, including cogeneration units; jet engines; and geothermal units. Calculations on exergy dissipation identify the processes and components where improvements will save energy resources. All the calculations reveal that combustion processes waste a great deal of exergy, leading to the conclusion that Direct Energy Conversion (DEC) devices, such as fuel cells, have significantly higher exergetic efficiencies and utilize fossil fuels in a more sustainable way. The exergy method is also applied to the photovoltaic (PV) cells and thermal solar systems that generate power as well as the solar collectors that deliver heat. Wind turbines are also analyzed using the exergy method: It is concluded that significant exergy destruction occurs because of Betz's limit and the wind turbine characteristics. A large number of examples in this chapter elucidate all the exergy calculations and provide guidance and resources for the application of the exergy methodology.

3.1 Heat Exchangers

A high fraction of the global energy conversion processes occurs in systems that fall under the generic term *heat exchangers* (HXs), systems where thermal energy is transferred from one fluid to another. Common types of HXs in the power production industry are *boilers, superheaters, reheaters, feed-water heaters,* and *condensers.* In vehicles the *radiators* are HXs that transfer heat from the cooling fluid of the engine to the air. A number of HXs in the *atmospheric distillation units* of refineries facilitate the transfer of the heat that evaporates the components of crude oil and then separates the fractions of these hydrocarbons in several condensers.

Heat exchangers are also ubiquitous in residences and office buildings as parts of heating, ventilation, and air-conditioning (HVAC) systems. The *radiators* that provide thermal comfort in buildings are essentially HXs that transfer the thermal energy of hot water or oil, which flows inside the radiator, to the air in the interior of the buildings. The *burners* convert the chemical energy of a fuel to thermal energy for the hot water needs of the residents and for the heating of buildings. All the air-conditioning (AC) systems include two HXs: one for the condensation of the refrigerant (process 2-3 in Figure 1.12) and the second for the evaporation of the refrigerant (process 4-1 in Figure 1.12). The latter process provides the "cooling effect" of the AC system.

Heat exchangers are open thermodynamic systems, where fluids – and sometimes fossil fuels – flow in and out. The most common types of HXs are described in this section and the exergy inputs and outputs are calculated. Immediate conclusions are derived for the minimization of exergy destruction and the more effective operation of these thermal systems.

3.1.1 Counter Flow and Cocurrent Flow Heat Exchangers

These HXs are used in most industrial applications where the objective is the transfer of heat from a hotter fluid (the primary fluid) to a colder fluid (the secondary fluid). The schematic diagrams of the operations of these HXs are shown in Figure 3.1, where the arrows denote the direction of heat flow. The temperatures of the two fluids are shown in the abscissae of the two diagrams; the ordinate, L, is a dimension along the heat exchange path. The subscripts p and s denote primary and secondary fluids; and the subscripts *in* and *out* denote the inlets and the outlets of the two fluids. For the spontaneous transfer of heat, at every point, L, along the heat exchange path, the temperature of the primary fluid must be higher than that of the secondary fluid.

It must be noted that, because the heat transfer path from the entrance to the exit of the heat exchanger is complex, in most heat exchangers the length, L, is not a physical distance, but a fictitious one that helps visualize the temperature differences in the HXs. In very simple and linear heat exchangers – such as concentric tubes – L is the actual distance from the entrance of the primary fluid.

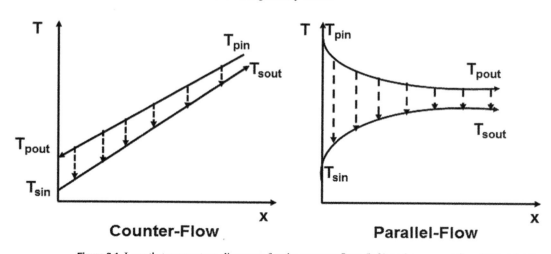

Figure 3.1 Length-temperature diagrams for the counter flow (left) and cocurrent flow HXs (right).

In the counter flow type, the primary and secondary fluids enter at opposite ends of the HX and leave at the other opposite ends. As a result of the configuration, it is possible (but not necessary) for the outlet temperature of the secondary fluid to be higher than that of the primary fluid, $T_{sout} > T_{pout}$, but not higher than the inlet temperature of the primary fluid, $T_{pin} > T_{sout}$. In the cocurrent flow HXs, the primary and secondary fluids enter and exit at the same sides. As a result, the temperature of the secondary fluid is lower than that of the primary fluid at all points within the HX as well as at the exit, $T_s < T_p$.

Both types of HXs are typically considered adiabatic systems because of two reasons:

1. Their exterior surface is well-insulated.
2. The low rate of heat lost to the surroundings is typically very low in comparison to the rates of enthalpy exchanged between the two fluids.

Therefore, the first law of thermodynamics for the HXs yields:

$$\dot{m}_p(h_{pin} - h_{pout}) + \dot{m}_s(h_{sin} - h_{sout}) = 0, \qquad (3.1)$$

The absolute value of each term in Eq. (3.1) represents the rate of heat exchanged between the two fluids. If the specific heat (at constant pressure) of the fluids is constant, the enthalpy differences may be given in terms of the temperature differences:

$$\dot{m}_p c_p(T_{pin} - T_{pout}) + \dot{m}_s c_s(T_{sin} - T_{sout}) \Rightarrow \frac{\dot{m}_s}{\dot{m}_p} = \frac{c_p(T_{pin} - T_{pout})}{c_s(T_{sout} - T_{sin})}. \qquad (3.2)$$

This expression is commonly used for the determination of the ratio of the flow rates or of one of the outlet temperatures.

The rate of lost work (rate of exergy destruction) is equal to the difference of the rates of exergies entering and leaving the HX streams. In the vast majority of HXs the exergy destruction due to the pressure loss is low and the rate of lost work is approximated by the temperature changes. Given that the temperature of the primary fluid is reduced from T_{pin} to T_{pout} and the temperature of the secondary fluid increases from T_{sin} to T_{sout}, the lost power – rate of exergy destruction – for both types of HXs is:

$$\dot{W}_{lost} \approx \dot{m}_s c_s \left[(T_{sin} - T_{sout}) - T_0 \ln \frac{T_{sin}}{T_{sout}} \right] + \dot{m}_p c_p \left[(T_{pin} - T_{pout}) - T_0 \ln \frac{T_{pin}}{T_{pout}} \right]. \qquad (3.3)$$

One term in the square brackets of Eq. (3.3) is positive and the other is negative. Their sum, the rate of lost work, is always greater than zero.

A figure of merit for HXs is the *effectiveness,* defined as the ratio of the actual rate of heat transfer, $\dot{Q}_{act}$, to the maximum possible rate of heat transfer, $\dot{Q}_{max}$:

$$\varepsilon = \frac{\dot{Q}_{act}}{\dot{Q}_{max}}. \qquad (3.4)$$

The maximum possible heat transfer is achieved when the fluid with the lower rate of specific heat capacity ($\dot{m}c$) attains the maximum possible temperature difference ($T_{pin} - T_{sin}$) [1] and this is achieved when the area of the HX is very large, actually infinite. When the rate of specific heat capacity of the primary fluid $(\dot{m}c)_p$ is lower, the effectiveness is defined as:

$$\varepsilon = \frac{\dot{Q}_{act}}{\dot{Q}_{max}} = \frac{\dot{m}_p c_p (T_{pin} - T_{pout})}{\dot{m}_p c_p (T_{pin} - T_{sin})} = \frac{T_{pin} - T_{pout}}{T_{pin} - T_{sin}}, \qquad (3.5)$$

When the rate of specific heat capacity of the secondary fluid $(\dot{m}c)_s$ is lower, the definition of the effectiveness yields:

$$\varepsilon = \frac{\dot{Q}_{act}}{\dot{Q}_{max}} = \frac{\dot{m}_s c_s (T_{sout} - T_{sin})}{\dot{m}_s c_s (T_{pin} - T_{sin})} = \frac{T_{sin} - T_{sout}}{T_{pin} - T_{sin}}. \qquad (3.6)$$

Counter current HXs typically have higher effectiveness and lower rates of exergy destruction.

Example 3.1: A counter flow HX is designed to reduce the temperature of alcohol after distillation from 78°C to 35°C. Cooling water enters the HX at 18°C. The mass flow rates of alcohol and water are 20 kg/s and 8 kg/s, and their specific heat capacities are 2.41 kJ/kg K and 4.18 kJ/kg K respectively. Determine the exit temperature of the water, the effectiveness of the HX, and the rate of exergy destruction during this process. The HX is adiabatic and the environmental temperature is 18°C (291 K).

Solution: The energy balance of the HX process, from Eq. (3.1) is written as:

$$\dot{m}_w c_w (T_{wout} - T_{win}) = \dot{m}_a c_a (T_{ain} - T_{aout}),$$

where the subscripts w and a indicate water and alcohol, respectively. This equation yields $T_{wout} = 62°C$ (335 K).

In this case water is the fluid with the lower rate of specific heat capacity $(\dot{m}c)$, which may attain the maximum possible temperature difference $(T_{ain} - T_{win})$ of 60°C. Hence the effectiveness of the counter flow HX is $\varepsilon = (62 - 18)/(78 - 18) = 73.3\%$.

The rate of exergy destruction (lost work) is obtained from Eq. (3.3): $8 * 4.18 * [(291 - 335) - 291 * \ln(291/335)] + 20 * 2.41 * [(351 - 308) - 291 * \ln(351/308)]$, or $\dot{W}_{lost} = 138.4$ kW.

Example 3.2: The cooling of alcohol, from 78°C to 35°C, in Example 3.1 accomplished in a cocurrent flow HX. The cooling water enters at 18°C (291 K) and exits at 32°C (305 K).[1] Determine the mass flow rate of the water, the effectiveness of the HX and the rate of exergy destruction. The HX is adiabatic and the environmental temperature is 18°C (291 K).

Solution: From the expression of the first law of thermodynamics, Eq. (3.1), the mass flow rate of water is 35.4 kg/s.

---

[1] In cocurrent flow heat exchangers, the exit temperature of the secondary fluid is less than that of the primary fluid as indicated in Figure 3.1b.

In this case, alcohol is the fluid with the lower rate of specific heat capacity ($\dot{m}c$) that may attain the maximum possible temperature difference ($T_{ain} - T_{win}$) of 60°C. Hence the effectiveness of the cocurrent flow HX is $\varepsilon = (78 - 35)/(78 - 18) = 71.7\%$.

The rate of exergy destruction in this HX is: $35.4 * 4.18 * [(291 - 305) - 291 * \ln(291/305)] + 20 * 2.41 * [(351 - 308) - 291 * \ln(351/308)]$ or $\dot{W}_{lost} = 191.3$ kW.

It is observed that for the same end result (the cooling of alcohol from 78°C to 35°C) the counter flow HX exhibits lower exergy destruction than the cocurrent flow HX. This is also reflected in the higher effectiveness of the counter flow HX.

It must be noted that an implicit assumption of both examples, as well as of Eq. (3.3), is that the exergy of the secondary fluid is utilized downstream. If the cooling water is dumped in the environment, as it happens with most cooling systems, additional exergy destruction follows when the hot water attains equilibrium with the environment at the temperature of 291 K.

3.1.2 Condensers

Condensers are essential equipment in steam/vapor power plants, heat pumps, and air-conditioning systems. On the one side of these HXs, the primary fluid is condensed and, in most cases, slightly subcooled. On the side of the secondary fluid the coolant – typically water or air – is heated up. A schematic diagram of the operation of a condenser is shown in Figure 3.2 with the inlet temperature of the condensing vapor, T_{pin}, equal to the saturation temperature, T_{sat}, that is, $T_{pin} = T_{sat}$. If the condensation of the primary fluid occurs isothermally, without subcooling, then $T_{pin} = T_{pout} = T_{sat}$. When subcooling occurs, the temperature difference $T_{sat} - T_{pout}$ is defined as the *subcooling range*. On the primary side, the specific exergy change during the isothermal condensation process is:

$$e_g - e_f = h_g - h_f - T_0(s_g - s_f) = h_g - h_f - T_0 \frac{(h_g - h_f)}{T_{sat}} = (h_g - h_f)\left(1 - \frac{T_0}{T_{sat}}\right). \quad (3.7)$$

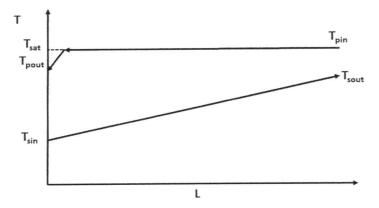

Figure 3.2 Length-temperature (L,T) diagram for a condenser.

It is apparent that, when condensation occurs at the environmental temperature T_0, there is zero exergy destruction. If the specific heats of the primary and secondary fluids are constant – a reasonable assumption because the subcooling range is very small – the total rate of exergy destruction in the condenser is:

$$\dot{W}_{lost} = \dot{m}_p \left[(h_g - h_f)\left(1 - \frac{T_0}{T_{sat}}\right) + c_p(T_{sat} - T_{pout}) - c_p T_0 \ln \frac{T_{sat}}{T_{pout}} \right] \\ - \dot{m}_s \left[c_s(T_{sout} - T_{sin}) - c_s T_0 \ln \frac{T_{sout}}{T_{sin}} \right]. \quad (3.8)$$

The first term in the last equation represents the exergy loss of the primary fluid that condenses and subcools and the second term is the exergy gain of the secondary fluid, the coolant.

Example 3.3: The condenser of a large power plant handles 340 kg/s of steam at pressure 6 kPa and 92% quality. The heat is removed by cooling water from a lake at 19°C and is returned back to the lake at 30°C. There is 1°C subcooling of the condensate, which is returned to the power cycle. Determine the mass flow rate of the cooling water, the exergy destruction in the condenser and the exergy destruction when the warmer cooling water is returned to the lake. The environmental temperature is that of the lake water, $T_0 = 19°C$ (292 K).

Solution: The temperature of the steam is the saturation temperature at 6 kPa, $T_{pin} = 36°C = 309$ K. The rate of heat removal for the isothermal condensation of the steam is: $0.92 * (2,567 - 152) * 340 = 755,412$ kW (only 92% of primary mass flow rate, the vapor part, condenses). Another $4.18 * 1 * 340 = 1,421$ kW of heat is removed for the subcooling of the condensate, for a total of 756,833 kW. The mass flow rate of the cooling water, which undergoes an 11°C temperature rise, is: $756,833/(11 * 4.18) = 16,460$ kg/s.

One notes that the mass flow rate of the cooling water is about 50 times more than that of the steam. This happens because the latent heat of steam is much higher than the sensible enthalpy of the cooling water ($h_{fg} \gg c\Delta T$) for the typical of cooling equipment.

The exergy destruction in the condenser is obtained from Eq. (3.8):
$755,412 * (1 - 292/309) + 340 * 4.18 * [309 - 308 - 292 * \ln(309/308)] - 16,460 * 4.18 * [303 - 292 - 292 * \ln(303/292)] = 27,729$ kW.

The exergy of the warm water as it exits the condenser is $16,460 * 4.18 * [303 - 292 - 292 * \ln(303/292)] = 13,907$ kW. All this exergy is destroyed in the environment, when the cooling water mixes with the lake water and cools to 19°C (292 K).

3.1.3 Boilers, Burners, Superheaters, Reheaters, Reactors

The function of this type of HXs is to increase the temperature and exergy of the working fluid and produce the high temperature fluid in power cycles. A fuel is typically consumed in this category of HXs: nuclear reactors consume nuclear fuel and transfer the generated heat to the reactor coolant; and fossil fuel equipment consumes coal, natural gas, or liquid hydrocarbons. The components and thermodynamic diagrams of the basic steam and air power cycles are depicted in Figures 1.7–1.10. In both types of cycles the heat addition occurs in the process denoted as 2-3. These heat addition processes determine the mass flow rate of the fuel needed for the process:

$$\dot{Q}_{2-3} = \dot{m}_p(h_3 - h_2) = \dot{m}_f \Delta H^0 \Rightarrow \dot{m}_f = \frac{\dot{m}_p(h_3 - h_2)}{\Delta H^0}, \tag{3.9}$$

where ΔH^0 is the heat of combustion of the fuel. Because on the fuel combustion side the burners, boilers, reheaters, and superheaters operate at atmospheric pressure, it is the heat of combustion at 1 atm that is typically used for the determination of the needed fuel.

Effectively, Eq. (3.9) yields the minimum amount of fuel needed in the cycle because there are energy losses in the equipment due to the following:

1. Incomplete combustion (this is minimized using excess air).
2. Flue gases and excess air exit at temperatures higher than the ambient.
3. Heat loss to the surroundings.

The mass flow rate of the fuel needed in actual heat addition equipment is usually 5–12% higher than what Eq. (3.9) indicates and depends on the efficiency of the combustion process.

The operation of burners, boilers, reheaters, and superheaters entails a high rate of exergy destruction, primarily caused during the following processes:

1. The combustion process, where the chemical exergy of the fuel is converted to the thermal energy of the combustion products. All chemical to thermal energy conversion processes entail high rates of exergy destruction.
2. The transfer of heat/enthalpy from the combustion products to the working fluid of the cycle. In the case of steam cycles this is the heat transferred to the water/steam of the cycles. In the case of gas burners this part is zero because the combustion products are directly fed to the gas turbine.
3. The combustion products (flue gases or exhaust gases) are rejected in the atmosphere at temperatures higher than the ambient temperature T_0.

During the heat addition to the working fluid the chemical exergy of the fuel is lost, while the exergy of the working fluid (the secondary fluid) increases. The total rate of power lost (rate of exergy destruction) during the process 2-3 in the combustion and heating equipment becomes:

$$\dot{W}_{lost} = \dot{m}_f \Delta G^0 + \dot{m}_s[(h_3 - h_2) - T_0(s_3 - s_2)]. \tag{3.10}$$

It must be noted that, when the mass flow rate of the fuel is measured in terms of mass units (e.g. kg/s) the corresponding Gibbs free energy must be in kJ/kg (not kJ/kmol as it is usually given in books and tables of properties). Table 3.1 shows the rate of heat supplied to the cycle; the flow rate of steam in the cycle; the rate of fuel consumed for the production of heat; the lost power in the boilers of typical steam power plant cycles

Table 3.1 Fuel consumption and lost power during the steam generation process in power plants. Property data from [5].

Power (MW)	Rate of heat (MW)	Rate of steam, $\dot{m}_s$ (kg/s)	Rate of fuel, $\dot{m}_f$ (kg/s)	Power lost, $\dot{W}_{lost}$ (MW)	Exergetic efficiency (η_{II})
Coal/carbon					
200	526	162.82	16.10	−284.02	0.46
400	1,053	325.64	32.19	−568.05	0.46
600	1,579	488.46	48.29	−852.07	0.46
Methane					
200	526	162.82	10.52	−295.36	0.46
400	1,053	325.64	21.04	−590.71	0.46
600	1,579	488.46	31.57	−886.07	0.46

with overall (first law) thermal efficiency 38%, when the fuel is coal and methane; and the exergetic efficiency of the boiler. The steam outlet temperature in all the cycles is 520°C; the pressure is 15 MPa; the ambient temperature is $T_0 = 27°C = 300$ K; and the efficiency of the turbo-generator is 0.82 (82%).

Table 3.1 shows that 54% of the exergy of the fuel is destroyed during the combustion and heat transfer processes. Given that the thermal efficiency of the cycle is 38% and that for the fuels coal and methane $\Delta H^0 \approx \Delta G^0$, the exergetic efficiency of the steam cycle is approximately 38%. It becomes apparent that most of the exergy loss in the cycle occurs in the boiler. Sections 3.2 and 3.3 include more details and calculations on the exergy destruction during the combustion processes.

3.1.4 Pinch-Point Temperature Difference

The second law of thermodynamics dictates that, for heat to be transferred from the primary to the secondary fluid, at all points of the HX surface the temperature of the primary fluid must be higher than the temperature of the secondary fluid. This presents a design problem with vapor generators (boilers) because of the sharp corners in the boiling curve. Figure 3.3 depicts the vapor generation process with a compressed liquid, the secondary fluid. Heat is supplied from gaseous combustion products, the primary fluid. Because of the sharp corner in the boiling curve, the minimum temperature difference between the two fluids occurs at the point in the HX where boiling starts. This is referred to as the *pinch point,* pp in Figure 3.3. Since $T_p > T_s$ at all points of the HX, the temperature difference at the pinch point, ΔT_{pp}, must always be positive. As it becomes apparent from an inspection of Figure 3.3, the positive temperature difference at the pinch point affects the temperature difference in all the other parts of the HX.

Because the effluents on the primary side are typically discarded in the environment as flue gases, for the maximization of heat transfer from the primary fluid and the minimization of the effluent's exergy, the temperature T_{pout} should be as close as possible to T_{sin}. Figure 3.3 shows that this is difficult to achieve when $\Delta T_{pp} > 0$ and that T_{pout}, as determined by the magnitude of ΔT_{pp}, can be significantly higher than T_{sin}. This implies that a great deal of exergy is lost with the effluents. The constraint $\Delta T_{pp} > 0$ and the

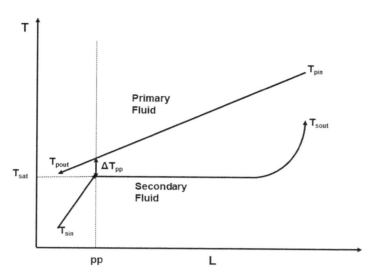

Figure 3.3 Vapor generation process in a boiler and the pinch point.

shape of the boiling curve indicate that two energy balances must be satisfied for the two HX parts separated by the vertical dashed line of Figure 3.3. The first is the energy balance before the pinch point of the adiabatic HX, to the right of Figure 3.3:

$$\dot{m}_p \left[h_{pin} - h_p(T_{sat} + \Delta T_{pp}) \right] = \dot{m}_s \left[h_s(T_{sout}) - h_{sf}(T_{sat}) \right], \qquad (3.11)$$

where h_{sf} denotes the enthalpy of the secondary fluid at the pinch point (liquid at the saturation pressure and temperature). The second equation is the energy balance after the pinch point, to the left of Figure 3.3:

$$\dot{m}_p \left[h_p(T_{sat} + \Delta T_{pp}) - h_p(T_{pout}) \right] = \dot{m}_s \left[h_{sf}(T_{sat}) - h_s(T_{sin}) \right]. \qquad (3.12)$$

In Eqs. (3.11) and (3.12) the parentheses denote the arguments of the enthalpy function and the square brackets multiplication. T_{sat} is the boiling temperature; and $h_{sf}(T_{sat})$ denotes the saturated liquid enthalpy of the secondary fluid. The addition of Eqs. (3.11) and (3.12) yields the following expression for the operation of the entire HX:

$$\dot{m}_p \left[h_p(T_{pin}) - h_p(T_{pout}) \right] = \dot{m}_s \left[h_s(T_{sout}) - h_s(T_{sin}) \right]. \qquad (3.13)$$

Eqs. (3.11), (3.12), and (3.13) are not independent and only two of them may be used to solve for two unknown variables. In typical calculations the first equation is used for the determination of the mass flow rates ratio and the last for the calculation of T_{sout}. The rate of exergy destruction (lost power, $\dot{W}_{lost}$) in the operation of a HX with a pinch point is:

$$\begin{aligned}\dot{W}_{lost} &= \dot{m}_p(e_{pin} - e_{pout}) + \dot{m}_s(e_{sin} - e_{sout}) = \\ &\dot{m}_p \left[h_{pin} - h_{pout} - T_0(s_{pin} - s_{pout}) \right] + \dot{m}_s \left[h_{sin} - h_{sout} - T_0(s_{sin} - s_{sout}) \right]\end{aligned} \qquad (3.14)$$

The first term in the last part of Eq. (3.14) represents the rate of exergy loss by the primary fluid and the second term is the rate of exergy gain by the secondary fluid. The sum of the two terms is always greater than zero.

Example 3.4: The exhaust gases of a gas turbine are used for steam generation. The mass flow rate of the gases is 168 kg/s and their exhaust temperature is 780 K. Liquid water at 37°C and 10 bar enters a countercurrent steam generator and exits as superheated steam at 320°C. Determine the rate of exergy destruction in this HX and the outlet temperature of the gases, when ΔT_{pp} is equal to 5°C, 10°C, and 30°C. $T_0 = 300$ K.

Solution: The HX is adiabatic and the exhaust gases are modeled as air. Since the pressure on the steam side is 10 bar, the boiling temperature is $T_B = T_{sat} = 180°C = 453$ K. The saturated liquid enthalpy at this temperature is $h_{sf}(T_{sat}) = 763$ kJ/kg; the enthalpy of the steam at 320°C is $h_s(T_{sout}) = 3,094$ kJ/kg; and the enthalpy of the pressurized liquid water entering the HX is $h_s(T_{sin}) = 155$ kJ/kg. The enthalpy of air at 780 K is $h_p(T_{pin}) = 926$ kJ/kg.

1. For pinch point temperature difference $\Delta T_{pp} = 5°C$, the temperature of the primary fluid (air) at the pinch point is 458 K and its enthalpy is $h_p(T_B + \Delta T_{pp}) = 586$ kJ/kg. From Eq. (3.11) the mass flow rate of the steam is: $168 * (926 - 586)/(3,094 - 763) = 24.5$ kg/s. The outlet enthalpy of the primary fluid is calculated from Eq. (3.13) as: $h_p(T_{pout}) = 369$ kJ/kg and this corresponds to $T_{pout} = 369$ K.

   The rate of exergy destruction with $\Delta T_{pp} = 5°C$ is: $32,901 - 23,024 = 9,877$ kW.

2. When $\Delta T_{pp} = 10°C$, the temperature of the primary fluid at the pinch point is 190°C (463 K) and the enthalpy $h_p(T_B + \Delta T_{pp}) = 591$ kJ/kg. The thermodynamic properties in the side of the secondary fluid remain the same. The mass flow rate of the secondary fluid is 24.1 kg/s. The outlet temperature of the exhaust gases is $T_{pout} = 377$ K and the rate of exergy destruction is: $32,668 - 22,685 = 9,983$ kW.

3. When $\Delta T_{pp} = 30°C$, the temperature of the exhaust gases at the pinch point is 210°C (483 K) and the enthalpy $h_p(T_B + \Delta T_{pp}) = 612$ kJ/kg. The mass flow rate of the secondary fluid is 22.63 kg/s. The outlet temperature of the exhaust gases is $T_{pout} = 403$ K and the rate of exergy destruction is: $31,661 - 21,265 = 10,396$ kW.

The following table summarizes the main results of this example:

	$\dot{m}_s$ (kg/s)	T_{pout} (K)	$\dot{W}_{lost}$ (kW)
$\Delta T_{pp} = 5°C$	24.50	369	9,877
$\Delta T_{pp} = 10°C$	24.14	377	9,983
$\Delta T_{pp} = 30°C$	22.63	403	10,396

It follows that, for the minimization of exergy losses ΔT_{pp} should be as low as possible. Lower ΔT_{pp} is achieved with higher heat exchange area, which is typically associated with higher investment cost for the HX. In the design of HXs with pinch point the engineers optimize the cost of exergy losses over the useful life of the HX with the added investment cost.

3.1.5 Feed-Water Heaters and Reheaters

Feed-water heaters (FWHs) are used in steam power plants to preheat liquid water before the evaporator section of the boiler. The preheater part of the water cycle supplies heat to water at the lower temperatures of the cycle. Instead of using the high-temperature heat from fuel combustion, the water is preheated using a fraction of the steam, which is extracted from the turbine. While the operation of the FWHs is irreversible and exergy is destroyed in their operation, the overall savings of high-exergy fuel generate exergy savings for the entire cycle. This is manifested in higher first-law and exergetic efficiencies for the cycle.

Let us consider the operation of an open FWH, which is part of a steam cycle and receives water from the pump at 40°C and 1.5 MPa. With $T_0 = 27°C$, the specific exergy of the pumped water is 2.6 kJ/kg. For the preheating part of the cycle, the liquid water temperature must increase to approximately 198°C, a process that entails enthalpy addition of 676.3 kJ/kg of water to be heated.[2] The specific exergy of the (almost saturated) water at 198°C is 100 kJ/kg. There are two ways to preheat the water:[3]

1. Using fuel (e.g. coal) for combustion.
2. Diverting a fraction of steam from the turbine and mixing this steam with the water.

In the first process the mass of carbon to be burned is: 676.3/32,700 = 0.0207 kg of carbon per kg of preheated water. This amount of fuel has exergy approximately equal to 676.3 kJ. Therefore, the exergy destruction in this case is: 676.3 − 1 ∗ (100 − 2.6) = 578.9 kJ/kg of preheated water.

In the second process one may extract from the turbine a fraction of the steam at 1.5 MPa and mix the steam with the low-temperature pressurized liquid water. The latent enthalpy of steam at this pressure is 1,946 kJ/kg. Assuming that the steam from the turbine is saturated (typical extraction pressures are very close to the vapor saturation line) in order to heat up the water from 40°C to 198°C, we need to extract from the turbine 676.3/1,946 = 0.347 kg of steam per kg of pressurized water. The extracted (bled) steam is mixed with the colder water from the pump and increases its temperature. The specific exergy of saturated steam at 1.5 MPa is 863 kJ/kg and this implies that the exergy of the 0.347 kg extracted steam from the turbine is 299 kJ.

In this case, the exergy destruction per kg of preheated water is: (863 − 100) ∗ 0.347 − 1 ∗ (100 − 2.6) = 167.4 kJ/kg of the preheated water. This is significantly less than the corresponding exergy destruction associated with coal combustion (578.9 kJ/kg).

Preheating with FWHs uses heat from the low-temperature and low exergy parts of the cycle (the "low quality" heat) to raise the water temperature and avoids the use of

---

[2] Modern coal and nuclear power plants include 2 or 3 FWHs with preheating ranges of 80–130°C for each preheating stage. Each FWH receives 10–20% of the steam from the turbine. An added advantage of using FWHs is that, because a fraction of steam is diverted, the size and cost of the low-pressure turbine are significantly reduced.

[3] An often advocated third way, preheating with solar irradiance, is not practical because solar energy is not always available.

high-temperature heat ("high quality" heat) to accomplish the preheating process. Thus, fuel exergy (or equivalently, a natural resource) is saved with the preheating process.

The function of the reheaters is to add heat to the cycles at relatively high temperatures. The steam that remains in the main part of the cycle, after a fraction is extracted for the FHW, may be further heated, usually at the upper temperature of the cycle, and then fed to the low-pressure turbine, where it produces additional work. Modern (and more efficient) steam power plants have two or three reheaters, with the operation of each reheater matching the operation of a corresponding FWH. The addition of heat at relatively higher temperature always improves the first law and exergetic efficiencies of power cycles.

3.1.6 Optimization of Heat Exchangers

The main source of irreversibility in all the types of HXs is the transfer of heat over the temperature difference of the primary and secondary fluids. Temperature differences are more pronounced in boilers, burners, and reheaters, where the primary fluid consists of the hot products of combustion processes. An additional source of irreversibility is related to the pressure drop and pumping power that maintains the flow of the primary and secondary fluids in the HXs. The latter is typically much smaller than the exergy destruction by heat transfer.

An analysis of heat transfer and mass separation processes by diffusion using the theory of nonequilibrium thermodynamics also concluded that the total rate of entropy production – and, by extent, the exergy destruction – is minimized when the entropy production rate is uniformly distributed along the heat or mass transfer area [2]. This has been called *equipartition of entropy production*. For heat exchangers, this uniform entropy production leads to the condition that the ratio $\frac{T_p - T_s}{T_p T_s}$, be uniform in all the segments of the heat exchanger.

An inspection of Figure 3.1 shows that the minimization of the overall exergy destruction in HXs is achieved when the temperature differences of the primary and secondary fluids is low. In the two diagrams of Figure 3.1, the minimum exergy destruction is achieved when the heating and cooling curves of the primary and secondary fluids approach as much as possible, subject to the constraint that the temperature of the hot (primary) fluid is everywhere higher than the temperature of the cold (secondary) fluid.

The overall heat transfer equation in a HX is [1]:

$$\dot{Q} = UA(LMTD) = UA \frac{(\Delta T)_{out} - (\Delta T)_{in}}{\ln\left[\frac{(\Delta T)_{out}}{(\Delta T)_{in}}\right]}, \qquad (3.15)$$

where U is the overall heat transfer coefficient of the HX; A is the heat transfer area; and the last term is called the *logarithmic mean temperature difference* (LMTD). The minimization of exergy destruction in all types of HXs invariably results in lower *LMTD*. Equation (3.15) dictates that, for the transfer of a given rate of heat in the HX, the lower *LMTD* must be compensated by an increase of the heat transfer area, which implies higher cost of equipment. The following example illustrates the trade-off between exergy destruction avoidance and higher heat exchange area.

Example 3.5: A counter-flow HX is designed to cool 42 kg/s of gasoline with specific heat 2.22 kJ/kg K from 65°C to 25°C. Cooling water at 18°C is available for this process. The design calculations show that the overall heat transfer coefficient of this HX is approximately $U = 140$ W/(m^2 K).[4] For the range of water mass flow rates 20–100 kg/s calculate the rate of exergy destruction and the required heat transfer area. $T_0 = 18°C = 291$ K.

Solution: The rate of heat transfer from the gasoline is: $42 * 2.22 * (65 - 25) = 3{,}730$ kW. The heat is transferred to the water, which enters the HX at 18°C. The exit temperature of water is related to the mass flow rate by the expression:

$$\dot{Q} = 3,730 \ kJ = 4.184 \dot{m}_w (T_{wout} - 18).$$

For a given value of the water mass flow rate, T_{wout} is first calculated; secondly, the LMTD of the HX is calculated and this specifies the area A using Eq. (3.15); finally, the rate of exergy destruction (lost power) is calculated from Eq. (3.3). The following table gives the calculated variables:

Water Mass Flow Rate (kg/s)	Water Outlet Temperature (°C)	LMTD (°C)	Heat transfer area (m^2)	Lost Power (kW)
20	62.6	4.3	5,964.2	52.7
22	58.5	6.7	3,803.7	74.3
26	52.3	9.6	2,651.0	108.3
30	47.7	11.4	2,209.9	133.8
40	40.3	14.0	1,749.9	176.3
50	35.8	15.5	1,546.7	202.4
60	32.9	16.5	1,424.9	220.1
80	29.1	17.7	1,275.1	242.6
100	26.9	18.4	1,178.3	256.2

The trade-off between lost power and heat transfer area (which determines the size and the cost) of the HX is apparent in this table: at 20 kg/s the rate of exergy destruction is low, at 52.7 kW, but the area required for the HX is very high, which implies high capital cost for the HX. At 100 kg/s, the rate of exergy destruction, at 256.2 kW, is significantly higher (signifying higher operational cost), but the required area for the heat transfer and the capital cost are lower. The engineer will typically choose the size of the HX and the mass flow rate of the cooling water by appropriately minimizing (e.g. using the present value method) the overall capital and operating cost over the life horizon of the HX.

---

[4] The overall heat transfer coefficient, U, has a weak dependence on the mass flow rates and slightly increases with the increase in the mass flow rate of water. In this example U is approximated as a constant.

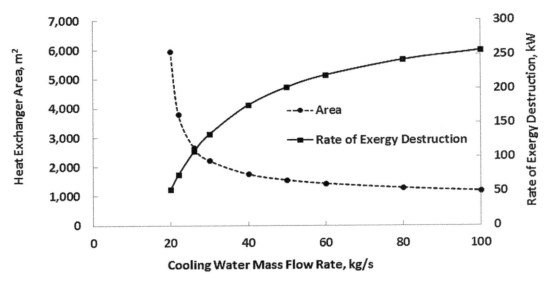

Figure 3.4 Surface area and rate of work lost in the HX of Example 3.5.

The two deviating variables that come in the optimization of HXs – exergy destruction and heat transfer area – are shown in Figure 3.4. Exergy destruction represents the variable (operational) energy cost during the lifetime of the HX, that is: $C_{vi} = f_1(W_{lost})$. The heat transfer area represents the capital investment cost of the HX, that is: $C_I = f_2(A)$. Hence, the total lifetime cost of the operation of the heat exchanger is represented by a function of the form: $f_2(A) + \Sigma f_1(W_{lost})$. It is apparent from the shape of the two curves in Figure 3.4 that the total lifetime cost of the HX has a minimum, which may be determined by an optimization process of the two variables. Several such optimization methods, including the present value method, are presented in Section 7.4.

3.2 Vapor Power Plants

More than 40% of the primary energy sources (TPES) globally is consumed for the production of electricity, an energy form that has become indispensable in contemporary human society. Approximately 52% (11,700 TWh) of the global electricity production is generated by nuclear, coal, and heavy oil power plants that utilize the Rankine cycle or one of its modifications [3, 4], with water being the working fluid for the vast majority of the cycles. Figure 3.5 shows the T, s diagram of such a modified Rankine cycle with superheat, one reheater, and one feed-water heater (FWH). The steam in the turbine is split in two parts, with the first directed to the FWH and the second to the condenser. The processes that make up this cycle are:

1. 1-2: Liquid water pressurization in a pump. Work is consumed in this part of the cycle and the exergy of the water increases.

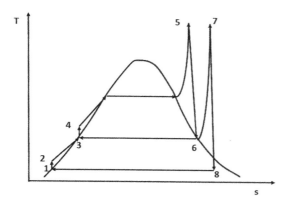

Figure 3.5 T, s diagram of a Rankine cycle with superheat, one reheat, and one FWH.

2. 2-3: Preheating the liquid water in a FWH. The liquid, pressurized water at state 2 is mixed with a fraction of the steam, denoted by y, at state 6. The effluent of the FWH is hotter liquid water at state 3.
3. 3-4: The water is pressurized in a second pump, to the upper pressure of the cycle.
4. 4-5: Liquid water is heated up to the saturation temperature and then boils. Further heating produces superheated steam at the highest temperature, T_5, of the cycle. This process is accomplished with the combustion of a fuel and entails significant work lost, as it may be seen in Table 3.1.
5. 5-6: Expansion of the steam in the high-pressure turbine. Work/power is produced in this part of the cycle and the exergy of the steam decreases.
6. 6-7: The reheating process, during which the temperature of the steam increases to approximately its highest level ($T_7 = T_5$).
7. 7-8: Expansion of the steam in the low-pressure turbine. The exit of this turbine is at subatmospheric pressure.
8. 8-1: The saturated steam/water mixture at the exit of the low-pressure turbine is condensed by cooling water and directed to the first pump to repeat the cycle.

Exergy calculations are conducted for a 400 MW coal-fired power plant that operates with the cycle of Figure 3.5. The conditions of the cycle are:

1. Condenser: $P_1 = 6$ kPa.
2. Feed-water heater at $P_2 = P_3 = P_6 = P_7 = 1.4$ MPa.
3. Steam at the entrance of the turbine: 15 MPa, 520°C.
4. Pumps' efficiencies 80%; turbines' efficiencies 83%.
5. The combustion efficiency of the boiler and superheater is 92% and the fuel (coal) may be modeled as carbon with LHV 32.7 MJ/kg.

Table 3.2 shows the variables of interest for the states 1–8 including the rate of exergy in each of the states. From the analysis of the cycle, the fraction of steam extracted at state 6 to preheat the liquid water at state 2 is $y = 0.2554$. The specific work production of the cycle is 1,196.5 kJ/kg and the specific heat input is 2,980.2 kJ/kg.

Table 3.2 Variables of interest for the Rankine cycle of Figure 3.5. $T_0 = 300$ K, $P_0 = 1.013$ bar.

State	P (MPa)	T (°C)	h (kJ/kg)	e (kJ/kg)	$\dot{E}$ (MW)
1	0.006	36.16	151.5	0.6	0.2
2	1.5	~36.16	153.4	1.9	0.5
3	1.5	198.3	845.0	155.8	41.8
4	16	~198.3	863.1	170.0	45.6
5	16	520	3,355.0	1,446.8	388.3
6	1.5	228	2,862.1	890.5	239.0
7	1.5	520	3,517.2	1,234.3	246.7
8	0.006	36.16	2,546.4	72.6	14.5

For the production of 400 MW, the mass flow rate of the working fluid (water/steam) is 334.3 kg/s; and the rate of heat input to the cycle is 996.3 MW. Since the combustion efficiency is 92%, the mass rate of carbon supplied for the delivery of the rate of heat is: $996.3/(0.92*32.8) = 33.02$ kg/s. The rate of exergy of the fuel (carbon), which is supplied to the boiler-superheater, is 1,085 MW.[5]

It is observed in Table 3.2 that the exergy of the working fluid significantly increases in the boiling/superheating (4-5) and the reheating (6-7) processes. Steam expansion in the high- and low-pressure turbines produces the power of the cycle and, simultaneously, reduces the specific exergy of the steam. Based on the values of Table 3.2, one may perform exergy balances to determine the lost power (rate of exergy destruction) in the equipment of this cycle.

1. First pump, process 1-2:

$$\dot{W}_{lost} = |\dot{W}_{12}| + \dot{E}_1 - \dot{E}_2 = \dot{m}_s(1-y)[(h_2 - h_1) + e_1 - e_2] = 0.15 \text{ MW}$$

2. FWH, with inputs from states 6 and 2 and output state 3:

$$\dot{W}_{lost} = y\dot{E}_6 + (1-y)\dot{E}_2 - \dot{E}_3 = \dot{m}_s[ye_6 + (1-y)e_2 - e_3] = 24.4 \text{ MW}$$

3. The second pump, process 3-4:

$$\dot{W}_{lost} = |\dot{W}_{34}| + \dot{E}_3 - \dot{E}_4 = \dot{m}_s[(h_4 - h_3) + e_3 - e_4] = 1.3 \text{ MW}$$

4. The boiler-superheater-reheater system receives the fuel; it receives water at state 4 and discharges steam at state 5; and also receives steam at state 6 and discharges it at state 7.

$$\dot{W}_{lost} = \left|\dot{m}_f \Delta G_f^0\right| + \dot{E}_4 - \dot{E}_5 + (1-y)(\dot{E}_6 - \dot{E}_7) = \left|\dot{m}_f \Delta G_f^0\right| + \dot{m}_s(e_4 - e_5)$$
$$+ (1-y)\dot{m}_s(e_6 - e_7) = 572.6 \text{ MW}$$

Since the combustion products of the fuel are exhausted in the atmosphere, zero exergy is assigned to them.

---

[5] In actual cycles a small pressure drop, on the order of a few kPa, occurs during the boiling and reheating processes. This pressure drop is not significant in the calculation of the enthalpy and exergy values of the working fluid.

5. The first turbine, process 5-6, receives steam at state 5, exhausts steam at state 6 and produces 164.8 MW of power.

$$\dot{W}_{lost} = \dot{E}_5 - \dot{E}_6 - \dot{W}_{56} = \dot{m}_s[e_5 - e_6 - (h_5 - h_6)] = 21.2 \ MW$$

6. The second turbine, process 7-8, receives steam at state 7, exhausts the steam at state 8 and produces 241.7 MW of power:

$$\dot{W}_{lost} = \dot{E}_7 - \dot{E}_8 - \dot{W}_{78} = \dot{m}_s(1-y)[e_7 - e_8 - (h_7 - h_8)] = 47.5 \ MW$$

7. Most of the exergy rate at state 8, (14.5 MW), is dissipated in the condenser.

The power lost in the several components is 681.5 MW and the net power produced, 400 MW. The 2 pumps consume 6.5 MW of power. The sum of the power lost and the net power is approximately equal to the exergy input of the fuel, 1,085 MW. The small discrepancy (0.32%) is due to numerical round-offs and the (low) uncertainty of the property values.

The most remarkable observation in this example is that the exergy destruction in the heating-by-combustion parts of the cycle – the boiler-superheater-reheater system – is higher than the total power produced. The rate of exergy destruction in the pumps is almost negligible, and the rates of exergy destruction in the FWH and the turbines are moderate. Even if the combustion efficiency were 100%, the rate of exergy destruction in the boiler-superheater-reheater would be very high. In total, 1,085 MW of fuel exergy are used to increase the exergy of water/steam by 342.7 MW in process 4-5 and by 85.6 MW in process 6-7. High exergy destruction occurs because a large fraction of the fuel's exergy is destroyed in the chemical-to-thermal energy transformation. A significant improvement in the utilization of fossil fuels would be the *Direct Energy Conversion (DEC)* of chemical energy into electric energy, a process accomplished in electrochemical systems, such as batteries and fuel cells. DEC devices do not utilize the heat-to-work conversion process and are not subjected to the Carnot efficiency limitation of Eq. (1-24).

3.3 Gas Turbines

Gas turbines operate with Brayton cycles or modifications of these cycles. The fuel of the gas turbines is typically natural gas or light hydrocarbons, which are injected in the combustion chamber as droplets. A few types of gas turbines are modified to be used with high quality pulverized coal. The four basic processes of the Brayton cycle, are depicted in Figure 3.6 and 1.10 (in the P,v and T,s diagrams) and may be summarized as follows:

1. 1-2: Atmospheric air enters the compressor (C) of the power plant and is compressed to a high pressure, typically in the range 5–30 bar.
2. 2-3: The compressed air enters the combustion chamber (CC), where it is mixed with the fuel. Combustion of the fuel produces a high-temperature mixture of the combustion products. Usually this process takes place with excess air and the products contain a fraction of oxygen and nitrogen together with carbon dioxide and water vapor. The combustion process entails significant work lost as the chemical exergy of the fuel is converted into the thermal exergy of the combustion products.

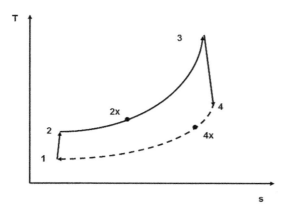

Figure 3.6 Thermodynamic diagram of the Brayton cycle. The states 2x and 4x pertain to regeneration.

3. 3-4: The expansion of the combustion products takes place in the turbine, which is connected to the generator and produces power. The products are exhausted to the environment at atmospheric pressure and, typically, at fairly high temperature.
4. 4-1: This is a process that takes place in the atmosphere: the high temperature exhaust gases are mixed with the atmospheric air, cool, and attain the atmospheric temperature, T_0. Since the mass of the atmosphere is by far higher than that of the exhaust gases, the atmosphere acts as a heat and mass reservoir that delivers to the compressor air at atmospheric pressure, temperature, and composition. Because this process does not occur within the equipment of the gas turbine, it is denoted with a broken line in Figure 3.6.

Let us consider a 50 MW gas turbine system that operates with a basic air standard Brayton cycle. The input to the compressor is air at 1 atm and 300 K. The high pressure of the cycle is 20 bar and the temperature at the entrance of the turbine is 2,000 K. The compressor efficiency is 80% and the turbine efficiency is 85%. Methane, CH_4, is the fuel of this cycle. The calculations show that the specific work produced by the turbine is 991 kJ/kg, of which 510 kJ/kg are consumed by the compressor leaving the net specific work of the cycle at 481 kJ/kg. For the production of 50 MW of power, the mass flow rate of air in the gas turbine is 104.0 kg/s. The specific heat input of the cycle is 1,443 kJ/kg, and the thermal (first law) efficiency of the cycle is approximately 33.3%.

The rate of heat input to the cycle is 150.0 MW and is generated by methane combustion. Since the LHV of this fuel is 50.020 MJ/kg, with complete combustion, approximately 2.999 kg/s of methane are consumed to produce the needed rate of heat. Table 3.3 shows the pressure, temperature, enthalpy, specific exergy, and rate of exergy at the four states of the gas turbine system. The properties of air were calculated with the REFPROP software [5].

The rates of exergy destruction in the three components of the gas turbine system are:

1. The compressor: $510 * 104$ (from the turbine) $+ 0 - 48,629 = 4,411$ kW.
2. The combustion chamber: $48,629 + 2.999 * 51,125$ (fuel exergy) $- 164,392 = 37,561$ kW.

Table 3.3 The state variables for the basic Brayton cycle.

State	Pressure (atm)	Temperature (K)	Specific Enthalpy (kJ/kg)	Specific Exergy (kJ/kg)	Rate of Exergy (kW)
1	1	300	426	0	0
2	20	788	936	468	48,629
3	20	2,000	2,379	1,581	164,392
4	1	1,202	1,388	522	54,296

3. The turbine: $164{,}392 - 510 * 104$ (to compressor) $- 54{,}296 - 50{,}000$ (power produced) $= 7{,}056$ kW.

It quickly becomes apparent that, as in the vapor cycle in Section 3.2, the irreversibility in the combustor by far exceeds the irreversibility in the other components. Combustion always produces high rates of exergy destruction.

Two significant observations in Table 3.3 are:

1. The temperature of the exhaust gases is higher than that of the air at the exit of the compressor.
2. The exergy of the exhaust gases, which are dispersed in the air, is significant. In a basic Brayton cycle this exergy is dissipated in the environment.

A moment's reflection proves that the higher temperature and exergy of the exhaust gases may be utilized to produce additional work and to reduce the lost power of the cycle. This happens with: the regeneration process; and the addition of a steam bottoming cycle.

3.3.1 Brayton Cycle with Regeneration

In addition to the states 1, 2, 3, and 4 of the basic Brayton cycle, Figure 3.6 also shows the states 2x and 4x, which are included in the modified Brayton cycle and utilize part of the high-enthalpy and high-exergy of the exhaust gases. In the modified cycle a counter-flow HX (the regenerator) is placed at the exit of the gas turbine to transfer part of the enthalpy of the exhaust gases to the compressed air from the compressor. The inlet to the hot side of this HX is state 4 (the exit of the turbine) and the outlet is denoted as state 4x, where the cooled gases are exhausted in the atmosphere. The inlet to the cold side is state 2 (the exit of the compressor) and the outlet is denoted as state 2x. From state 2x the compressed and heated air is directed to the combustion chamber. The effectiveness of the HX defines the temperature of the state 2x:

$$\varepsilon = \frac{T_{2x} - T_2}{T_4 - T_2}. \tag{3.16}$$

In this case the maximum amount of heat is transferred when the temperature of the compressed air, the secondary fluid at the exit of the HX, is equal to T_4. Because part of the heat input of the cycle is supplied by the regenerator, and $T_{2x} > T_2$, a lesser amount of heat and a lesser amount of fuel are needed to produce the combustion products at T_3. Let us include a regenerator with 84% effectiveness in the gas turbine system of the last

Table 3.4 The state variables for the Brayton cycle with a regenerator.

State	Pressure (atm)	Temperature (K)	Specific Enthalpy (kJ/kg)	Specific Exergy (kJ/kg)	Rate of Exergy (kW)
1	1	300	426	0	0
2	20	788	936	468	48,629
2x	20	1,136	1,331	738	76,747
3	20	2,000	2,379	1,581	164,392
4	1	1,186	1,388	522	54,296
4x	1	840	993	245	25,506

section. Table 3.4 shows the pressure, temperature, specific enthalpy, specific exergy, and rate of exergy at the states of the modified cycle [5].

The addition of the regenerator does not affect the net specific work and the air mass flow rate through the system. It affects the heat supplied by the fuel, because the rate of heat needed in the burner is less, approximately 109.0 MW. Therefore, the rate of fuel supplied for the combustion is reduced to 2.179 kg/s, which corresponds to a 111.4 MW rate of exergy. Since the system produces 50 MW of electric power, its thermal efficiency becomes 44.9%, a significant improvement compared to the basic Brayton cycle case (33.3%).

The numerical values of the lost power in the four components of this system are:

1. In the compressor: $510 * 104$ (from the turbine) $+ 0 - 48{,}629 = 4{,}411$ kW.
2. In the combustion chamber: $76{,}747 + 2.179 * 51{,}125$ (fuel exergy) $- 164{,}392 = 23{,}756$ kW.
3. In the turbine: $164{,}392 - 510 * 104$ (to compressor) $- 54{,}296 - 50{,}000$ (power produced) $= 7{,}056$ kW.
4. In the regenerator: $54{,}296 - 25{,}506 + 48{,}629 - 76{,}747 = 672$ MW.

It is observed that the exergy destruction in the combustion chamber is significantly reduced, primarily because of the reduction of the consumed fuel. The exergy destruction in the regenerator, which is absent in the basic cycle, is very low. It is also observed that, even though a fraction of the enthalpy and the exergy of the exhaust gases is utilized in the cycle to reduce the flow rate of the fuel, the exhaust gases are still at high temperature, 840 K, and carry 25.5 MW rate of exergy. This exergy may be utilized in a combined cycle to produce additional power.

3.3.2 The Combined/Bottoming Cycle

The combined cycle entails the addition of a vapor (usually steam) cycle that receives heat from the exhaust gases of the Brayton cycle. The vapor cycle produces additional power, which is added to the power of the gas cycle. Because the vapor cycle utilizes the waste heat of the gas cycle, its temperatures are lower and it has been referred to as the *bottoming cycle*. A schematic diagram of the entire combined cycle is depicted in Figure 3.7, with the gas cycle at the top and the vapor cycle at the bottom. The counter

3.3 Gas Turbines

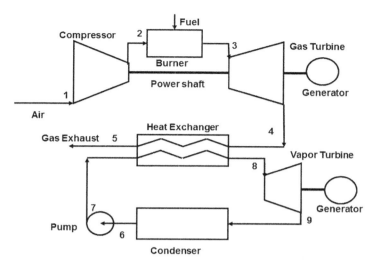

Figure 3.7 The combined Brayton-Rankine cycle.

flow HX in the middle of Figure 3.7 connects thermally the two cycles: The exhaust gases at the exit of the gas turbine are directed to this HX, which is sometimes referred to as *steam generator,* because it produces the steam for the bottoming cycle. This HX is the boiler of the vapor cycle and does not need fuel to operate.

The vapor is the secondary fluid in the HX; it evaporates at an intermediate pressure; it is usually superheated; and it is directed to the vapor turbine. At the exit of the turbine the vapor condenses in the condenser; it is pressurized by the pump; and is directed back to the vapor generator to complete the Rankine cycle.

In most combined cycle systems, water/steam is used in the Rankine cycle, but any other suitable fluid (e.g. a refrigerant, ammonia, or a hydrocarbon) may be used. As it is shown in Figure 3.7, the combined cycle utilizes two turbine-generator systems and produces two streams of power, one from the gas turbine and the second from the vapor turbine. The system uses fuel only for the gas combustion in the upper cycle. The result is an advanced power system with lower lost power; significantly better exergy utilization; and higher efficiency than other conventional electricity generation units. Overall thermal (first law) efficiencies in the range 60–64% have been achieved with combined cycles [6].

Let us consider that a bottoming cycle with a counter current steam generator is added to the system analyzed in the last section with the variables on the gas side as shown in Table 3.4. The bottoming cycle generates steam at 15 bar and 360°C.[6] The specific enthalpy difference of the steam cycle after the pinch point is $(3,169 - 845)$ kJ/kg; with a 5°C pinch-point temperature difference, the temperature of the gas at the pinch point is approximately 477 K (204°C) and its enthalpy 606 kJ/kg. The energy balance in this part of the HX determines the mass flow rate of the steam, 17.32 kg/s. With the addition

---

[6] In all practical designs of combined cycles, the pressure and temperature of the bottoming cycle are optimized to produce maximum total power.

Table 3.5 Parameters of interest of the Brayton cycle and its improvements.

	Basic Brayton	With Regenerator	Combined Cycle
Power produced (MW)	50	50	63.928
Methane consumed (kg/s)	2.999	2.179	2.179
Thermal efficiency (%)	33.3	45.2	58.7
Exergetic efficiency (%)	32.6	44.9	57.4
Heat input (MW)	150	109	109
Rate of exergy destruction (MW)	103.3	61.4	47.5
Rate of exergy in exhaust gases (MW)	54.3	25.5	0.5

of the bottoming cycle the exhaust gases exit the combined cycle at 357 K and carry only 509 kW rate of exergy.

The specific exergy of the steam at the entrance of the steam turbine is 1,034 kJ/kg. A 78% efficient steam turbine produces 806 kJ/kg of work, of which approximately 1.8 kJ/kg is consumed by the pump of the bottoming cycle. Therefore, the bottoming cycle would produce an additional $(806 - 1.8) * 17.32 = 13,928$ kW of power raising the total for the combined cycle to 63,928 MW. With the rate of heat input to the combined cycle at 109.0 MW the thermal efficiency of the combined cycle increases to 58.7%.

The lost work in the upper part of the combined cycle is the same as in the Brayton cycle with the regenerator. The addition of the bottoming cycle utilizes the rate of exergy of the exhaust gases, which in this case is reduced from 25.5 MW to 0.5 MW. This is a small amount of waste heat that may be recovered in low-temperature applications.

Table 3.5 shows the parameters of interest pertaining to the basic Brayton cycle and the improvements with the regenerator and bottoming cycle. It is observed that the installation of the regenerator and the addition of the bottoming cycle almost double the thermal efficiency of the basic Brayton cycle. This is achieved by utilizing the rate of exergy of the exhaust gases, which becomes almost zero with the introduction of the bottoming cycle.

3.4 Cogeneration

Several industrial, domestic, and commercial applications are performed using processes that require more than one energy form. For example, a steel mill utilizes high temperature heat for the production of steel from iron ore and coal, as well as electricity for the operation of the presses and other machinery that form the steel plates. Food pasteurization plants (for milk, juices, tomato paste, etc.) require large quantities of heat at moderate temperatures (85–90°C) in addition to the electric power for the circulation pumps and the bottling equipment. Large supermarkets use natural gas for heating as well as electricity for lights and the several refrigerators and freezers in the displays. Most households use electricity for lighting and air-conditioning as well as natural gas for cooking and hot water.

3.4 Cogeneration

The production of heat and electricity in separate processes consumes high quantities of primary energy and entail high rates of exergy destruction. It is technically feasible to combine two or more of the energetic processes and produce the same outcomes using a lesser quantity of primary energy, thus saving exergy and primary energy sources. In such cases two or more outcomes are accomplished by the same system. When the desired outcomes are the production of electric power and heat, the two outcomes may be combined in a single system, the *cogeneration* system [7] that generates both electric power and heat using a single system.

In the case of milk and juice pasteurization a heat source is needed at approximately 85°C. The conventional method to pasteurize is to burn fuel (e.g. liquid or gaseous hydrocarbons) and heat up water at a temperature slightly higher (e.g. 90°C) than that required for the pasteurized product; then use a HX to heat up the food product(s) to the desired temperature by cooling the water. When cogeneration is used, a system is designed that utilizes a vapor cycle to produce electric power as well as the heat needed for the pasteurization process. The needed rate of heat comes from the vapor cycle by one of two processes:

1. Using a condenser at approximately 0.57 bar pressure, the saturation pressure of 85°C. Effectively the condenser becomes the heater for the food product and supplies heat at 85°C.
2. Extracting a fraction of the steam from the turbine at 85°C (or at a slightly higher temperature) and feeding the steam to a HX. The heat exchange process raises the temperature of the food product.

Figure 3.8 is a schematic diagram of the first method, where the steam discharged by the turbine is at the desired saturation pressure. The steam enters the HX that serves as the condenser for the cycle as well as the "heater" of the milk. The power producing cycle is a basic Rankine steam cycle with the condensate extracted at the pressure that corresponds to the desired saturation temperature, 0.57 bar and 85°C in this case.

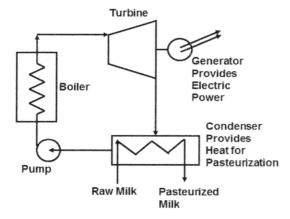

Figure 3.8 Hot water production for a pasteurization process using the condenser of a power unit.

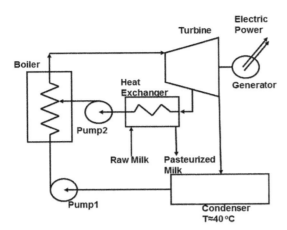

Figure 3.9 Hot water production for a pasteurization process using a FWH system.

The Rankine cycle produces electric power as well as the needed heat for the pasteurization process.

The second method is depicted in Figure 3.9. A fraction of the steam from the Rankine cycle is extracted from the turbine and diverted to the HX as in the case of a feed-water heater. The steam condenses and delivers the heat of condensation to the food product for pasteurization, with the rest of the cycle producing electric power. The exergetic analysis of this system is the same as the analysis of the FWH in Section 3.1.5.

The exergetic need of the pasteurization process is low, because the temperature of the food product is raised to moderate temperatures, typically less than 100°C. In the case of conventional pasteurization methods, fuel is consumed for the production of low-temperature heat, which implies that a large fraction of the exergy of the fuel is destroyed. In both cogeneration methods the Rankine cycle uses a (typically higher) quantity of fuel to supply the cycle with the required heat at the higher temperature. A fraction of the fuel exergy produces the electric energy and a smaller fraction is used for the pasteurization process. The combination of the two processes results in higher exergetic efficiencies and lower exergy destruction (lost work) per unit mass of the fuel consumed.

Example 3.6: A food-processing plant uses 58 L/s of hot water at 82°C. The hot water is supplied by an old burner, which consumes natural gas and has 85% combustion efficiency. The temperature of the water inlet to the burner is 20°C. It is proposed to substitute this outdated system with a small power plant operating with a simple Rankine cycle that supplies the needed heat at its condenser. The steam inlet to the turbine is at 400°C and 60 bar; the condenser is at 90°C (70.14 kPa); the turbine efficiency is 82% and the pump efficiency is 75%. Calculate the exergy destruction in all the processes of the old and new systems and determine their exergetic efficiencies. $T_0 = 293$ K.

Solution: The mass flow rate of the hot water is approximately 58 kg/s and the rate of heat required for the needed temperature increase of 62°C is:

$58 * 4.184 * 62 = 15{,}046$ kW. The specific exergy of hot water at 82°C and 1 atm is 24.1 kJ/kg and the rate of exergy of the hot water produced is $24.1 * 58 = 1{,}398$ kW.

The LHV of natural gas is $LHV = 1{,}070$ kJ/scf (37,787 kJ/m^3) and its exergy is $\Delta G^o = 1{,}120$ kJ/scf (39,558 kJ/m^3).

With the old burner the needed rate of heat is supplied from the combustion of natural gas at the expense of $15{,}046/(1{,}070 * 0.85) = 16.54$ scf/s (0.468 m^3/s). This corresponds to exergy rate of 18,525 kW. Therefore, the exergetic efficiency of the burner is: $1{,}398/18{,}525 = 7.5\%$ and the rate of exergy destruction in the old system is 17,127 kW.

With the new system the specific enthalpy, entropy and exergy at the four states of the Rankine cycle are:

h1 = 377 kJ/kg	s1 = 1.1925 kJ/kg K	e1 = 30.5 kJ/kg
h2 = 385 kJ/kg	s2 = 1.1979 kJ/kg K	e2 = 36.9 kJ/kg
h3 = 3,177 kJ/kg	s3 = 6.5408 kJ/kg K	e3 = 1,263.5 kJ/kg
h4 = 2,473 kJ/kg	s4 = 6.9559 kJ/kg K	e4 = 438.3 kJ/kg

The heat from the condenser is solely used for the production of hot water. Therefore, the required mass flow rate of the steam in the Rankine cycle is: $15{,}046/(2{,}473 - 377) = 7.2$ kg/s. Hence, the rate of heat input to the cycle is 20,102 kW and this is supplied by the combustion of $20{,}102/1{,}070 = 18.8$ scf/s of natural gas corresponding to a rate of exergy for the natural gas supplied of 21,056 kW. The power output of the cycle is 5,069 kW.

Therefore, the exergetic efficiency of the new system is: $(1{,}398 + 5{,}069)/21{,}056 = 30.7\%$. The rate of exergy destruction in the new system is: $21{,}056 - 1{,}398 - 5{,}069 = 14{,}589$ kW.

It is observed that the substitution of the old burner with the new system results in significantly higher exergetic efficiency and lower rate of exergy destruction, which implies that primary energy resources are preserved. It must be noted that this substitution would require capital investment by the owners of the food processing plant.

Cogeneration of electric power and heat is also accomplished with gas cycles, using the turbine exhaust gases. The gas turbine exhaust is always at high temperature and, hence, the waste heat of the gas cycle may be used for the supply of lower-temperature heat. Cogeneration of heat and electric power by a single system requires less primary energy/fuel input than two systems that produce separately the electric energy and the heat [3, 7]. Because heat and electric power cogeneration systems save a great deal of primary energy; have higher (both first-law and exergetic) efficiencies; and reduce the cost of the heating processes, they have become very popular since the 1980s and are now widely used in commercial establishments, large residential complexes, and office buildings.

One of the regulatory measures that has significantly promoted the widespread use of cogeneration, is the deregulation of the electric power industry in most OECD countries and the associated provision that requires electricity transmission and sale corporations to purchase the excess electric power produced by smaller producers (e.g. by the owners of cogeneration plants) at the prevailing wholesale price. Thus, entrepreneurs who invest in the substitution of older heating systems with cogeneration facilities are assured that there is a market for the electric power they produce. Example 3.7, below, illustrates the exergetic advantage for the installation of a cogeneration system in an industrial pasteurization process.

Example 3.7: A large milk pasteurization plant uses hot water at 85°C at a rate 1.65 m^3/min. The water is supplied at an average temperature 18°C and is currently heated by natural gas in an old burner with 80% efficiency. It is proposed to buy a 10 MW gas turbine, which has 41% thermal efficiency, use the waste heat of the turbine for the supply of hot water, and sell the produced electricity. Determine: a. the amount of natural gas consumed daily by the two systems; b. The exergetic efficiencies of the old and new system; and c. the daily exergy destruction in the old and the new system. $T_0 = 18°C = 291$ K.

Solution: a. The volumetric flow rate of the water is 1.65/60 m^3 = 0.0275 m^3/s and the mass flow rate is approximately 27.5 kg/s. The rate of heat required for the water temperature increase from 18°C to 85°C is: $27.5 * 4.184 * (85 - 18) = 7,709$ kW. Since the heat supplied by natural gas is 1,070 kJ/scf (37,787 kJ/m^3) and the burner has 80% efficiency, the pasteurization plant consumes $7,709/(0.8 * 1,070) = 9.01$ scf/s of natural gas (0.255 m^3/s). The daily amount of natural gas consumed is: 778,464 scf (22,052 m^3).

The 10 MW gas turbine requires heat input $10/0.41 = 24.4$ MW and its waste heat is 14.4 MW, more than sufficient to supply the demand of 7,709 kW for the pasteurization process. The daily heat input to the gas turbine is $2.11 * 10^9$ kJ and the daily amount of natural gas consumed is: 1,970,243 scf (55,812 m^3). The increase in the daily consumption of natural gas is mainly due to the production of the electric power.

b. The exergy of 1 scf of natural gas is approximately $-\Delta G^0 = 1,120$ kJ/scf (39,558 kJ/m^3). Therefore, the daily exergy input to the old burner is: $871.88 * 10^6$ kJ and that to the new system is $2,207 * 10^6$ kJ.

The exergy of the hot water (an incompressible liquid) produced daily is obtained from Eq. (2.21): $27.5 * 24 * 60 * 60 * [4.184 * 67 - 291 * 4.184 * \ln(358/291) = 66.62 * 10^6$ kJ. The gas turbine produces daily $864.26 * 10^6$ kJ of electric energy in addition to the hot water.

The exergetic efficiency of the old system is: $66.62 * 10^6/871.88 * 10^6 = 7.6\%$ and of the new system $(66.62 * 10^6 + 864.26 * 10^6)/2,207 * 10^6 = 42.2\%$. This is a significant improvement!

c. From the above values of exergy production and consumption, the daily exergy destruction in the old system is $871.88 * 10^6 - 66.62 * 10^6 = 805.26 * 10^6$ kJ; and that in the new system: $2,207 * 10^6 - 66.62 * 10^6 - 864.26 * 10^6 = 1,276 * 10^6$ kJ. The higher exergy destruction in the new system is due to the additional production of 10 MW electric power. Had this quantity of electricity been produced in a separate facility, with the old burner operating, the combined exergy destruction would have been significantly higher than $1,276 * 10^6$ kJ.

3.5 Jet Engines

One of the most important inventions of the 20th century, the jet engine, generates high thrust to power aircrafts at high subsonic and supersonic speeds. Modern and powerful jet engines have enabled the development of large aircraft, which may carry almost 800 passengers from one continent to another.

The function of an aircraft engine is not to produce net power, but to provide high thrust, and this is achieved by the high-velocity air jet at the exit of the engine. The engine utilizes a variance of the Brayton cycle, where the turbine produces sufficient power to drive the compressor and the (very small) electric power requirements of the aircraft. The remainder of the energy and exergy of the gases at the exit of the turbine is spent for the acceleration of the working fluid in a nozzle, which provides the needed thrust to propel the aircraft. Figure 3.10 depicts the T,s diagram and the components of the engine, with labels for the several states of the working fluid, which is typically modeled as air. The parts of Figure 3.10 in broken lines correspond to the *afterburner*, which is an optional component that significantly increases the thrust of the engine and is primarily used with supersonic military aircraft. The main components of jet engines, their functions, and the thermodynamic processes that take place in the engines are as follows:

1. The *diffuser*, process 1-2, converts the velocity of the incoming air to higher pressure.
2. The *compressor*, process 2-3, further increases the pressure of the air.
3. The *burner*, process 3-4, receives the fuel – kerosene – which reacts with oxygen in the air and increases the temperature of all the reaction products.
4. The *turbine*, process 4-5, provides sufficient power for the compressor and the electric power requirements of the aircraft.

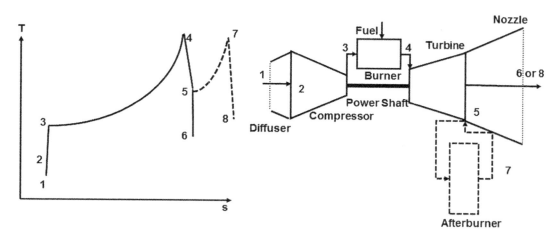

Figure 3.10 Components and T,s diagram of a jet engine.

Table 3.6 Thermodynamic properties and specific exergy at states 1-6 of the jet engine.

State	T (K)	P (MPa)	h (kJ/kg)	s (kJ/kg K)	e (kJ/kg)
1	222.0	0.0220	0.0	0.0000	0.00
2	258.1	0.0366	36.2	0.0047	35.14
3	664.1	0.7325	453.2	0.1102	428.72
4	1,400.0	0.7325	1,294.0	0.9562	1,081.75
5	1,038.7	0.1769	868.7	1.0134	643.76
6	620.4	0.0220	406.8	1.0449	174.79

5. The *nozzle*, process 5-6 (or 7-8 with an afterburner), accelerates the hot, pressurized gas from the turbine exhaust to very high velocity that produces the thrust.
6. The *afterburner*, process 5-7, is a second burner and acts as a reheater that increases the temperature of the gases at the exit of the turbine to produce higher thrust. The afterburner is an optional component and exhausts in the nozzle.

Let us examine the operation and determine the thrust from a commercial jet engine (without an afterburner) attached to an aircraft, which is cruising at 35,000 feet (10,668 m) with speed 970 km/hr (269 m/s). The atmospheric pressure at this level is $P_0 = 22.6$ kPa and the ambient temperature is $T_0 = 222$ K. The inlet diameter of the engine is 1.4 m. The compressor pressure ratio is $P_3/P_2 = 20$; the temperature of the gases at the entrance of the turbine is $T_4 = 1,400$ K; and the nozzle exhausts to the atmosphere. The electric power need for the aircraft is 2% of the power delivered to the compressor. The fuel is kerosene with LHV, $-\Delta H^0 = 43,200$ kJ/kg and the equivalent Gibbs' free energy of the combustion reaction is $-\Delta G^0 = 41,850$ kJ/kg. The isentropic efficiencies of the diffuser and the nozzle are 96%; the isentropic efficiencies of the compressor and the turbine are 84% and 88% respectively; and the combustion efficiency is 98%.

The properties of the working fluid are obtained from [5]. Table 3.6 shows the thermodynamic properties as well as the specific exergy of air at all the states of this jet engine. The datum for the enthalpy and entropy is at the environmental conditions, $P_0 = 22.6$ kPa, and $T_0 = 222$ K.

The ambient air density at this altitude is 0.3454 kg/m³ and the mass flow rate of the air in the engine is calculated to be: $3.14 * (1.4/2)^2 * 0.3454 * 269 = 142.9$ kg/s. The rate of heat input to the jet engine is $142.9 * (1,294 - 453.2) = 120,150$ kW. With LHV for the fuel 43,200 kJ/kg and 98% combustion efficiency, the mass flow rate of the fuel is: 2.838 kg/s. The exergy rate of the fuel is: 118,770 kW. From the enthalpy difference between the inlet and outlet of the nozzle, the exit velocity of the air relative to the jet engine is 961.2 m/s. Therefore, the thrust generated by the engine is: $(142.9 + 2.838) * 961.2 - 142.9 * 269 = 101,643$ N.

One may calculate the exergy flow and exergy destruction in the six components of this jet engine to determine components where significant improvements may be made:

Table 3.7 Properties and specific exergy at all states of the jet engine with an afterburner

State	T (K)	P (MPa)	h (kJ/kg)	s (kJ/kg K)	e (kJ/kg)
1	222.0	0.0220	0.0	0.0000	0.00
2	258.1	0.0366	36.2	0.0047	35.14
3	664.1	0.7325	453.2	0.1102	428.72
4	1400.0	0.7325	1294.0	0.9562	1081.75
5	1038.7	0.1769	868.7	1.0134	643.76
7	1400	0.1769	1293.6	1.3641	990.80
8	845.6	0.0220	650.6	1.3797	344.28

1. In the diffuser the rate of exergy destruction is: $142.9 * (269^2/2{,}000 - 35.1) = 148.7$ kW.
2. In the compressor, which receives $417 * 142.9 = 59{,}589$ kW of power from the turbine, the rate of exergy destruction is: $59{,}589 - 142.9 * (428.72 - 35.14) = 3{,}348$ kW.
3. In the burner the rate of exergy destruction is: $118{,}770 + 142.9 * 428.72 - 145.7 * 1{,}081.75 = 22{,}423$ kW.
4. In the turbine the rate of exergy destruction is: $145.7 * (1{,}081.75 - 643.76) - 59{,}589 * 1.02 = 3{,}036$ kW (the factor 1.02 includes the electric power need of the aircraft).
5. In the nozzle the rate of exergy destruction is: $145.7 * [643.76 - 174.79 - (961.2)^2/2{,}000] = 1{,}022$ kW.

3.5.1 Jet Engines with Afterburner

The afterburner reheats the gases at the exit of the turbine and before they enter the nozzle. The higher enthalpy of the gases generates higher nozzle exit velocity and higher thrust. Following the calculations in Section 3.5, let us consider that an afterburner with 98% combustion efficiency increases the temperature of the gases to 1,400 K – the same temperature as at the turbine inlet. Table 3.7 shows the thermodynamic properties and specific exergy of the states that are pertinent to the operation of the jet engine with an afterburner.

The heat input to the afterburner requires the combustion of an additional 1.433 kg/s of fuel. When the afterburner is engaged, the exit velocity of the gases increases to 1,134 m/s and the produced thrust increases from 101,643 N to $(142.9 + 2.838 + 1.433) * 1{,}134 - 142.9 * 269 = 128{,}371$ N.

In the actual operation of the aircraft, the engagement of the afterburner starts a transient process when thrust and aircraft velocity increase until a new equilibrium velocity is reached: The increased thrust increases the aircraft velocity and this also causes the increase of the air mass flow rate intake in the diffuser. The higher mass flow rate in the entire engine generates an even higher thrust and higher aircraft velocity. This sequence continues until the aircraft reaches a mechanical state of equilibrium, where the drag force on the aircraft is equal to the applied thrust. In most cases, the new state of equilibrium is at supersonic velocity for the aircraft. At the new equilibrium cruising velocity, the air input to the engine is higher than 142.9 kg/s, the burner fuel consumption is higher and the generated thrust should be calculated based on the new aircraft velocity and mass flow rates.

With the use of the afterburner in the engine, the states 1-5 and the exergy destruction are the same as in the list 1–5 of Section 3.5. The rate of exergy destruction in the afterburner and the nozzle are as follows at the commencement of the afterburner process:

6. In the afterburner/reheater the rate of exergy destruction at the commencement of the afterburning process is: $1.433 * 41{,}850 + 145.7 * 643.76 - 147.1 * 990.80 = 8{,}020$ kW.
7. In the nozzle the rate of exergy destruction is: $147.1 * [990.80 - 334.28 - (1{,}134)^2/2{,}000] = 1{,}992$ kW.

In addition, in both cases of jet engines with and without afterburners, the exergy of the hot high-velocity gases at the exit of the nozzle is dissipated in the environment.

It is apparent that most of the exergy destruction in the jet engines occurs in the burners and afterburners, where fuel combustion occurs. As with the power plants this is due to the chemical-to-thermal energy conversion, a process where a great deal of exergy is destroyed. While in the terrestrial power plants we may employ DEC devices, bottoming cycles, and other exergy minimization equipment, because of weight and reliability limitations, we are very much limited on what modifications we may do to the jet engines that operate at 10,000 m above ground.

3.6 Geothermal Power Plants

Geothermal power plants utilize resources at moderate to low temperatures that rarely exceed 200°C. All geothermal resources – dry steam, two-phase fluid, liquid water at low temperature, liquid water at high pressure, and dry rock – are low-exergy resources. The thermodynamic properties of the geothermal fluids are approximated by the properties of salty water [8] and very often are modeled as pure water. Figure 3.11 depicts the specific exergy of saturated water in the liquid and vapor states as well as of water-steam mixtures as a function of temperature, and with the dryness fraction x as a parameter. The water and steam properties were obtained from [5], $T_0 = 300$ K and $P_0 = 100$ kPa. It is observed that, the exergy of the dry steam resources is 5–10 times higher than the temperature of saturated liquid water. Dry steam resources (e.g. those in the Geysers, California and Larderelo, Tuscany) have more specific exergy than other resources.

It is also observed in Figure 3.11 that, because of the prevailing lower temperatures, the exergy of geothermal resources is significantly lower than the exergy of the steam supplied to the turbines of conventional and nuclear power plants. For this reason, geothermal power plants are constructed differently than fossil fuel and nuclear units, with an emphasis on higher exergetic efficiencies and lower lost power. The geothermal resource that supplies the power plant with thermal energy dictates to a large extent the type of power plant that is built.

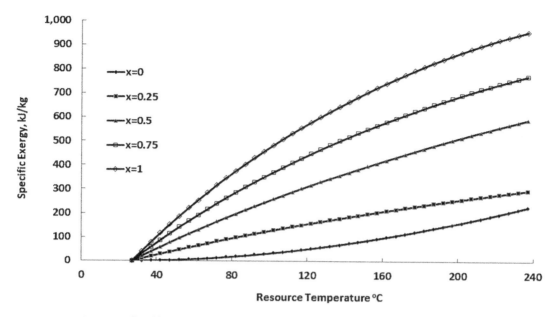

Figure 3.11 Specific exergy of saturated water and of water-steam mixtures. Property data from [5], $T_0 = 300$ K, $P_0 = 100$ kPa.

3.6.1 Dry-Steam Units

The dry-steam power plants are the simplest of all geothermal units and utilize the few dry steam resources, which have the highest specific exergy as shown in Figure 3.11. Steam from the geothermal well passes through a *steam dryer* or *separator* where water droplets are removed. It is then fed to a turbine, which drives the electric generator, produces power, and finally exhausts in the condenser. When the mass flow rate of the steam produced is denoted by $\dot{m}$ and the specific exergy of the steam by e_1, the maximum power from the dry steam power plant is:

$$\dot{W}_{max} = \dot{m} e_1 = \dot{m}[h_1 - h_0 - T_0(s_1 - s_0)]. \qquad (3.17)$$

With a condenser temperature close to T_0 and a turbine efficiency η_T, the actual power produced by dry steam units, when the condenser temperature is at T_0, is:

$$\dot{W} \approx \dot{m} \eta_T [h_1 - h_0 - T_0(s_1 - s_0)]. \qquad (3.18)$$

Example 3.8: Five geothermal wells provide 46 kg/s of dry saturated steam to a geothermal power plant at 210°C. What is the maximum power this plant may deliver and what is the power delivered when the efficiency of the turbine is 78%? $T_0 = 300$ K, $P_0 = 1$ atm.

Solution: The specific exergy of saturated steam at 210°C is approximately 894 kJ/kg. Therefore, the maximum power that may be produced by this power plant is $46 * 896 = 41{,}216$ kW. Assuming that the condenser of the power plant is maintained at the environmental temperature, $T_0 = 300$ K, the power produced with a 78% efficient turbine is 32,148 kW.

3.6.2 Single-Flashing Units

Figure 3.12 shows the schematic diagram of a single-flashing geothermal unit and the corresponding thermodynamic T,s diagram. The geothermal fluid from the well, at state *1*, is a liquid or two-phase mixture composed of saturated steam and liquid water. The fluid enters a flashing chamber, where its pressure is reduced to that of state 2. At the end of the flashing chamber and the steam drier, steam and water are separated. In the T, s diagram the separate states are denoted with the prime symbol (2') for the liquid and the double-prime symbol (2") for the vapor. Flashing chambers are well-insulated and the process 1-2 is at constant enthalpy. The fraction of steam produced in the flashing process is obtained from the enthalpy balance:

$$h_1 = h_2 = (1 - x_2)h_{2'} + x_2 h_{2''} \Rightarrow x_2 = \frac{h_1 - h_{2'}}{h_{2''} - h_{2'}}. \quad (3.19)$$

The flashing process is highly irreversible and entails a great deal of lost work. Typically, 10–25% of the geothermal fluid is converted to steam in the flashing chamber. When the total mass flow rate of the geothermal fluid from the well is denoted by $\dot{m}$, the amount of steam directed to the turbine is $\dot{m}x_2$. The maximum power that may be produced by the turbine is equal to the product of the exergy of the vapor phase and the mass flow rate of this vapor:

$$\dot{W}_{\max} = \dot{m} x_2 e_{2''}. \quad (3.20)$$

The states 2, 2', and 2" are at the same pressure and temperature. The flashing pressure at state 2 – or, equivalently, the saturation temperature $T_2 = T_{sat}(P_2)$ that corresponds to P_2 – is an adjustable parameter that may be optimized to yield maximum power. A quick analysis reveals that when the temperature T_2 is very close to the geothermal fluid temperature T_1 the exergy of the produced steam is high but the

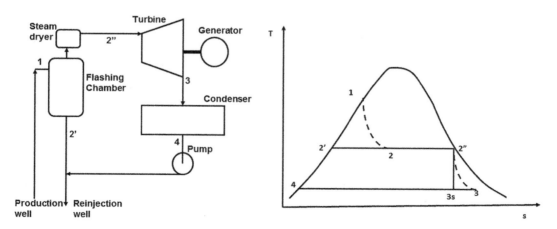

Figure 3.12 Schematic diagram of a single-flashing geothermal unit and the corresponding T, s diagram.

3.6 Geothermal Power Plants

fraction of steam produced, x_2, almost vanishes and the product $x_2 e_{2''}$ is also close to zero. At the other extreme, when the temperature T_2 is close to that of the environment, which is also the condenser temperature, the exergy of the produced steam almost vanishes and the product $x_2 e_{2''}$ again is close to zero. Between the two extremes there is an optimum value where this product and the power denoted by Eq. (3.20) are maximized. The maximum value of the product $x_2 e_{2''}$ is obtained by numerical optimization, with T_2 being the optimization parameter. Figure 3.13 shows the results of the optimization process when the geothermal fluid is saturated liquid water at 210°C and the condenser is at the environmental temperature of 27°C. It is observed that the product $x_2 e_{2''}$ attains a maximum value when T_2 is approximately equal to the average temperatures of the geothermal fluid and the environment, $T_{2opt} = (T_1 + T_0)/2$ [9, 10]. If the condenser temperature, T_3, is higher than T_0 then T_{2opt} is approximately equal to the average of the condenser and the geothermal fluid temperatures:

$$T_{2opt} = \frac{T_1 + T_3}{2} \quad \text{and} \quad P_{2opt} = P_{sat}(T_{2opt}). \tag{3.21}$$

The actual power produced by a turbine with efficiency η_T is:

$$\dot{W}_{act} = \dot{m} x_2 \eta_T (h_{2''} - h_{3s}) = \dot{m} x_2 (h_{2''} - h_3), \tag{3.22}$$

where η_T is the turbine efficiency and state 3 is the thermodynamic state of the turbine effluent.

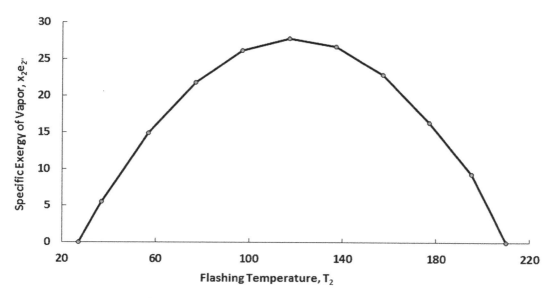

Figure 3.13 Maximum power in single-flash geothermal power plants is obtained when the flashing temperature is the average of the condenser and the resource temperatures.

Example 3.9: Determine the maximum specific work, in kJ/kg of the geothermal resource, the thermal efficiency and the exergetic efficiency of single flashing-units, which utilize geothermal resources that produce saturated liquid water at $T_R = 120, 140, 160, 180, 200$ and $220°C$. $T_0 = 30°C = 303$ K.

Solution: The maximum specific work is obtained from Eq. (3.20) as: $x_2 e_{2''}$, with x_2 determined from Eqs. (3.19) and (3.21). The thermal (first law) efficiency is based on the specific heat, q_R, which may be obtained when the geothermal resource is cooled to T_0: $q_R \approx c(T_R - T_0)$. The exergetic efficiency is based on the exergy of the resource, e_R. The following table gives the pertinent parameters and efficiencies for this example.

Resource Temp. T_R (°C)	Resource Exergy e_R (kJ/kg)	Dryness Fraction x_2	Exergy of Vapor, $e_{2''}$ (kJ/kg)	Specific Work (kJ/kg)	Thermal Efficiency, $\eta_{th} (=\eta_I)$ (%)	Exergetic Efficiency, η_{II} (%)
120	47.1	0.0818	312.6	25.6	6.8	54.3
140	68.4	0.1016	371.2	37.7	8.2	55.1
160	93.0	0.1222	426.5	52.1	9.6	56.0
180	121.0	0.1439	478.6	68.9	11.0	56.9
200	152.3	0.1668	527.9	88.0	12.4	57.8
220	186.8	0.1912	574.5	109.8	13.8	58.8

An important observation in this table is that the thermal (first law) efficiencies are very low. This is a consequence of the low temperatures of most geothermal resources. The exergetic efficiencies are fairly high – and invariably higher than the first law efficiencies. This signifies that geothermal resources are rather well utilized with single-flashing power plants.

3.6.3 Dual-Flashing Units

In an optimized single-flashing unit only 10–25% of the geothermal fluid is converted to steam during the flashing process. This leaves 75–90% of the total mass flow rate from the geothermal wells to be discarded or reinjected at the intermediate temperature T_2, thus, wasting a large fraction of the geothermal resource's exergy. The discarded fluid may be utilized to produce more power, a process accomplished in the dual-flashing geothermal power plants.

The dual-flashing units utilize a second flashing chamber, which admits the liquid effluent from the first flashing chamber. The reduced pressure and temperature in the second chamber produces additional steam at the reduced pressure P_3 with corresponding saturation temperature $T_3 = T_{sat}(P_3)$. The second quantity of steam also passes through a *steam drier/separator* to remove any droplets and is directed to a low-pressure turbine, where it produces an additional amount of power. The schematic diagram of the dual flashing geothermal unit and the corresponding

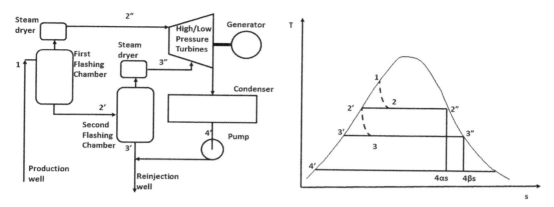

Figure 3.14 Schematic diagram of a dual flashing geothermal power plant and the corresponding T, s diagram.

thermodynamic diagram are shown in Figure 3.14. As with the single flashing units, the fraction of steam, x_2, produced in the first flashing chamber is calculated using Eq. (3.19). The additional dryness fraction at state 3 is calculated from the similar expression:

$$h_{2'} = h_3 = (1 - x_3)h_{3'} + x_3 h_{3''} \Rightarrow x_3 = \frac{h_{2'} - h_{3'}}{h_{3''} - h_{3'}}. \qquad (3.23)$$

The mass flow rate of the effluent from the first flashing chamber is not $\dot{m}$, but $\dot{m}(1 - x_2)$. Therefore, the vapor mass flow rate from the second flashing chamber to the turbine is: $\dot{m}(1 - x_2)x_3$. The total maximum power produced by the dual-flashing units is derived from the two streams of steam and is given by the expression:

$$\dot{W}_{max} = \dot{m}[x_2 e_{2''} + (1 - x_2)x_3 e_{3''}]. \qquad (3.24)$$

The actual work considers the actual temperature/pressure of the condenser, state 4, and the efficiencies of the high- and low-pressure turbines.

As in the case of the single flashing power plants, the pressures and corresponding temperatures of the two flashing chambers are free parameters to be optimized. When the geothermal well produces compressed liquid or saturated liquid water, the optimum temperatures for the flashing chambers are approximately given when the available temperature range is equally divided by the temperatures T_2 and T_3:

$$T_2 = T_4 + \frac{2(T_1 - T_4)}{3} \quad \text{and} \quad T_3 = T_4 + \frac{T_1 - T_4}{3}. \qquad (3.25)$$

The corresponding pressures of the two flashing chambers are the saturation pressures of the last two temperatures: $P_2 = P_{sat}(T_2)$ and $P_3 = P_{sat}(T_3)$. In actual geothermal power plants the lowest flashing pressure is designed to be higher than the atmospheric pressure ($P_3 > P_0$) in order to avoid air in-leakage in the flashing chamber.

3.6.4 A Useful Exercise in Exergy Analysis

Dual flashing units utilize a higher fraction of the geothermal fluid exergy than single flashing units. However, still the effluent of the second flashing chamber is liquid water at a temperature higher than T_0. A third and then a fourth flashing processes may be added to extract higher fractions of the geothermal fluid exergy, but this does not practically happen because of the added cost. The isenthalpic flashing processes are irreversible and produce significant entropy and lost work. The latter shrinks with the magnitude of the flashing temperature drop and vanishes as the flashing temperature drop diminishes. Adding more flashing processes reduces the total lost work by decreasing both the temperature of the liquid water effluent and the lost work in each flashing process. Therefore, the addition of several flashing chambers improves both the first-law and the exergetic efficiency of geothermal power plants.

It becomes apparent that, when the state of the geothermal resource is fixed, the conversion of the maximum amount of the exergy of the resource to electric energy is achieved by two improvements: 1. the minimization of the temperature difference between flashing processes; and 2. the maximization of the efficiency of the turbines that produce power. This offers an interesting exercise in the theory of thermo-mechanical energy conversion: Consider a multi-flash system, of n flashing stages – 1-2, 2-3, ..., k-(k+1), (k+1)-(k+2), ..., n-(n+1) – part of which is shown in Figure 3.15: The vapor from the kth flashing process is fed to the kth turbine and the liquid is fed to the $(k+1)$th flashing chamber. The vapor from the last chamber is fed to the $(k+1)$th turbine, the liquid to the $(k+2)$th flashing

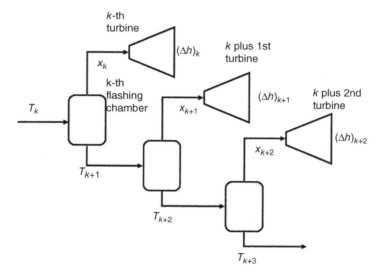

Figure 3.15 A multi-flash system, with n flashing stages will extract all the exergy of the geothermal fluid as $n \to \infty$.

chamber etc. If a total of n flashing chambers is used, the total power produced by the geothermal power plant is:

$$\dot{W}_{tot} = \sum_{k=1}^{n} \dot{m}_k (\Delta h'')_k \eta_k, \qquad (3.26)$$

where $\dot{m}_k$ is the mass flow rate of vapor produced in the (k)th flashing chamber; $(\Delta h'')_k$ is the isentropic enthalpy drop in the (k)th turbine, which is equal to the exergy of the vapor at the temperature T_k, that is $(\Delta h'')_k = e''_k$; and η_k is the efficiency of the (k)th turbine. Since the flashing processes are isenthalpic, the mass flow rate of steam produced in the (k)th chamber is given by the expression:

$$\dot{m}''_k = \dot{m}'_{k-1} \frac{h'_{k-1} - h'_k}{h_{fg}} = \dot{m}(1 - x_1)(1 - x_2)\ldots(1 - x_{k-1})x_k. \qquad (3.27)$$

In Eq. (3.27) $\dot{m}$ is the mass flow rate supplied by the geothermal wells to the first flashing chamber; the prime symbol $(')$ denotes the saturated liquid state; and h_{fg} is the latent heat of vaporization, which is almost constant over small ranges of the water/steam temperature. Hence, the total power from the entire series of flashing and expansions may be written as follows:

$$\dot{W}_{tot} = \dot{m} \sum_{k=2}^{n+1} \left[\prod_{i=2}^{k} (1 - x_{i-1}) \right] x_k (\Delta h)_k \eta_k. \qquad (3.28)$$

When the number of the flashing processes is large, a very small amount of steam is produced in each flashing chamber. The enthalpy drop on the liquid side may be approximated in terms of the specific heat of saturated liquid water, c'_p. Assuming that the properties h_{fg}, and c'_p, are constant, the following approximation is derived for the total power produced by the multiple flashing systems, in terms of the flashing temperatures, $T_1, T_2, T_3, \ldots, T_k$:

$$\dot{W}_{tot} \approx \frac{\dot{m} c'_p}{h_{fg}} \sum_{k=2}^{n+1} \left[\prod_{i=2}^{k} (1 - x_{i-1}) \right] (T_{k-1} - T_k) e''_k \eta_k. \qquad (3.29)$$

Using the theory of Lagrange undetermined multipliers for the maximization of total power, one deduces that – as with the single- and the dual-flashing units – maximum power is produced when the available range of temperatures is divided into n equal parts:

$$T_1 - T_2 = T_2 - T_3 = \ldots = T_{k-1} - T_k = \ldots T_{n-1} - T_n = T_n - T_0, \qquad (3.30)$$

Equations (3.21) and (3.25) are special cases of this general expression, derived for $n = 2$ and $n = 3$.

At the limit, $n \to \infty$, which is practically approximated by introducing a very large number of flashing chambers and turbines, and with $\eta_k = 100\%$ the total power generated is equal to the theoretical maximum power: the product of the mass flow rate from the wells and the exergy of the geothermal fluid at state 1:

$$\lim_{n \to \infty} (\dot{W}_{tot}) = \dot{W}_{max} = \dot{m}[e(T_1) - e(T_0)]. \tag{3.31}$$

This maximum power is delivered, in principle, by an isentropic *two-phase expander*, which makes possible the gradual expansion of hot liquids. Although there is considerable engineering research in this area, the actual development of a two-phase expander has been, so far, elusive.

In practice, there are diminishing returns in the marginal amount of power that may be extracted by having more than two flashing chambers and more than two turbines. Currently, the cost of the additional equipment beyond two flashing chambers does not justify the small amounts of generated power. Instead, binary power plants are used, which typically utilize a higher fraction of the geothermal fluid exergy and, as a consequence, have higher exergetic efficiencies.

3.6.5 Binary Units

When the geothermal fluid at the wellhead has low enthalpy and exergy (e.g. when it is liquid water at temperatures less than 150°C) the flashing process produces very small quantities of steam that has low exergy. For such resources it is advantageous to use a heat exchanger to transfer the thermal energy from the geothermal resource to a *secondary fluid* or *working fluid*, which produces vapor and may expand in a turbine. Effectively, the HX becomes the boiler of a basic Rankine cycle that uses the secondary fluid for the production of power. These cycles are called *Organic Rankine Cycles (ORCs)*. A schematic diagram of a binary geothermal unit is shown in Figure 3.16. The geothermal fluid – the primary fluid – enters the HX from the production well and exits at the reinjection well. Heat is transferred in the HX to the secondary fluid, the working fluid of the Rankine cycle. The latter exits the HX as saturated or superheated vapor at sufficiently high pressure and temperature, and imparts its energy to the turbine, which drives the electric generator. The Rankine cycle of the secondary fluid is completed with the condenser and the condensate pump.

Suitable choices for the secondary fluid are substances with high saturation pressures and relatively low saturation temperatures. Candidate fluids are most of the refrigerants, including ammonia, and several hydrocarbons, such as propane, butane, iso-butane, and pentane. Figure 3.17 depicts the *L-T* diagram for the operation of the HX: The primary geothermal fluid supplies with heat the secondary fluid and cools, following the upper curve and going through the states 1-4. The secondary fluid preheats in process 4-3, evaporates in process 3-2 and becomes superheated vapor in process 2-1. If $\dot{m}_p$ denotes the mass flow rate of the primary fluid and $\dot{m}_s$ the mass flow rate of the secondary fluid, the enthalpy balances in the three parts of the HX of Figure 3.17 yield the following equations for the three parts of the HX:

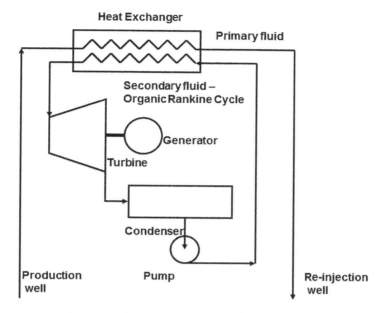

Figure 3.16 Schematic diagram of a binary geothermal power plant

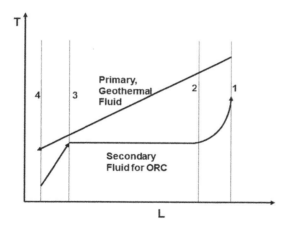

Figure 3.17 Length-Temperature diagram of the binary unit's HX. The pinch point (at 3) is apparent.

$$\dot{m}_p(h_{p1} - h_{p2}) = \dot{m}_s(h_{s1} - h_{s2})$$
$$\dot{m}_p(h_{p2} - h_{p3}) = \dot{m}_s(h_{s2} - h_{s3}). \quad (3.32)$$
$$\dot{m}_p(h_{p3} - h_{p4}) = \dot{m}_s(h_{s3} - h_{s4})$$

The primary fluid remains in the liquid state throughout the cooling process. Therefore, its enthalpy difference may be approximated by the products of the (assumed constant) specific heat capacity and the temperature differences. The same approximation may be used for the preheating (process 4-3) of the secondary fluid. During the evaporation process (3-2) the specific enthalpy of the secondary fluid is the latent heat, h_{fg}. Hence, the set of Eq. (3.32) simplifies to:

$$\dot{m}_p c_{pp}(T_{p1} - T_{p2}) = \dot{m}_s(h_{s1} - h_{s2})$$
$$\dot{m}_p c_{pp}(T_{p2} - T_{p3}) = \dot{m}_s h_{fg} \qquad , \qquad (3.33)$$
$$\dot{m}_p c_{pp}(T_{p3} - T_{p4}) = \dot{m}_s c_{ps}(T_{s3} - T_{s4})$$

where c_{pp} and c_{ps} are the specific heat capacities of the liquid state of the primary and the secondary fluid, respectively.

The mass flow rate, $\dot{m}_p$, and the primary fluid inlet temperature, T_{p1}, of geothermal binary units are imposed by the state of the geothermal resource. The evaporation temperature of the secondary fluid, T_{s2}, is decided by optimization, as described in Example 3.10. HX design considerations impose a 3–10°C difference at the *pinch point* of the HX, which occurs at state 3 in Figure 3.17. Therefore, the difference $T_{p3} - T_{s3}$ is in the range 3–10°C. These conditions and the set of equations in Eq. (3.32) or Eq. (3.33) are sufficient to determine the mass flow rate of the secondary fluid; $\dot{m}_s$, the exhaust temperature of the secondary fluid, T_{s1}; and the exhaust temperature of the primary fluid, T_{p4}. In an optimized binary geothermal unit, the exhaust temperature of the geothermal resource, T_{p4}, is very close to T_0, and this implies that almost all of the exergy in the geothermal resource is used in the ORC. In general, binary geothermal power plants utilize a higher percentage of the exergy of the geothermal fluid than flashing units.

Example 3.10: Six geothermal wells produce 420 kg/s of water at 10 bar and 120°C. It has been decided to produce power with a simple ORC cycle without superheat and with isobutane as the working fluid. The pinch-point temperature difference is 5°C. Cooling water is available at 30°C. Design a cycle for the production of maximum power. $T_0 = 30°C = 303$ K.

Solution: The specific exergy of water at 10 bar and 120°C (with $T_0 = 30°C = 303$ K) is 47.1 kJ/kg and, therefore, the maximum power that may be obtained from this geothermal resource is 19,782 kW.

Given a state 2 for the isobutane exiting the HX in the ORC (defined by the boiling temperature, T_b, and $x = 1.0$), the maximum power of the cycle is equal to $\dot{W}_{max} = \dot{m}_s \Delta h_s$, where Δh_s is the specific enthalpy difference in an isentropic expansion. State 2 may be optimized to produce maximum power using a numerical iteration. The pertinent set of equations in this case is obtained from Eq. (3–33):

$$\dot{m}_w c_w(120 - T_b - 5) = \dot{m}_i h_{ifg}$$
$$\dot{m}_w c_w(T_b + 5 - T_{wout}) = \dot{m}_i(h_{i3} - h_{i4}) ,$$

where T_b is the boiling temperature of isobutane, which is denoted by the subscript *i*. The subscript *w* denotes the water. Using T_b as a parameter, the following table is obtained:

Boiling Temperature, T_b (°C)	Spec. Enth. Difference, Δh_s (kJ/kg)	Iso-butane mass flow rate, $\dot{m}_i$ (kg/s)	Power, $\dot{W}_{max}$ (kW)	Temperature T_{wout} (°C)	Exergetic efficiency, η_{II} (%)
30	0.0	461.7	0	31.8	0.0
40	10.6	422.8	4,478	36.0	22.6
50	20.6	382.1	7,871	41.3	39.8
60	30.1	339.0	10,211	47.7	51.6
70	37.5	293.1	10,984	55.5	55.5
75	41.8	268.7	11,235	59.9	56.8
80	44.8	243.2	10,890	64.8	55.0
90	53.4	188.0	10,045	76.2	50.8
100	62.0	124.7	7,726	90.3	39.1
110	69.4	47.5	3,302	108.5	16.7
115	75.3	0.0	0	120.0	0.0

It is observed in this table that maximum power of 11,235 kW is produced when the isobutane boiling temperature is approximately 75°C. At this optimum point the exergetic efficiency of the binary unit is 56.8%. Comparing these results with those of Example 3.9 for the resource temperature of 120°C and 420 kg/s geothermal fluid input (10,752 kW equivalent power and 54.3% exergetic efficiency), it becomes apparent that the maximum power and the corresponding exergetic efficiency are higher in an optimized binary unit than in a single-flashing unit that utilizes the same resource. This happens primarily because flashing entails more irreversibilities than the heat transfer process in ORCs.

3.7 Fuel Cells

All the exergetic calculations with vapor and gas power plants show that the combustion processes destroy a high fraction of the chemical exergy of the fuels. The conversion of chemical energy to thermal energy (to be subsequently converted to mechanical and electric energy) is inherently highly irreversible, entails a great deal of lost work, and is characterized by low exegetic efficiency. One avoids the irreversibilities of such energy transformations by employing *Direct Energy Conversion* (*DEC*) systems, where the conversion of the chemical energy of the fuel to electricity takes place directly without the intermediate production of heat. With DEC systems the Carnot limitation on the energy conversion processes do not apply and the systems may achieve very high first-law and exergetic efficiencies. Currently available DEC systems and processes are convenient, clean, and typically have significantly higher efficiencies than thermal energy conversion systems. The *fuel cells* are DEC systems, similar to batteries. Unlike the batteries, fuel cells are open thermodynamic systems that receive a fuel and an oxidant (e.g. hydrogen and air), exhaust the reaction products (water), and continuously produce electric power.

The first oxygen-hydrogen fuel cells were developed by Sir Humphrey Davy in 1802 (almost one century before the internal combustion engine) and were used in tractors as early as 1839. However, later in the 19th century their use in vehicles shrank with the development of the more convenient internal combustion engines and the discovery of petroleum resources. Among the advantages of fuel cells are:

1. Their thermal and exergetic efficiencies can be very high. In principle, the exergetic efficiency of fuel cells can be 100%.
2. Their operation is clean. Fuel cells produce extremely small amounts of nitrogen oxides (NOX).
3. Their operation is noiseless.
4. Their starting time is very short and their operation may be interrupted or ended at will.
5. Unlike the batteries – another type of DEC system – fuel cells may operate continuously without the need to recharge and regenerate.

The main disadvantage of fuel cells is that the generated voltage is very low, on the order of 1 V. High current compensates for the low voltage and the production of the needed power. The high current in engines causes higher electric energy dissipation and exergy destruction.

Figure 3.18 depicts the schematic diagram of a hydrogen-oxygen fuel cell. Gaseous hydrogen is supplied to the cell on the side of the anode, the negative terminal, and air is supplied on the side of the cathode, the positive terminal. The electrodes are made of a porous material for the gases to diffuse and are coated with a catalyst, usually platinum or palladium that makes the reactions possible at the operating temperatures. A basic solution between the electrodes, typically potassium hydroxide (KOH), dissociates to produce K^+ and OH^- ions. The hydrogen is ionized at the anode, thus releasing two

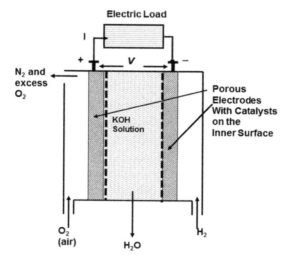

Figure 3.18 Schematic diagram of the hydrogen-oxygen fuel cell.

electrons, and diffuses through the porous electrode towards the cathode where it combines with the hydroxyl ions (OH$^-$) to form water molecules, according to the reaction:

$$H_2 + 2HO^- \rightarrow 2H_2O + 2e^-. \tag{3.34}$$

The anode of the fuel cell has a surplus of electrons and is negatively charged. On the side of the cathode, oxygen gas diffuses into the pores of the electrode and combines with water molecules and electrons to form hydroxyl ions (OH$^-$):

$$\frac{1}{2}O_2 + H_2O + 2e^- \rightarrow 2HO^-. \tag{3.35}$$

The cathode has a deficiency of electrons and is positively charged. The potential difference between the anode and the cathode induces an electric current to pass through the external circuit – the electric load in Figure 3.18.

The overall reaction in this fuel cell is the formation of water:

$$H_2 + \frac{1}{2}O_2 \rightarrow H_2O. \tag{3.36}$$

Hydrogen and air are continuously supplied to their respective compartments in the fuel cell and the produced water is continuously removed via a drain. This allows the fuel cell to operate continuously for long periods of time without the need of recharging, a significant limitation of chemical batteries. In actual fuel cells the production of water dilutes the basic KOH solution. For this reason, part of the diluted solution is periodically drained and concentrated KOH solution is added to restore the desired concentration of the basis.

Because the fuel cell is a DEC device, the maximum work it may produce is equal to the change of the Gibbs' free energy of the reaction, 236,100 kJ/kmol in the case of the hydrogen-oxygen reaction. The maximum voltage that may be developed in the hydrogen-oxygen fuel cell is obtained from Faraday's equation:

$$V_{max} = \frac{-\Delta G^o}{\zeta F} = \frac{236,100}{2 * 96,500} = 1.223\,V. \tag{3.37}$$

where ζ is the number of electrons deposited in the anode ($\zeta = 2$ in this case) and F is Faraday's constant, 96,500,000 Cb/kmol. This is a low voltage, which is typical of fuel cells and batteries. Practical fuel cell systems are connected in series (in stacks) that produce higher voltage.

3.7.1 Practical Types of Fuel Cells

In the first two decades of the 21st century there is a great deal of research and development on practical fuel cells that have high energy conversion efficiency (approaching their theoretical efficiency) are durable, and produce power reliably. Several types of fuels cells have been developed to generate power for engineering systems, such as automobiles and space crafts, and to produce household and industrial

Table 3.8 Fuel cell reactions, Gibbs' free energy, and maximum voltage at 298 K.

Fuel	Reaction	$-\Delta G^o$ (kJ/kmol)	V_{max} (V)
H_2	$H_2 + 1/2 O_2 \rightarrow H_2O$	236,100	1.22
CO	$CO + 1/2 O_2 \rightarrow CO_2$	275,100	1.43
CH_4	$CH_4 + 2O_2 \rightarrow CO_2 + 2H_2O$	831,650	1.08
CH_3OH	$CH_3OH + 3/2 O_2 \rightarrow CO_2 + 2H_2O$	718,000	1.24
C_2H_5OH	$C_2H_5OH + 3O_2 \rightarrow 2CO_2 + 3H_2O$	1,357,700	1.17
NH_3	$4NH_3 + 3O_2 \rightarrow 2N_2 + 6H_2O$	337,900	1.17

power. Practical fuel cells are characterized and classified by the type of electrolytes and porous electrodes [11]:

1. *Proton Exchange Membrane Fuel Cells (PEMFCs)* use polymeric membranes (e.g. Nafion™ 117) and water-KOH solutions. They operate at temperatures below 100°C, where the water remains liquid. The catalysts in the electrodes are noble metals – gold, silver, palladium, and platinum.
2. *Alkaline Fuel Cells (AFCs)* have powered the Apollo missions to the moon and the space shuttles. They also use KOH solution as the electrolyte, they operate in the temperature range 60–250°C and use platinum, chromium, and nickel as catalysts.
3. *Phosphoric Acid Fuel Cells (PAFCs)* use the phosphoric acid, H_3PO_4, as their electrolyte and operate in the range 160–210°C. Porous silicon carbide is used for the diffusion of ions and their electrodes are made of nobble metals (ruthenium, rhodium, palladium, silver, osmium, iridium, platinum, and gold), which are not eroded by the acid, and catalyze the reactions.
4. *Molten Carbonate Fuels Cells (MCFCs)* utilize molten lithium potassium carbonate salt as the electrolyte in a porous matrix of lithium-aluminum oxide and with nickel electrodes. Because the salt electrolyte melts at high temperatures, the range of their operating temperatures is 600–800°C.
5. *Solid Oxide Fuel Cells (SOFCs)* use a dense layer of ceramic material that conducts the ions. Lanthanum-strontium-manganite (LSM) is a material that is frequently used for the cathode and yttrium-stabilized zirconia mixed with nickel metal for high conductivity is a material commonly used as the anode of SOFCs. These fuel cells may operate at temperatures up to 1,100°C.

Advances in chemistry and materials science continuously produce new materials with desirable properties that are adopted for the development of more efficient, more powerful, and more reliable fuel cells.

One of the inherent disadvantages of hydrogen fuel cells is the low density of the gas. Hydrogen fuel cells require high volumes for the tanks that supply the gas. High temperature fuel cells allow the use of other fuels with complex molecules – hydrocarbons, alcohols, and natural gas, principally composed of methane and carbon monoxide – to be used instead of hydrogen. A list of several fuels that have been used in practical fuel cells, their Gibbs' free energies [12], and the maximum voltage they may deliver is in Table 3.8.

Methanol (CH_3OH) and ethanol (C_2H_5OH) are liquids under standard conditions with significantly higher densities than the other fuels. Even though the specific energy (in kJ/kg) of the liquid fuels is lower than that of hydrogen and other gases, the high energy density (in kJ/m^3) is a significant advantage of the liquid fuels because fuel cells and fuel storage systems do not need to occupy high volumes or to be under very high pressures. Ammonia is also a promising fuel because it is stable, relatively inexpensive, and may be stored and transported easier than hydrogen.

3.7.2 Irreversibilities in Fuel Cells

Ions from the anode and cathode must travel through the porous electrodes by diffusion, an extremely slow transport process. The ion transport and the reactions on the surface of the catalysts and the porous media are irreversible processes that entail entropy production and lost work due to the following mechanisms:

1. Ohmic losses: The flow of ions is an electric current and is resisted by the other molecules within the porous medium. Ohmic losses are similar to the resistance losses in electric conductors.
2. Concentration losses: High current densities limit the mass transfer in the porous media and cause the build-up of high concentration gradients.
3. Activation losses: These are caused by the higher-than-reversible reaction activation potentials on the surface of the catalysts and the finite rate of the chemical reactions.
4. Fuel crossover losses: Liquid hydrocarbon molecules that are water soluble are dragged with the H^+ ions through the pores to the cathode by electro-osmosis. When they reach the cathode, the hydrocarbon molecules combine with the oxygen ions (and effectively undergo combustion) without producing electrons [13].

All the irreversibilities associated with the practical operation of fuel cells are lumped together in the concept of the fuel cell efficiencies. The *voltage efficiency*, the *electric efficiency*, and the exergetic efficiency are three figures of merit commonly used for fuel cells:

$$\eta_V = \frac{V}{V_{max}}, \quad \eta = \frac{E}{-\Delta H^o}, \quad \text{and} \quad \eta_{II} = \frac{E}{-\Delta G^o}, \qquad (3.38)$$

where V is the actual voltage produced and V_{max} the maximum voltage obtained from Eq. (3.37); E is the actual electric energy produced by the fuel cell per kmol of fuel; and $-\Delta H^o$ and $-\Delta G^o$ are the molar enthalpy and Gibbs' free energy of the reaction, respectively. Because for most reactions used with fuel cells $\Delta H^o \approx \Delta G^o$, the last two figures of merit are almost equal. Electric efficiencies of commercial fuel cells are in the range 50–60% [14] with experimental fuel cells approaching 75%. The higher range of these efficiencies is by far greater than the exergetic efficiencies of fossil-fuel thermal engines.

Example 3.11: A cylindrical tank with 1.2 m diameter and 3 m height contains hydrogen at 100 bar and 300K. The hydrogen in the tank may be used to produce electric energy by combustion in a gas turbine with 40% overall thermal efficiency or in a fuel cell with 63% efficiency. Determine the amount of electric energy produced using the two methods and their exergetic efficiency.

Solution: The volume of the cylinder is: $V = \pi D^2 H/4 = 3.39$ m^3. At 100 bar, hydrogen may be modeled as an ideal gas. The mass of the hydrogen in the cylinder is: $m = PV/RT = 100 * 10^5 * 3.39/(4{,}157 * 300) = 27.2$ kg (if, instead of the ideal gas equation, REFPROP [5] is used the mass is 26.4 kg, a 3% difference).

The heat of combustion of hydrogen (LHV) is 119,950 kJ/kg. Hence, the combustion of 27.2 kg of hydrogen produces $3.26 * 10^6$ kJ of heat. When this heat is converted in a thermal engine with 40% thermal efficiency, it produces $1.31 * 10^6$ kJ or 362.5 kWh of electric energy.

If a fuel cell is used, its efficiency is based on $-\Delta G^o$, which is 118,050 kJ/kg. With 63% efficiency, the fuel cell produces $118{,}050 * 27.2 * 0.63 = 2.02 * 10^6$ kJ = 562 kWh.

The corresponding exergetic efficiencies of the two electric energy production methods are 40.1% and 63%. It is apparent that the DEC conversion via fuel cell produces 55% more electric energy than the gas turbine and it is clearly the thermodynamically superior energy conversion method.

Example 3.12: A car has a gasoline-operated IC engine with 22% efficiency and a 74 L gasoline tank. This engine is substituted with a hydrogen-air PEMFC that has 58% efficiency. The electric motor of the new car is 95% efficient. Determine the mass of hydrogen required in a hydrogen tank for this car to have the same distance range. If the hydrogen is kept at 400 bar, and behaves as an ideal gas, how much is the volume of this tank if the car operates at ambient temperatures close to 300 K?

Solution: The actual work that becomes available as the motive power of the car must be first calculated. The density of gasoline is 0.719 kg/L, and the mass of the gasoline in the tank is 53.2 kg. Since the heat of combustion (LHV) of gasoline (modeled as octane) is 44,430 kJ/kg, the IC-engine vehicle with a full tank has available $74 * 0.719 * 44{,}430 = 2.36 * 10^6$ kJ of thermal energy. With a 22% efficient engine, the car has available $0.52 * 10^6$ kJ of mechanical work, which is used for the propulsion of the car.

Because the fuel cell engine must have the same distance range, with a 95% electric motor efficiency and 58% fuel cell efficiency, there must be enough hydrogen in the tank to supply the same quantity of work. Therefore the exergy of the hydrogen in the tank must be $0.52 * 10^6/(0.95 * 0.58) = 0.94 * 10^6$ kJ.

From Table 2.2: for hydrogen $-\Delta G^o = 236{,}100$ kJ/kmol and, hence, $0.94 * 10^6/236{,}100 = 4.00$ kmols (8 kg) of hydrogen need to be stored in the hydrogen tank at 400 bar ($400 * 10^5$ Pa). At this pressure, using the ideal gas approximation for hydrogen

($PV = nRT$), yields: $V = 4.00 * 8{,}314 * 300/400 * 10^5 = 0.249$ m^3 = 249 L. A more accurate value for the density of hydrogen at the 400 bar pressure and 300 K temperature is 25.815 kg/m^3 [5]. This yields a more accurate value for the volume of the tank: 310 L.

It is observed in this example that, because of the low density of hydrogen, the tank of the fuel cell driven car must be significantly larger than that of the gasoline driven car.

Example 3.13: Instead of a hydrogen-oxygen fuel cell, an ethanol-air fuel cell with 53% electric efficiency is developed to be used with the car of Example 3.12. Determine the mass of ethanol to be stored in the car and the volume of the ethanol tank. The density of ethanol is 789 kg/m^3.

Solution: The fuel cell in this case must supply the same quantity of work and the exergy of ethanol in the tank must be $0.52 * 10^6/(0.95 * 0.53) = 1.03 * 10^6$ kJ. Therefore, $1.03 * 10^6/1{,}357{,}700 = 0.759$ kmol (34.9 kg) of ethanol need to be stored in the tank. Since the density of ethanol is 789 kg/m^3, the volume of the tank is 44.2 L.

It is observed that the use of a liquid fuel in the fuel cell results in significant reduction of the volume of the fuel tank. An additional benefit (for safety) is that the pressure of the fuel is atmospheric and not in the hundreds of atmospheres.

3.7.3 Combined Heat and Power Production with Fuel Cells – Cogeneration

The fraction of the chemical energy, $-\Delta G^o$, that is not converted to electricity is dissipated as heat in the fuel cell stack. Because of this, fuel cell systems generate high heat fluxes that must be removed by cooling systems, typically by cooling water. All or part of the heat produced may be used for the heating needs of buildings and industrial processes. The combined generation of electric power and heat would utilize up to 85% of the total exergy of the fuel [14].

Electric power cogeneration with the combination of fuel cells and a Rankine cycle may be used in systems where heat is not needed. Cogeneration is particularly advantageous with high temperature fuel cells, for example the MCFC and SOFC types that operate at temperatures in the range 600–1,000°C. The following example illustrates the rates of exergy flow and the electric power produced by such a system.

Example 3.14: Twenty solid oxide fuel cells operate with liquid methanol (CH_3OH) and produce 10 MW of electric power (0.5 MW each). The exergetic efficiency of the fuel cells is 52% and their operational temperature is 780°C (1,053 K). A simple Rankine cycle is proposed to remove this heat and produce additional power. The lower temperature of the cycle is 36°C, and the steam is supplied to the turbine at 40 bar and 500°C. The efficiencies of the steam turbine and the pump are 82% and 80%

respectively. Determine the methanol rate consumed, the power produced, the overall exergetic efficiency of the combined system, and the lost power in the components of this system. $T_0 = 300$ K.

Solution: The fuel cells produce 10 MW of power with efficiency 52%. Therefore, the equivalent exergy input to the 20 fuel cells is 10/0.52 = 19.231 MW. From Table 3.8 the Gibbs' free energy of methanol, $-\Delta G^o$, is 718 MJ/kmol = 22.44 MJ/kg. Therefore, 0.857 kg/s of methanol are used by the system (equivalent to 0.027 kmol/s).

The rate of heat available for the steam cycle is the energy dissipated in the fuel cells: 19.231 − 10 = 9.231 MW. The analysis of the proposed Rankine cycle shows that the specific pump work is 5 kJ/kg of water/steam; the specific turbine work is 1,036 kJ/kg of water/steam; and the specific heat input is 3,289 kJ/kg of water/steam. Therefore, the mass flow rate of the steam in the Rankine cycle is 2.807 kg/s and the additional power produced is (1,036 − 5) ∗ 2.807 = 2,894 kW (2.894 MW) for a total power production by the fuel cell-Rankine cycle system of 12.894 MW.

The exergetic efficiency of the combined system is: 12.894/19.231 = 67.04%.

The rate of exergy entering the entire system is: 19.231 MW. The fuel cells produce 10 MW of power and 9.231 MW of heat at 1,053 K. The latter corresponds to a rate of exergy equal to 9.231 ∗ (1 − 300/1,053) = 6.601 MW. Hence, the lost power in the fuel cells is 2.630 MW.

During the heat transfer process from the fuel cells to the cycle, 6.601 MW of exergy is received for the production of 2.807 kg/s of steam with specific exergy 1,323 kJ/kg. The exergy rate out of the HX is 3.714 MW and the lost power is 2.887 MW.

The rest of the steam cycle receives rate of exergy 3.714 MW and produces 2.894 MW of power. The lost power in the rest of the cycle (which includes the heat dissipation in the condenser) is: 0.820 MW.

It is observed that the highest rates of lost work (and entropy production) occur in the operation of the fuels cells and the HX. This indicates that engineers should strive to improve the operations of the two subsystems, for example by using fuel cells with higher efficiency and by optimizing the pressure and temperature of the generated steam.

It must be noted that the exergetic efficiency of 67.04% achieved in the combined system with the SOFC is by far higher than the efficiency of other thermal engines, where fuel combustion takes place. This happens because most of the power produced by this combined system (10 MW) is generated by DEC in the fuel cell and not by combustion in the boiler.

3.8 Photovoltaics Systems

The total power from the sun that reaches the outer limit of the earth's atmosphere is $1.73 * 10^{14}$ kW, by far higher than the global demand for power. The solar power – often called *incident solar radiation* and shortened to *insolation* – contributes annually total energy of $5.46 * 10^{21}$ MJ, more than 100 million times the annual energy used by the earthlings. Solar energy is abundant, free of charge, almost uniformly distributed in most populated areas, and available to all nations on the planet. However, only a very small fraction of the insolation is currently converted to useful energy. Photovoltaic (PV) cells and thermal solar power plants contribute a small fraction of the globally produced electricity, while passive solar heating systems provide with space heating and hot water a small fraction of the domestic heating needs. Another small fraction of the insolation is used in the photosynthesis process for the production of food and biomass.

3.8.1 The Solar Spectrum and Insolation in the Terrestrial Environment

The radiation emitted by the sun is not monochromatic and has the characteristics of black-body radiation emitted in the range 5,800–5,900 K. The full spectrum of solar power is transmitted to the outer limit of the terrestrial atmosphere. When the radiation enters the atmosphere, gases preferentially absorb parts of the spectrum: Ozone absorbs in the ultraviolet range; diatomic oxygen, water vapor, and carbon dioxide absorb at discrete wavelengths in the visible and infrared parts. By the time the insolation reaches the surface of the earth, part of the radiation intensity has already been absorbed. As a consequence, the spectrum received by energy conversion on the earth's surface is not smooth and continuous, but exhibits several gaps due to the preferential absorption by the atmospheric gases. The three graphs of Figure 3.19 depict the energy density (in W/m² nm) of the spectrum at the edge of the terrestrial atmosphere; the approximation of this spectrum as black body radiation emitted at T = 5,900 K; and the energy density of the solar spectrum received on a horizontal surface at sea level. The area under each graph represents the total irradiance (in W/m²) of the spectrum. The parameter m denotes the thickness (mass) of the atmosphere through which the solar energy traverses, with $m = 1$ denoting that the solar rays are perpendicular to the earth's surface. The energy density decreases with increasing m because a higher fraction of the solar radiation is absorbed with the increased thickness of the gaseous atmospheric layer. On a given location at the earth's surface, m is a minimum at the solar noon.

The sun may be approximated as a *black-body*, whose radiation has several convenient properties. Among these are: 1. the power emitted or absorbed is homogeneous and constant in all directions; and 2. the energy density, I_λ, is solely a function of the temperature of the black body. It must be noted in Figure 3.19 that approximately 32% of the solar insolation does not reach the surface of the earth and is retained or reflected by the gases in the atmosphere.

At the edge of the terrestrial atmosphere the energy density of solar radiation may be given by Plank's expression:

$$I_\lambda = 6.85 * 10^{-5} \frac{2hc^2}{\lambda^5 \left(\exp\left({hc}/{\lambda k_B T} \right) - 1 \right)}, \tag{3.39}$$

where λ is the radiation wavelength; T is the absolute temperature of the radiation source; c is the speed of light, $3 * 10^8$ m/s; k_B is the Boltzmann constant, $1.380 * 10^{-23}$

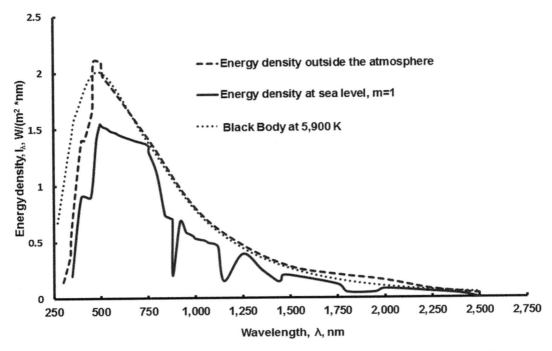

Figure 3.19 The energy density as a function of the wavelength for a black body at 5,900 K; the solar spectrum at the edge of the atmosphere; and the solar spectrum on the surface of the earth. Data from [10] and www.nrel.gov/gis/solar.html.

J/K; and h is Plank's constant, $6.62*10^{-34}$ Js. The constant $6.85*10^{-5}$ is a geometric factor that accounts for the distance from the sun to the earth. Since the speed of light is equal to the product of the frequency of radiation ν and the wavelength:

$$c = \lambda \nu, \tag{3.40}$$

the energy density may also be written in terms of the frequency of radiation, I_ν:

$$I_\nu = 6.85*10^{-5} \frac{2h\nu^3}{c^2 \left(\exp\left(^{h\nu}/_{k_B T}\right) - 1 \right)}. \tag{3.41}$$

It is apparent from Eqs. (3.39) to (3.41) that the energy density of the solar spectrum is approximately zero at the two extremities of the spectrum – where ν and λ are zero and very large respectively – and that between these values there is a maximum. The maximum of the energy density, λ_{max}, is given by *Wien's Law*:

$$\lambda_{max} T = 0.0029 \ mK. \tag{3.42}$$

The total power (insolation), $\dot{W}_{rad}$, emitted by a black body is calculated by integrating the energy density over wavelengths or frequencies:

$$\dot{W}_{rad} = A \int_0^\infty I_\lambda d\lambda = A \int_0^\infty I_\nu d\nu = A\sigma T^4, \tag{3.43}$$

where σ is the Stephan–Boltzmann constant, $5.67 * 10^{-8}$ W/(m² K⁴). The total insolation in a given location on the earth's surface is determined by integrating the local solar spectrum. Given the characteristics of the spectrum on the earth's surface, as shown in Figure 3.19, this may only be accomplished numerically.

In practice the local irradiance is determined experimentally by radiometers. Based on measurements and modeling, extensive data sets have been developed and published in most countries that contain the hourly values of irradiance at several locations [15]. Since the insolation on the surface of the earth is variable with spatial and temporal dependence, the total incident energy during a time period $0 - t$ is:

$$E = \int_0^t \dot{W}_{rad} dt. \tag{3.44}$$

An alternative way to analyze the solar radiation – one that assists with the operation of PV cells – is to consider that the radiation is emitted and received as a flux of photons with distinct wavelengths, frequencies, and energies. The relationship between the wavelength and frequency is expressed by Eq. (3.40) and the energy of the photons is:

$$E_{ph} = \frac{hc}{\lambda} = h\nu. \tag{3.45}$$

Figure 3.20 depicts the frequency and wavelength of photons as a function of their energy in eV using Eq. (3.45). Energetic photons have the ability to dislodge electrons from their ordinary orbits, convert them to free electrons and generate a voltage in PV cells. Photons with low energy do not dislodge electrons and their energy is dissipated as heat.

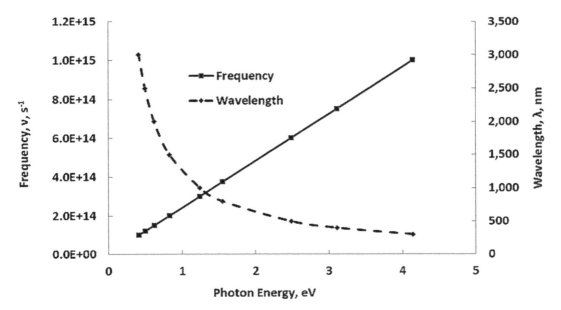

Figure 3.20 Frequency and wavelength of photons as a function of their energy.

3.8.2 The p-n Junctions and Photovoltaic Cells

Atomic nuclei are surrounded by electrons that revolve in well-defined orbits. The electrons have distinct energy levels and are subjected to the *Pauli Exclusion Principle*, which stipulates that two electrons may not exist in the same atom, unless their states are different. Electrons in isolated atoms occupy distinct *levels of energy*, which are sharply defined. In groups of atoms, where the electrons interact with several nuclei, the energy levels are not sharp but resemble *energy bands* that are significantly wider for the outer electrons of an atom. The Pauli principle applies and excludes orbits of electrons between the bands, energy levels that are sometimes called *forbidden bands*. At the upper end of the electron energies, there is a continuous band where electrons exist as *free electrons*, move within the matrix of the several atoms within the material, and are "shared" by several nuclei. When a voltage is applied, the free electrons generate an electric current.

The energy gap between the free electron band and the immediately lower band is narrow, typically between 0.5 and 1.4 eV, in semiconductor materials. The *doping* process, which adds very low amounts (a few parts per million) of an *impurity* in semiconductors, creates a separate *allowed energy* level within the last forbidden band, very close to the free electron band. With the addition of relatively low energy from photons, electrons from this allowed energy level "jump" into the free electron band, thus creating a surplus of electrons. Such semiconductors are called *n-type* and the new energy levels, the *donor levels*. In the *p-type* semiconductors, electrons from the last valence band "jump" onto the new energy level, which is called an *acceptor level* and has a deficit of electrons (sometimes referred to as a surplus of "holes"). The processes that create the n-type and p-type of semiconductors are depicted schematically in Figure 3.21, where the symbols + and − denote the electric charges generated by the migration of electrons between bands.

High purity crystalline silicon (Si) is typical of semiconductor base materials. Common doping materials are: boron for the creation of the n-type; and arsenic for the creation of the p-type semiconductors. When a thin layer of a p-type semiconductor

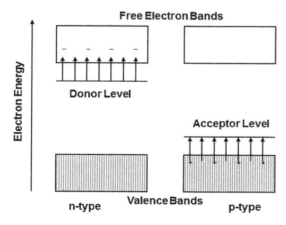

Figure 3.21 Electron excitation from the donor and acceptor levels.

comes in contact with a layer of an n-type semiconductor, a voltage difference is created across their interface, which is called the *p-n junction*. If the opposite ends of the p-n junction are connected to an external circuit, electric current flows until the electric balance of the valence and free electron bands is restored. The continuous excitation of electrons by incident photons maintains the voltage difference across the p-n junction and continuously generates electric current and power in the *PV cells*.

By operating with this mechanism, PV cells directly and continuously convert the insolation to electric energy without the intermediate step of heat generation. They are DEC devices and are not subjected to the Carnot limitations of thermal engines.

3.8.3 Exergy Conversion in PV Cells

The first limitation of the PV cells is related to the process of electron excitation to the acceptor bands in the p-type and the free electron bands in the n-type of semiconductors [16]. As shown in Figure 3.21 the migration of electrons to these bands requires finite energy input, which is provided by the photons – *the photoelectric effect*. Electrons with low energies – at the high wavelength part of the spectrum – do not possess the electron excitation energy and cannot cause electron migration. The energy of these photons is dissipated as heat in the p-n junction. On the other extreme of the spectrum highly energetic photons interact with only one electron; they impart to this electron the excitation energy; the electron migrates; but the remaining photon energy is dissipated in the p-n junction. Figure 3.22 shows this effect on the energy intensity captured by two PV cells with energy gaps 0.7 eV (corresponding to a maximum photon wavelength

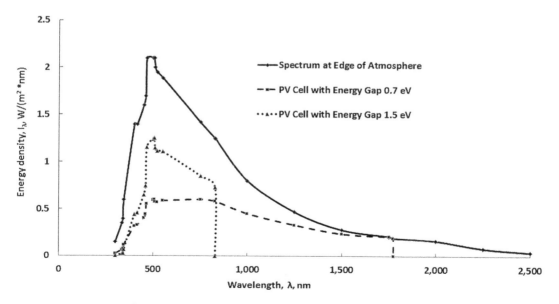

Figure 3.22 Energy spectrum at the outer limit of the atmosphere (data from www.nrel.gov/gis/solar.html) and light intensity captured by two PV cells with energy gaps 0.7 eV and 1.5 eV.

1,773 nm) and 1.5 eV (corresponding to a maximum photon wavelength 828 nm). The incident radiation has the characteristics of the solar spectrum at the edge of the terrestrial atmosphere (i.e. the cells generate power for a satellite). The electric power produced is represented by the areas under their respective curves, which have the units W/m^2. The remaining energy is dissipated and represents exergy destruction. With incident solar irradiance 1,353 W/m^2 the rate of exergy destruction in the PV cell of 0.7 nm energy gap is 782 W/m^2; and the exergy destruction in the PV cell of 1.5 eV energy gap is 943 W/m^2.

The exergy destruction and the exergetic efficiency of PV cells depend on the respective energy gap of the PV materials. When the energy gap corresponds to photon wavelengths at either of the two ends of the incident spectrum, the power of the PV cell vanishes. This implies that there is a maximum power – corresponding to the minimum exergy destruction – at an intermediate energy gap, which corresponds to a spectrum wavelength λ_m, and energy $E_{phm} = hc/\lambda_m$. For the radiation spectra at the surface of the earth and at the outer limit of the terrestrial atmosphere, which are shown in Figure 3.19, the wavelength λ_m and the corresponding energy gap E_{phm} are determined by trial and error. For the continuous black-body radiation spectrum, defined by Eq. (3.39), the exergetic efficiency of PV cells with energy gap E_g may be calculated analytically:

$$\eta_{PV} = \frac{15 E_g}{\pi^4 kT} \int_{E_g/kT}^{\infty} \frac{(hc/\lambda kT)^2}{\exp(hc/\lambda kT) - 1} d(hc/\lambda kT), \qquad (3.46)$$

where $E_g = hc/\lambda_g$, with λ_g being the wavelength corresponding to the energy gap of the PV cell; and T is the temperature of the radiating black body. The remainder of the incident radiation is dissipated in the PV cell and contributes to the destruction of exergy. Figure 3.23 shows the fraction of exergy, which is destroyed in a PV cell and the efficiency such a PV cell may achieve as a function of the energy gap, E_g. The incident radiation is emitted by a black body at 5,900 K, which approximates the solar spectrum at the edge of the atmosphere. The two graphs imply that the maximum theoretical conversion efficiency is approximately 54% and may be achieved with PV cells that have energy gap 1.27 eV.

The second limitation for the conversion of sunlight to electricity in PV cells is related to the external circuit. The power delivered to the external circuit is equal to the product of voltage and current, VI, while the power produced by the PV cells is a monotonically increasing function of the insolation. Figure 3.24 shows the typical relationship of the voltage and current produced by a PV cell for two values of the insolation. An inspection of Figure 3.24 proves that there is a single point in the voltage-current curve, where maximum power is produced and the slope of the tangent to the curve is equal to -1. Since the resistance of the external circuit (the *external load*, R) relates V and I, ($V = RI$), the power produced by the cell is determined by the external load and the insolation. Figure 3.24 also shows the voltage-current characteristics for four values of the external load, $R_1 < R_2 < R_3 < R_4$. When the insolation and external resistance are prescribed, the power produced is not necessarily a maximum. The

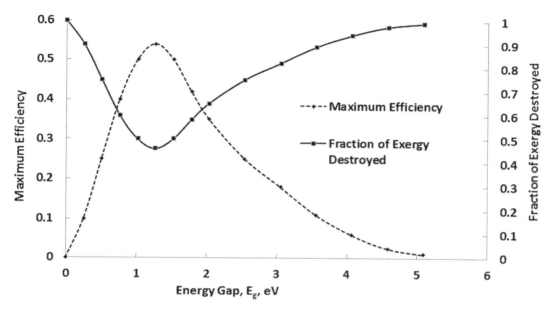

Figure 3.23 Fraction of exergy destroyed in PV cells and maximum efficiency, when the radiation is emitted by a black body at 5,900 K

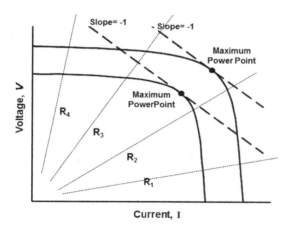

Figure 3.24 Relationship between the voltage and current produced by PV cells for two values of the insolation.

difference between the maximum power and the actual power produced contributes additional exergy destruction.

Commercial PV cell systems include a *Maximum Power Point Tracker (MPPT)* that ensures power is produced at the maximum power point when subjected to a given insolation. The MPPT includes a variable resistance that matches the external load to the resistance corresponding to the maximum power. Naturally, there is a small amount of power dissipation in this additional resistance.

Other, of lesser significance, sources of dissipation and exergy destruction in PV cells are:

1. Reflection of irradiance at the surface of the cell. These losses may be reduced to almost zero with transparent coatings.
2. Free electrons that are created far from the junction – outside the diffusion length for the electrons. These free electrons do not cross the junction and do not contribute to the electric field and the power production. This source of dissipation is minimized by creating very thin junctions.
3. Photons that interact with the nuclei and do not excite the electrons.

The first cause of exergy destruction – the mismatch between the energy carried by photons and the energy gap of the p-n junction – contributes by far the highest amount of exergy loss in PV cells. This limitation is not inherent in the nature of the energy resource (the insolation), but is caused by the type of the device used for the conversion of solar energy (the PV cell). To minimize the effect of this source of exergy destruction one may develop multilayer (multi-junction) PV cells, where each layer has a different energy gap, E_g, with the energy gap decreasing in each successive layer. A schematic representation of such a PV cell with four layers is shown in Figure 3.25, where the vertical lines represent the PV layers – each layer is a separate p-n junction – and the length of the arrows denotes the energy of the incident photons. The layers are arranged in order of decreasing energy gap, $E_{g1} > E_{g2} > E_{g3} > E_{g4}$. The first layer intercepts the highest energy photons with energy gap, E_{g1} and produces electric energy. This layer is transparent to the remaining photons, which pass to be

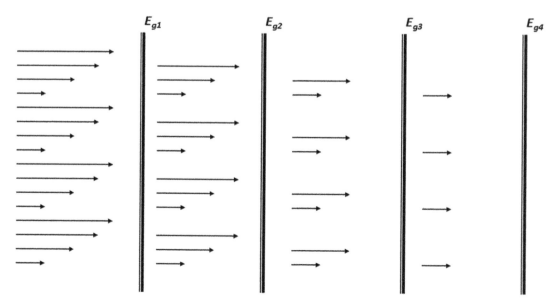

Figure 3.25 A multilayer PV cell with energy gaps, $E_{g1} > E_{g2} > E_{g3} > E_{g4}$ produces higher power. The lengths of the arrows represent the magnitudes of photon energies.

intercepted by the other layers. The second layer intercepts the photons with energies between E_{g1} and E_{g2} and adds to the production of electric energy. The third layer produces electric energy by intercepting photons in the energy range E_{g2} to E_{g3} and the fourth layer does the same by intercepting photons in the range E_{g3} to E_{g4}. By intercepting the photons at energy levels close to their own energies, the multi-junction PV cells generate significantly higher power (and have significantly lower lost work) than monolayer cells. The maximum efficiency of a four-layer PV cell would increase from the 54% of Figure 3.23 to 82%. A stack of infinite layers/junctions operating with concentrated sunlight would achieve the efficiency limit denoted by Eq. (2.37) [17, 18].

3.8.4 Efficiency of Solar Cells

The International Electrotechnical Commission (IEC) defines international standards for all electrical and electronic devices. The commission has defined the efficiency of PV cells as the ratio of the power produced by the cell and the power received at the *standard terrestrial conditions (STC)*. The latter is defined as: irradiance of 1 kW/m^2; spectral distribution close to that of the solar radiation through air mass of m = 1.5; and cell temperature 25°C.

$$\eta = \frac{\dot{W}_{act}}{\dot{W}_{STC}}, \quad (3.47)$$

where the numerator is the actual power produced by the PV cells and the denominator denotes the power received under STC. Typically, PV cell efficiencies drop by approximately 1% for every 4–6°C temperature above 25°C [19].

While this is considered as the first-law efficiency of solar cells, the definition takes into account the "environment" of solar energy conversion, the terrestrial atmosphere. Given that the STC irradiance in Eq. (3.47) denotes the heat supplied by the solar radiation and that this heat corresponds to an exergy as in Eq. (2.37), the exergetic efficiency of the solar cell is:

$$\eta_{II} = \frac{\dot{W}_{act}}{\dot{W}_{STC}(1 - T_0/T)} = \frac{\eta}{(1 - T_0/T)}, \quad (3.48)$$

where T is the sun's temperature, approximately 5,800 K. The denominator of the last term is approximately equal to 0.95 and this yields the approximate relationship: $\eta_{II} \approx 1.05\eta$, which also signifies that the numerical values of the first-law and the exergetic efficiencies of solar cells are very close.

A comprehensive pictorial compilation of the evolution of PV cell first-law efficiencies since 1975 has been accomplished by the *National Renewable Energy Laboratory (NREL)* in the USA and is shown in Figure 3.26 [20]. The captions in Figure 3.26 also indicate the type of technology used in the PV cell; the laboratory of manufacture of the PV cell; and information on the number of junctions. Two principal conclusions may be deduced from Figure 3.26:

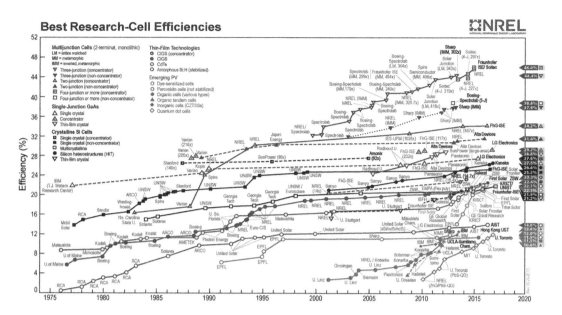

Figure 3.26 The evolution of PV cell efficiencies.
From [20] by permission of the National Renewable Energy Laboratory (NREL) in USA

1. The efficiencies of PV cells based on a particular technology improve with time, but reach a plateau, when the technology matures and few improvements can be made.
2. The efficiencies of multi-junction cells are significantly higher than those of single-junction cells.

An interesting fact about practical PV cells is that their efficiency, as defined in Eq. (3.47) in terms of the irradiance in their environment, is lower in space, where the environment receives solar irradiance unfiltered by the terrestrial atmosphere. The local solar irradiance outside the terrestrial atmosphere is higher, and the higher denominator in Eq. (3.47) results in lower first-law efficiencies, typically by 1–2%. Nevertheless, a given PV cell generates higher power at the outer limit of the atmosphere, despite the apparent lower efficiency.

3.9 Solar Thermal Systems

Solar thermal systems capture solar radiation as heat in a working fluid and use the heat for the production of electric power, for space heating, for water heating, and for industrial processes.

3.9.1 Power Production

Rankine cycles, similar to those depicted in Figures 1.7, 1.8, and 3.5, are typically used for the electricity production from solar energy. A *central receiver,* absorbs the reflected radiation from a large number of *solar reflectors,* and replaces the boiler. The working fluid removes the heat from the central receiver and evaporates. The produced vapor drives a turbine-generator system. When the working fluid is water, steam is generated in the central receiver and is fed directly to the turbine. When a molten salt is used for thermal storage, steam is produced in a HX and is subsequently fed to the turbine. Organic fluids may also be used in solar Rankine cycles, but so far, they are absent from solar thermal installations. Because the reflectors concentrate the solar power in the central receiver, these systems are widely known as *Concentrated Solar Power* (CSP) systems.

Figure 3.27 is an aerial photograph of *Solar Two* – a 10 MW thermal power plant that was built near Barstow, California. The central receiver is at the top of the receiving tower. Solar irradiance reflects on the 1,818 surrounding heliostats (each with 40 m^2 reflecting area) and is directed to the receiver, where molten salts (60% $NaNO_3$ and 40% KNO_3) absorb the solar energy and produce superheated steam in a HX that is fed to the steam turbine. The molten salts pool, the main HX, the cooling system of the small power plant and other auxiliary facilities are also observed in Figure 3.27. Solar-Two was decommissioned in 1999 and was replaced by *Solar Tres*, a scaled up 15 MW pilot project with thermal storage capacity. Solar Tres is located in Andalucía, Spain, where it now operates under the name *Gemasol*.

The exergy losses in the solar Rankine cycle are very similar to the losses encountered in typical vapor power plants, as explained in Section 3.2. In addition, thermal solar power plants include exergy losses caused by the following [10]:

Figure 3.27 An aerial photograph of Solar-Two, a 10 MW solar thermal plant near Barstow California, decommissioned in 1999. (Courtesy of Sandia National Laboratories)

1. *Cosine losses:* The function of the heliostats is to reflect the insolation to the central receiver. Because of this they are not perpendicular to the sun rays and the insolation is reduced by the cosine of the angle between the sun's rays and the perpendicular direction to the reflecting surfaces.
2. *Blocking and Shadowing:* Shadows are cast by the central receiver and the buildings of the power plant. In addition, part of the sunlight is blocked by heliostats, especially during the early morning and the late afternoon hours when the sun is close to the horizon.
3. *Reflective losses:* A perfectly reflective surface, with 100% reflectivity, does not exist. The currently used heliostat surfaces reflect 90–95% of the incident solar energy.
4. *Attenuation:* Losses between the heliostats and the central receiver that are caused by water vapor, particulates, and smog.
5. *Reflection at the central receiver:* The receiver has a finite absorptivity and part of the irradiance from the heliostats is reflected at its surface. Absorptive paints and the use of cavities minimize the reflection losses.
6. *Spillage:* Position control errors in the tracking system of the heliostats and high winds cause geometric distortions on the reflective surfaces. As a result, part of the irradiance from the heliostats is "spilled" around the receiver and is lost.
7. *Receiver Heat Losses:* By radiation conduction and convection. Because the central receiver must absorb as much as possible of the reflected irradiance from the heliostats, there is no thermal insulation in the receiver, where heat loss is significant.

Example 3.15 A proposed thermal solar power plant will produce 15 MW of electric power for 16 hours during the summer days, when air-conditioning is needed. Molten salts are used for energy storage. The 24-hour averaged irradiance during the summer days in the region is 310 W/m². A simple Rankine steam cycle is used to produce power: The steam enters the turbine at 540°C and 4 MPa and the condenser temperature is 35°C. The isentropic turbine efficiency is 84% and the pump efficiency is 80%. Other (averaged during the day) losses are: cosine losses 9%; heliostat reflecting losses 7%; blocking, shadowing, attenuation, and spillage losses 6%; receiver reflection and heat losses 20%. Determine the area of the heliostats required for the production of the 15 MW power and the exergy destruction at every step of the energy conversion process. $T_0 = 300$ K.

Solution: At first we need to analyze the Rankine cycle and determine the daily amount of heat needed for the production of 15 MW for 16 hours a day (240 MWh). From the steam tables, the specific enthalpy, entropy, and exergy of the water and steam in the four states of the Rankine cycle as depicted in Figure 1.8 are:

$h_1 = 147$ kJ/kg	$s_1 = 0.5053$ kJ/kg K	$e_1 = 1$ kJ/kg
$h_2 = 152$ kJ/kg	$s_2 = 0.5095$ kJ/kg K	$e_2 = 5$ kJ/kg
$h_3 = 3{,}537$ kJ/kg	$s_3 = 7.2071$ kJ/kg K	$e_3 = 1{,}380$ kJ/kg
$h_4 = 2{,}423$ kJ/kg	$s_4 = 7.8959$ kJ/kg K	$e_4 = 60$ kJ/kg

The specific net work of the cycle is $w_{net} = (1{,}114 - 5) = 1{,}109$ kJ/kg; the specific heat input is $q_{in} = 3{,}385$ kJ/kg; and the thermal efficiency of this simple Rankine cycle is 32.7%. For the production of 15 MW power, $15/0.327 = 45.87$ MW of heat rate input is needed; and the mass flow rate in the steam cycle is: $15{,}000/1{,}109 = 13.53$ kg/s.

Because the thermal solar power plant operates for 16 hours per day (and produces a total of 240 MWh) the daily heat input is $Q_{in} = 734$ MWh or $2.642 * 10^9$ kJ/day.

The daily insolation in the area of this power plant is $310 * 24$ Wh/m^2/day or 7.44 kWh/m^2/day. Accounting for the losses at the heliostats and the central receiver, the daily energy received by the molten salts to be transferred to the steam cycle is: $7.44 * (1 - 0.09) * (1 - 0.07) * (1 - 0.06) * (1 - 0.20) = 4.73$ kWh/m^2/day $= 17{,}046$ kJ/m^2/day. Therefore, $2.642 * 10^9/17{,}046$, or approximately 155,000 m^2 of heliostat area must be erected to supply the cycle with the needed heat input.

Based on the 310 W/m^2 average irradiance, the total incident energy associated with the total area of the heliostats is: $4.152 * 10^9$ kJ/day. From Eq. (2.38) the exergy of the total incident radiation is: $E_{inc} = 3.932 * 10^9$ kJ/day. Since the electrical energy production of the power plant is $0.864 * 10^9$ kJ/day, the first law efficiency of the plant is 20.8% and the exergetic efficiency is 21.9%.

Calculation of exergy destruction:

The exergy destruction associated with the cosine loss is: $3.932 * 10^9 * 0.09 = 0.357 * 10^9$ kJ/day; and the exergy transmitted by the heliostats is: $3.574 * 10^9$ kJ/day.

From the heliostat-receiver system, the exergy imparted to the molten salts is: $3.574 * 10^9 * 0.93 * 0.94 * 0.80$ kJ/day $= 2.500 * 10^9$ kJ/day. Therefore, the exergy destruction in this system is $1.074 * 10^9$ kJ/day.

During the 16 hours of the steam cycle operation, the exergy transfer from the molten salts to the steam cycle is: $13.53 * 16 * 60 * 60 * 1{,}380 = 1.075 * 10^9$ kJ/day. Therefore, the exergy destruction associated with the molten salts HX is: $2.500 * 10^9 - 1.075 * 10^9 = 1.425 * 10^9$ kJ/day.

The exergy destruction in the turbine, which produces 15,000 kW, for the 16 hours daily it operates and exhausts steam with specific exergy $e_4 = 60$ kJ/kg, is:
$1.075 * 10^9 - (15{,}000 + 13.55 * 60) * 16 * 60 * 60 = 0.164 * 10^9$ kJ/day.

It is observed in this example that the highest rate of exergy destruction occurs in the central receiver-steam generator system, where the molten salts receive radiation exergy and transfer it as thermal exergy to the steam cycle. The exergy destruction in the heliostat-receiver system is also significant.

3.9.2 Passive Solar Heating – Solar Collectors

The most widespread use of solar energy globally is the passive heating of space and water. Passive heating by the sun helps avoid the use of fossil fuels or electric power and conserves primary energy resources. *Solar collectors* (or simply *collectors*) are stationary systems, facing the prevailing direction of the sun. The collectors are essentially dark enclosures covered by a glass surface that "traps" the incident radiation. Their temperatures typically reach 20–60°C higher than the ambient. Circulating water picks up the heat from the solar collectors and dissipates it in HXs (commonly called *radiators*) to achieve a comfortable air temperature in buildings. Alternatively, the water may be stored in insulated tanks to supply the hot water needs of the building. Because solar collectors trap both direct and diffuse solar radiation, they operate well even in the cloudy autumn and winter days. Energy and exergy losses in solar collectors are higher than those in heliostats, which follow the disk of the sun, and often exceed 50% of the incident radiation.

Example 3.16: A solar collector is designed to meet the hot water needs of a household, which are 650 L (0.65 m^3) of water per day at 50°C. Water is supplied to the household at 18°C. The 24-hour averaged irradiance in the region is 202.5 W/m^2. The total energy losses in the collector are 48% of the incident solar energy. Calculate the area of the collector and determine its first law and exergetic efficiencies. $T_0 = 18°C = 291$ K.

Solution: For the water to be heated up from 18°C to 50°C, the total daily heat input is

$Q = V\rho c_P(T_1 - T_0) = 0.65 * 1,000 * 4.184 * 32 = 87,027$ kJ.

During a day the collector "collects" heat per unit area $q_{col} = (1 - 0.48) * 202.5 * 24 * 60 * 60 = 9,098$ kJ/m^2. Therefore the area of the collector is $A = 9.566$ m^2 and the incident solar energy is 167,359 kJ.

The first law efficiency of this collector is based on the heat absorbed and is equal to: $87,027/(167,359) = 52\%$.

The exergy of the incident radiation on the collector is: $E_{inc} = (1 - 291/5,800) * 167,359 = 158,962$ kJ. The exergy of the 650 kg of the incompressible hot water may be calculated from Eq. (2.21): $E_{wat} = 650 * 4.184 * [323 - 291 - 291 * \ln(323/291)] = 4,461$ kJ. Therefore, the exergetic efficiency of the solar collector is 2.8%.

This is an indication that significant exergy destruction occurs in passive solar collectors. Despite this, one has to consider that, if the energy and exergy from the sun were not "collected" they would have dissipated in the environment, anyway.

Also, that solar energy is free of cost and, thus, the owners of the solar collector do not have to pay for "fuel."

Example 3.17: In order to improve the exergetic efficiency of hot water production by solar energy, it has been suggested that a heat pump driven by PV cells be used instead of the collector. For the conditions of Example 3.16 calculate: a. the area of the PV cells; b. the exergy destruction; and c. the exergetic efficiency of the new system if the COP of the heat pump is 4.8 and the efficiency of the PV cells is 20%.

Solution: From Example 3.16, the daily amount of heat to be supplied to the water is $Q = 87{,}027$ kJ. With COP $= 4.8$, the daily amount of electric energy that would produce this quantity of heat in a heat pump system is: $W = 18{,}131$ kJ/day.

a. With 0.2025 kW/m² insolation, the needed area of the PV cells is: $A = 18{,}131/(0.2025 * 24 * 60 * 60 * 0.2) = 5.181$ m².
b. The incident exergy of the irradiance on the PV cells is: $E_{inc} = (1 - 291/5{,}800) * 5.181 * (0.2025 * 24 * 60 * 60) = 86{,}099$ kJ and the exergy of the 650 kg hot water produced is 4,461 kJ. Hence, the exergy destruction is 81,638 kJ.
c. The exergetic efficiency of the PV-with-heat-pump system is: $4{,}461/86{,}099 = 5.2\%$.

It must be noted that, even though the exergetic efficiency almost doubles with the new system, the added cost of the PV cells and the heat pump may not justify the investment for the new system. Passive solar collectors are simple and economical systems; they require very low maintenance; and they operate without the consumption of fuel and are free of cost for their owners. For these reasons they are widely used in many regions of the globe.

3.10 Wind Turbines

The rate of kinetic energy in a stream of air (the wind) through an engine of diameter D was calculated in Section 2.8.4 to be:

$$\dot{W}_{max} = \dot{E} = \frac{1}{2}\dot{m}V_i^2 = \frac{\pi}{8}\rho D^2 V_i^3, \tag{3.49}$$

where V_i is the incoming velocity of the wind; ρ is the density of the air, approximately 1.2 kg/m³; and $\dot{m}$ is the mass flow rate, which is intercepted by the power producing engine – the wind turbine. However, for an engine to continuously produce power the fluid must be discharged from the engine at finite velocity and this limits the maximum power to be generated from wind turbines to the so-called *Betz's limit* of Eq. (2.43) [3, 21]:

$$\dot{W}_{max}^{tur} = \frac{8}{27}\rho A V_i^3 = \frac{2\pi}{27}\rho D^2 V_i^3 = \frac{16}{27}\dot{E}. \tag{3.50}$$

Wind turbines are engines with very long blades rotating on a vertical plane that is almost perpendicular to the wind direction. The larger and most commonly used turbines operate with two or three long blades that are pivoted on a horizontal axis and are commonly referred to as *horizontal axis turbines*. The generation of power by the wind turbines does not exactly follow the wind velocity. Figure 3.28 (obtained from data of the V90–3.0 MW turbine by Vestas Inc.) depicts the typical relationship between the power produced by an actual wind turbine and the prevailing wind velocity [22]. Four distinct velocity ranges are apparent in Figure 3.28:

1. From zero wind velocity to 3.5 m/s – the *cut-in velocity* of the turbine – the turbine does not produce any power (its blades do not rotate).
2. From 3.5 m/s to 15.0 m/s the power produced by the turbine increases following closely the cubic dependence $\dot{W} \sim V^3$:

$$\dot{W} = \frac{1}{2}\eta_T A \rho V^3 = \frac{\pi}{8}\eta_T D^2 \rho V^3. \qquad (3.51)$$

3. The velocity 15.0 m/s is the *rated velocity* for this turbine. The power produced in the range $15 < V < 25$ m/s is almost constant, called the *rated power* of the turbine, which is 3.0 MW in this case.
4. The range beyond the *cut-out velocity* – 25 m/s for this turbine – is the upper limit where the engine may operate safely and generate electric power. Once the wind velocity exceeds the cut-out velocity, the engine is shut off to avoid damage (the turbine is *feathered*) and the generated power vanishes.

When the wind velocity increases beyond the rated velocity (a design parameter) the turbine does not convert the increased kinetic energy of the wind to generate more power. A large fraction of the rate of exergy from the wind is not harnessed in this range, resulting in significant lost power. The variability of the wind velocity and the design characteristics of wind turbines have a significant impact on the time-averaged power and the energy produced by wind turbines over long time periods. Table 3.9

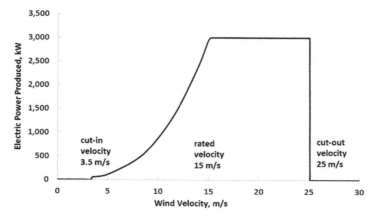

Figure 3.28 Typical relationship between the power produced by an actual wind turbine and the prevailing wind velocity.
Data adapted from [22]

Table 3.9 Parameters of wind exergy and wind power utilization on March 22, 2011. Wind velocity data courtesy of the Wind Energy Laboratory, West Texas A&M University.

Hour	Wind Velocity, V (m/s)	Wind Exergy, $\dot{W}_{max}$ (MW)	Betz' Limit, $\dot{W}_{max}^{tur}$ (MW)	Power Production (MW)	Lost Power, $\dot{W}_{lost}$ (MW)
0:00:00	15.1	13.14	7.78	3.00	10.14
1:00:00	15.3	13.66	8.10	3.00	10.66
2:00:00	14.5	11.63	6.89	2.71	8.92
3:00:00	14.1	10.69	6.34	2.49	8.20
4:00:00	13.7	9.81	5.81	2.29	7.52
5:00:00	13.1	8.58	5.08	2.00	6.58
6:00:00	12.9	8.19	4.85	1.91	6.28
7:00:00	12.1	6.76	4.01	1.57	5.18
8:00:00	13.2	8.77	5.20	2.04	6.73
9:00:00	18.3	23.38	13.86	3.00	20.38
10:00:00	19.9	30.07	17.82	3.00	27.07
11:00:00	18.4	23.77	14.08	3.00	20.77
12:00:00	18.1	22.62	13.41	3.00	19.62
13:00:00	17.4	20.10	11.91	3.00	17.10
14:00:00	17.5	20.45	12.12	3.00	17.45
15:00:00	18.2	23.00	13.63	3.00	20.00
16:00:00	17.7	21.16	12.54	3.00	18.16
17:00:00	15.7	14.76	8.75	3.00	11.76
18:00:00	13.2	8.77	5.20	2.04	6.73
19:00:00	8.4	2.26	1.34	0.53	1.73
20:00:00	6.8	1.20	0.71	0.28	0.92
21:00:00	8	1.95	1.16	0.46	1.50
22:00:00	9.2	2.97	1.76	0.69	2.28
23:00:00	8.9	2.69	1.59	0.63	2.06
Daily Totals, MWh		310.38	183.93	52.64	257.74
Daily Averages, MW		12.93	7.66	2.19	10.74

shows the hourly-averaged wind velocity measurements on March 22, 2011 (a windy day) outside the city of Canyon, Texas. It also shows the wind exergy and the Betz's limit that correspond to the averaged wind velocity; the power generated by a turbine with the same characteristics as the one in Figure 3.28; and the lost power, defined as the difference between the wind exergy and the actual power generated.

The following observations, which are typical of all wind turbine generation systems, are made from the data of Table 3.9:

1. During every hour the power produced is always less than the exergy of the wind and less than the Betz's limit.
2. Because the power production curve of the turbine flattens at $V = 15$ m/s, the wind turbine does not take advantage of the significantly higher exergy of the wind at the higher velocities range.

3. While the rated power of the wind turbine is 3 MW, because of the hourly variation of the wind power, the daily average power is 2.19 MW.
4. The average exergetic efficiency of the turbine is $2.19/12.93 = 16.9\%$ (28.6% if based on the Betz's limit). However, this is not a benefit/cost ratio similar to the ones used for fossil fuel power plants because the "cost of fuel" (the wind) is zero.

Problems

(Unless otherwise stated consider the environment at 300 K and 0.1 MPa.)

1. A counter flow HX receives the exhaust of a gas turbine at 458°C to heat up water from 18°C to 85°C for pasteurization. The mass flow rate of the exhaust gas is 5.6 kg/s and the HX has a minimum temperature difference 15°C. Determine the mass flow rate of the water, the effectiveness of this heat exchanger and the rate of exergy destruction.

2. A steam turbine receives 150 kg/s of steam at 520°C and 5 MPa. The steam enters the condenser at 40°C and $x = 0.96$. Cooling water at 28°C is supplied to the condenser and leaves the condenser at 35°C. There is 1°C subcooling in the condenser. Determine the flow rate of cooling water and the rates of exergy destruction in the turbine and the condenser.

3. A steam turbine receives 150 kg/s of steam at 520°C and 5 MPa. Because of lack of cooling water, a dry cooling system is used. The steam from the turbine enters the condenser at 55°C and $x = 1.0$. Air enters the condenser at 28°C and leaves at 50°C. There is 1°C subcooling in the condenser. Determine the flow rate of cooling air and the rates of exergy destruction in the turbine and the condenser.

4. The FWH of a power plant receives 200 kg/s of water from the pump at 38°C and 2.8 MPa. The liquid water temperature increases to 130°C using steam at 135°C and 95% dryness fraction. Determine the flow rate of the steam and the rate of exergy destruction in the FWH. What would be the rate of exergy destruction if, instead of the steam, methane was used for the preheating of the water?

5. A jet engine is cruising at 39,000 feet (11,887 m) with speed 980 km/hr. The atmospheric pressure at this level is $P_0 = 21.5$ kPa and the ambient temperature is $T_0 = 218$ K. The inlet diameter of the engine is 1.6 m; the compressor pressure ratio is $P_3/P_2 = 20$; the temperature of the gases at the entrance of the turbine is $T_4 = 1,500$ K; and the nozzle exhausts to the atmosphere. The electric power need for the aircraft is 1.8% of the power delivered to the compressor. The fuel is kerosene with LHV 43,200 kJ/kg and the equivalent Gibbs' free energy of the combustion reaction is 41,850 kJ/kg. The isentropic efficiencies of the diffuser and the nozzle are 96%; the isentropic efficiencies of the compressor and the turbine are 82% and 85% respectively; and the combustion efficiency is 98%. Determine the thrust of the engine and the exergy destruction in all the components of this engine.

6. A geothermal well produces 62 kg/s of water at 180°C and 2.5 MPa. An entrepreneur claims that his (proprietary) energy conversion system will produce 9 MW of electric power. Would you invest in this project?

7. A geothermal well produces 55 kg/s of water and steam at 200°C with dryness fraction $x = 0.12$. The steam is separated and expanded in a turbine to 40°C. The remaining liquid water is flashed to 120°C and expanded also to 40°C. Both turbines have 80% efficiencies. What is the maximum power that may be extracted from this resource, what is the power that is actually obtained and what is the exergetic efficiency of this geothermal unit?

8. A geothermal well produces 68 kg/s of water at 180°C and 1.8 MPa. The water is used in a heat exchanger to vaporize the butane of an ORC. The butane exits the heat exchanger as saturated vapor at 90°C; the turbine efficiency is 80%; the condenser of the ORC is at 38°C; and the pump is 75% efficient. There is 5°C temperature difference at the pinch point. Determine the power produced by the ORC and the exergetic efficiency of the entire system.

9. A hydrogen-oxygen fuel cell consumes 0.0012 kmol/hr of hydrogen and supplies voltage of 1.1 V and current of 35 A. Determine the voltage efficiency and the exergetic efficiency of this fuel cell.

10. A car has a gasoline IC engine with 25% efficiency and a 60 L gasoline tank. This engine is substituted with a hydrogen-air PEMFC that has 65% efficiency. The motor of the new car is 95% efficient. Determine the mass of hydrogen required in a hydrogen tank for this car to have the same distance range. If the hydrogen is kept at 200 bar, and behaves as an ideal gas, how much is the volume of this tank if the car operates at ambient temperatures close to 300 K?

11. A thermal solar collector produces daily 2,250 L (2.25 m^3) of hot water at 50°C. Water is supplied to the collector at 18°C. The 24-hour averaged irradiance in the region of the collector is 240 W/m^2 and the energy losses in the collector are 36% of the incident solar energy. Calculate the area of the collector and determine its first law and exergetic efficiencies. $T_0 = 18°C = 291$ K.

12. A wind turbine has 40 m long blades and 1,500 kW rated power. The cut-in velocity of the turbine is 2.0 m/s, the rated velocity 11 m/s and the cut-out velocity 25 m/s. The following are the average wind characteristics in the region where the turbine is deployed:

$0 < V < 2$ m/s	10% of the time.
$2 < V < 5$ m/s	16% of the time.
$5 < V < 8$ m/s	21% of the time.
$8 < V < 11$ m/s	14% of the time.
$11 < V < 18$ m/s	25% of the time.
$18 < V < 25$ m/s	12% of time
$25 < V$ m/s	2% of the time.

Determine the annual total energy this turbine produces and the average exergetic efficiency of the turbine. The air density is 1.18 kg/m^3 and the turbine efficiency is constant at 38%.

13. Obtain the average operating data of an actual steam power plant in your area, perform an exergetic analysis of all the components, and suggest ways to improve the exergetic efficiency of the power plant.

14. Obtain the average operating data of an actual PV power plant. Using the local irradiance in the region calculate the exergetic efficiency of this power plant. (Hint: the websites of most PV power units provide ample information on the energy produced and the websites of National Laboratories – the NREL in the USA – provide data on the irradiance.)

References

[1] F. P. Incropera, and D. P. DeWitt, *Fundamentals of Heat and Mass Transfer* (New York: Wiley, 2000).
[2] D. Tondeur, and E. Kvaalenf, Equipartition of Entropy Production. An Optimality Criterion for Transfer and Separation Processes. *Industrial & Engineering Chemical Research*, **26** (1987), 50–6.
[3] E. E. Michaelides, *Energy, the Environment, and Sustainability* (Boca Raton, FL: CRC Press, 2018).
[4] International Energy Agency, *Key World Statistics* (Paris: IEA, 2017).
[5] REFPROP, *Reference Thermodynamic and Fluid Transport Properties* (Boulder, CO: National Institute of Standards and Technology, 2013).
[6] L. S. Langston, Anticipated but Unwelcome. *Mechanical Engineering*, **140** (2018), 36–41.
[7] J. H. Horlock, *Cogeneration – Combined Heat and Power* (Malabar, FL: Krieger, 1997).
[8] E. E. Michaelides, Thermodynamic Properties of Geothermal Fluids. *Geothermal Resources Council - Transactions*, **13** (1981), 361–64.
[9] J. Kestin, ed., *Sourcebook of Geothermal Energy* (Washington DC: US Department of Energy, 1980).
[10] E. E. Michaelides, *Alternative Energy Sources* (Berlin: Springer, 2012).
[11] J. O'M. Bockris The Origin of Ideas on a Hydrogen Economy and Its Solution to the Decay of the Environment. *International Journal of Hydrogen Energy*, **27**, (2002) 731–40.
[12] M. J. Moran, and H. N. Shapiro, *Fundamentals of Engineering Thermodynamics*, 6th ed. (New York: Wiley, 2004).
[13] V. Neburchilov, J. Martin, H. Wang, and J. Zhang, A Review of Polymer Electrolyte Membranes for Direct Methanol Fuel Cells. *Journal of Power Sources*, **169** (2007), 221–38.
[14] *Fuel Cells*, US-DOE, Fuel Cell Technology Office, Washington DC (2015).
[15] S. Wilcox, *National Solar Radiation Database 1991–2010 Update: User's Manual*, Technical Report NREL/TP-5500-54824 (August 2012).
[16] W. Shockley, and H. J. Queisser, Detailed Balance Limit of Efficiency of p-n Junction Solar Cells. *Journal of Applied Physics*, **32** (1961), 510–19.
[17] A. De Vos, and H. Pauwels, On the Thermodynamic Limit of Photovoltaic Energy Conversion. *Applied Physics*, **25** (1981), 119–25.
[18] A. De Vos, *Thermodynamics of Solar Energy Conversion* (Weinheim: Wiley-VCH, 2008).
[19] S. Dubey, N. J. Sarvaiya, and B. Sheshadri, Temperature Dependent Photovoltaic (PV) Efficiency and Its Effect on PV Production in the World – A Review. *Energy Procedia*, **33**, (2013), 311–21.
[20] S. Kurtz S., and D. Levi, 2017, *Best Research-Cell Efficiencies*, www.nrel.gov/pv/cell-efficiency.html, last accessed October 2019.
[21] A. Betz. Schraubenpropeller mit Geringstem Energieverlust, (mit einem Zusatz von L. Prandtl) *Nachrichten von der Gesellschaft der Wissenschaften zu Göttingen, Mathematisch-Physikalische Klasse* (1919) 193–217.
[22] Vestas (2019) www.vestas.com/en/products/#! last visited October 2019.

4 Exergy Consumption and Conservation

Summary

Our society does not need energy *per se*. We use energy in its various forms to accomplish desired actions – commuting to work, keeping the interior of our homes at comfortable temperatures, and producing industrial goods. It is proven that the so called "minimum energy" needed for the accomplishment of desired actions is actually a thermodynamic maximum, defined by the exergy concept. The application of the exergy method determines the *benchmark* for the minimum energy resources required for the performance of all desired actions and tasks. Using the exergy methodology, the minimum energy requirements (benchmarks) are determined for a variety of processes including: natural gas transportation, refrigeration, liquefaction, drying, water desalination, and petroleum refining. The minimum energy requirements for the lighting, heating, and air-conditioning of buildings are calculated by the application of the exergy method as well as the minimum energy needed for the transportation of people and commercial goods. Given their importance for the transition to renewable energy forms, the exergy method is applied to energy storage systems to determine the best way to store energy. Several examples in this chapter offer assistance and resources for the application of the exergy methodology to energy-consuming systems and processes.

4.1 Energy Conservation or Exergy Conservation?

"Energy conservation" is a misnomer because the total energy (the mass-energy when nuclear reactions are considered) is naturally conserved as articulated in the first law of thermodynamics. What is meant by this colloquial expression is that lesser energy is spent for the accomplishment of an action desired by humans. Let us consider the desired action to maintain the interior of a building in Berlin at a comfortable 22°C on a cold winter day: Heat is transferred from the warm building to the colder environment and at the same time energy/heat is supplied to the building. When the rate of heat transferred to the environment is equal to the rate of heat input to the building, the interior temperature is maintained at the desired comfortable level for its occupants. During these energy exchanges, the total "energy" of the building remains the same and the occupants are content. However, there is significant exergy destruction, due to the transfer of heat to the building and the dissipation of the same quantity of heat in the environment.

Now, let us consider that the owners of the building decided to improve its insulation. As a result, the rate of heat transferred to the environment is reduced by 30%. As a consequence, the rate of heat needed to maintain the interior of the building at the same temperature of 22°C decreases by 30% too; and (if everything else remains the same) the rate of fuel to the burner that supplies this rate of heat is reduced by the same fraction. The addition of the insulation is touted as "an energy conservation measure." However, the energy in the building and the environment is always conserved, regardless of the insulation. In this case it is important to realize that, while energy is always conserved, the effect of the insulation is that lesser energy resources (fuel for a heater or electricity for a heat pump) are spent to maintain the comfortable temperature in the building's interior. Because the energy resources are inextricably tied to exergy, one may deduce that, with the addition of the insulation, the exergy destruction for maintaining the interior of the building at the desired temperature is reduced.

The colloquial "energy conservation" expression may be more precisely articulated in terms of thermodynamic variables, such as: *exergy conservation, lesser exergy destruction,* or *lesser entropy production.* Articulating the colloquial "energy conservation" in terms of thermodynamic variables introduces a quantitative and precise method to analyze all "energy conservation" projects and processes using optimization tools. The application of exergetic calculations introduces a quantitative method for the lesser consumption of primary energy resources, and by extent, for the conservation of the natural energy resources – the total primary energy sources (TPESs) in the earth.

4.1.1 Desired Actions, Natural Resource Consumption, Conservation and Higher Efficiency

While Chapter 3 primarily deals with the supply side of energy, the subject of "energy conservation" is related to the demand side. By reducing the exergy destruction – or, equivalently, by reducing the work lost and entropy production – in systems and processes, the human society, as a whole, uses less of the TPESs that supply the global energy demand. The use of *natural resources* that supply the global population with energy is reduced and the natural resources are conserved for future use.

From the beginning one must realize that the human society uses energy resources, not because it needs energy *per se*, but in order to accomplish *desired actions* or to fulfill needed *societal tasks*. These actions or tasks are, in general, performed by individuals and groups and are typically accomplished using "energy" in one or more of its several forms. Examples of such actions or tasks are [1, 2]:

- Maintain a warm comfortable temperature for the residents of a building in Berlin during the winter months.
- Maintain a cool comfortable temperature for the residents of a building in Miami during the summer months.
- Provide adequate lighting in the classrooms of a high school in Vietnam for the students to read books comfortably.
- Cook 0.5 kg of pasta to produce a meal in a household.

- Transport three hundred passengers from London to Oxford.
- Produce 5,000 gallons of gasoline from crude oil.
- Produce 80 tons of cement to be used in the construction of a building.
- Manufacture 600 sections, each of 6" diameter and 8 m length, of a galvanized pipe.

It is the performance of these actions or the fulfillment of the pertinent tasks that is desired by the human society, not necessarily the consumption of energy or exergy. The various forms of energy – natural gas, electricity, gasoline – are consumed as a consequence of accomplishing the desired actions. If the same actions and tasks could be performed reliably and conveniently using lesser quantities of energy or exergy, the consumers are indifferent. Actually, if the actions could be performed using reduced energy or less expensive forms of energy, the consumers would welcome the outcome. Simply put, the amount of energy consumed is of no consequence to the consumers as long as the task is reliably (and oftentimes comfortably) fulfilled.[1]

The vast majority of the desired actions may be accomplished by several methods that use different forms and different amounts of exergy. Let us consider again the desired action of maintaining a comfortable environment for the residents of the building in Berlin. This may be accomplished by several alternative methods including the following:

1. Use the existing/old natural gas burner of the building and maintain the interior temperature at 22°C, as it was done in the last thirty years.
2. Use the existing burner of the building, reduce the interior temperature at 20°C, and ask the residents to wear an extra layer of clothing (a sweater) for their comfort. Less fuel and less exergy will be used for the heating of the building.
3. Replace the existing burner with an efficient heat pump and maintain the temperature at 22°C. Because of the higher COP, and after accounting for the natural gas conversion to electricity, less overall exergy will be consumed for the fulfillment of the desired task.

The first alternative is the *status quo:* do nothing and continue with the same method the task was fulfilled in the past. For the second alternative, it is necessary that the residents change their normal behavior by putting on an extra sweater. For this alternative to be accomplished it is necessary that the buildings' residents cooperate or somehow assist with the fulfillment of the task. This is a typical *energy conservation* project, because an action of cooperation or assistance is necessary for lesser exergy to be used, without the installation of new equipment and without the expense of capital for the investment. The third alternative involves the replacement of the older system with a new one. This alternative requires investment, but it does not require any action by the residents, for whom the change in the method of heating is not even noticeable.

---

[1] On the contrary, if these tasks are not fulfilled (e.g. because of lack of sufficient energy supply or because energy has become too costly), there is widespread discontent and, sometimes, public demonstrations and rage as it happened in the years 2007–8 in India, Indonesia, Pakistan, the Philippines, and several other developing nations because of lack of affordable energy; in January 2017 in Mexico as a result of the rise of the gasoline price; in late 2018 in France (yellow vests); and in Iran in the autumn of 2019.

This is typical of an *energy efficiency* project. Lesser exergy (and lesser primary energy) is used with the higher efficiency of the heat pump without any discomfort and additional action by the residents.

In summary, *energy conservation* typically implies some kind of action or cooperation, by the community: switch off lights; drive less miles; use carpools; use bicycles rather than cars; reduce the building temperature in the winter and raise it in the summer, etc. *Energy efficiency* is typically accomplished by the alteration or replacement of machinery, equipment, or processes and requires capital investment. Energy efficiency usually does not require an action or cooperation and is typically transparent by those affected. Table 4.1 offers a few examples of *energy conservation* and *energy efficiency* activities. All the activities ultimately result in natural resource conservation and lesser exergy destruction.

A third category of activities associated with the popular "energy conservation" concept is *energy substitution,* where another form of energy, typically a renewable form, substitutes for the energy derived from fossil fuels to accomplish a task or a desired action. The substitution of a gas burner with a solar collector to supply hot water is an example of energy substitution. Using a clothesline rather than an electric dryer to dry the clothes is another example. Energy substitution does not necessarily imply lesser exergy destruction: for example, calculations show that more solar exergy is destroyed with the operation of the solar collector than chemical exergy in the burner. However, the solar exergy would have been, anyway, dissipated in the environment if it were not used by the solar collector. The vast majority of energy substitution projects utilize the exergy of renewable energy sources, which dissipate if they are not utilized, in order to consume lesser quantities of other primary energy resources, most notably fossil fuels. This justifies the widespread use of energy substitution projects as "energy conservation" projects.

Table 4.1 Activities that lead to energy conservation and energy efficiency. Regardless of the category, all activities result in to lesser exergy destruction.

Energy Conservation	Energy Efficiency
Switch off lights when out of a room.	Replace the fluorescent lights with LEDs.
Carpool with the neighbors for the daily commute.	Replace the *BMW-318i* you currently commute with a *Ford Escort* .
Increase the thermostat temperature from 20°C to 25°C in the summer. Do the opposite in the winter and adjust the dress code appropriately.	Use an efficient ground source heat pump (GSHP) to replace the older air-conditioning and heating systems.
Implement the daylight energy savings time.	Increase the thermal insulation in the roof of buildings.
Mandate a maximum speed limit of 88 km/hr on the highways.	Install an additional feed-water heater in the electric power plant.
Recycle materials.	Replace the old refrigerator with one that has higher COP.

4.2 Maximum Negative Work – The "Minimum Work"

As individuals fulfill several desired actions (e.g. commuting to work; air-conditioning the house during the summer; cooking meals) and strive to consume the least amount of primary energy sources for the completion of these actions, it is useful to know what the *minimum energy and exergy* for the fulfillment of every action is. For example, what is the minimum quantity of exergy that will cook 0.5 kg of pasta? Or, what is the minimum amount of electricity/exergy that will keep a household's temperature at 25°C during a hot summer day? The minimum exergy for the accomplishment of a given task becomes the *benchmark* for the process/action that requires exergy input. The quantitative knowledge of this benchmark allows engineers and scientists to design equipment that aim to approach this benchmark as closely as possible. Naturally such designs involve processes with lesser irreversibilities.

The concept of exergy is the best quantitative variable for the determination of such benchmarks and the setting of goals. It was shown in Chapter 2 that exergy is the measure of the maximum mechanical work an energy resource may produce. Equations (2.7), (2.17), and (2.26) are readily used to determine the maximum mechanical work (or power) the energy resources may generate and offer the benchmark for the produced mechanical work. The concept of exergy may be similarly used to determine *the minimum amount of mechanical work required for a work-using process, the desired action*: Following the thermodynamic convention, which is depicted in Figure 1.1, the mechanical work, W, produced by thermodynamic systems is positive and the mechanical work consumed by thermodynamic systems is negative. Let us consider the annual amount of work (electricity) consumed by three refrigerators, A, B, and C, that perform the same task (e.g. keep the food at 4°C) and consume annually $W_A = 1,200$ kWh, $W_B = 1,050$ kWh, and $W_C = 1,800$ kWh. There is no doubt that refrigerator B is the most efficient because it consumes the "minimum amount of work," 1050 kWh. However, when we apply the thermodynamic convention, the annual work associated with the three refrigerators is: $W_A = -1,200$ kWh, $W_B = -1,050$ kWh, and $W_C = -1,800$ kWh. Clearly, the work of the most efficient refrigerator, B, corresponds to the maximum algebraic number, because $-1,050 > -1,200 > -1,800$.

The example of the three refrigerators shows that, what is colloquially referred to as "the minimum amount of work consumed," is actually an algebraic maximum according to standard thermodynamic convention. Figure 4.1 schematically shows this notion with work consumption by a process as a function of an arbitrary variable x. The algebraic maximum work corresponds identically to the absolute value of what would colloquially be called "minimum work."

The exergy analyses of systems and processes always determine the maximum work and maximum power associated with equipment and processes. In addition, the exergy analyses do not make a distinction between positive and negative values of the work. As a consequence, the concept of exergy and exergy analyses may be suitably used for the determination of the "minimum work," the benchmark in work-consuming processes and systems, where the algebraic value of the mechanical work is negative.

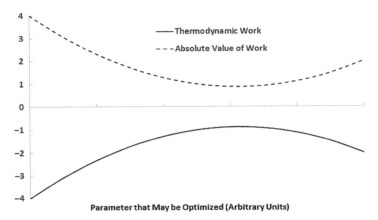

Figure 4.1 The "minimum work" of work-consuming processes is, actually, a thermodynamic maximum.

4.2.1 Application of the Exergy Method to Gas Compression

Let us consider that the desired action is the continuous compression of air in an open system from the ambient conditions ($P_0 = 1$ atm, $T_0 = 300$ K) to 30 atm. A typical compressor, with 80% isentropic efficiency would consume 614 kJ/kg to fulfill this task ($w = -614$ kJ/kg, according to the thermodynamic convention). During the compression process the pressure of the air increases to 30 atm and, simultaneously, its temperature increases to 883 K. The temperature rise occurs because of the chosen compression process, not because the desired action requires it. Even if the compression process were isentropic (100% compressor efficiency) the exit temperature of the compressor would have been 772 K and the specific isentropic work, $w_s = -491$ kJ/kg.

The concept of exergy assists in the determination of the "minimum work" or "minimum power" to be spent for the continuous pressurization of air from 1 to 30 atm in a compressor – an open thermodynamic system. This open system has one inlet and one outlet. The conditions at the inlet are: $P_0 = 1$ atm, $T_0 = 300$ K; and the desired condition at the outlet is $P_1 = 30$ atm. The outlet temperature is irrelevant to the desired action, which is simply the pressurization of air, and not necessarily the heating of air. Eqs. (2.19) and (2.20) may be used for the calculation of the maximum work and the exergy difference between the two states 0-1:

$$w_{max} = e_0 - e_1 = h_0 - h_1 - T_0(s_0 - s_1) \approx c_P(T_0 - T_1) - c_P T_0 \ln \frac{T_0}{T_1} + RT_0 \ln \frac{P_0}{P_1}.$$
(4.1)

One observes that, of the three terms in the last part of Eq. (4.1), only the term $RT_0 ln(P_0/P_1)$ is relevant to the desired action, the pressurization of air. This term is always negative since $P_0 < P_1$, signifying that work must be spent for the pressurization. The other two terms in the last part of Eq. (4.1), $c_P(T_0 - T_1) - c_P T_0 \ln(T_0/T_1)$, pertain to the initial and final temperatures of the system and are irrelevant to the desired action – to provide pressurized air and not hotter pressurized air.

It may be easily proven by elementary calculus that, for all values of T_0 and T_1:

$$c_P T_0 \left(1 - \frac{T_1}{T_0}\right) - c_P T_0 \ln\left(\frac{T_0}{T_1}\right) = c_P T_0 \left(\left[1 - \frac{T_1}{T_0} + \ln\left(\frac{T_1}{T_0}\right)\right]\right) \leq 0. \quad (4.2)$$

Since this part of Eq. (4.1) is negative (or zero for $T_0 = T_1$), the two terms related to the temperature in the equation always make the exergy difference more negative. This implies that additional mechanical work is spent in the compressor for the increase of the air temperature during the compression process. The temperature terms in Eq. (4.1) always increase the absolute mechanical work required for gas compression, even though the final effect – the temperature increase – does not directly pertain to the desired action, the compression of the gas. A glance at Eqs. (4.1) and (4.2) proves that, if the compression occurs isothermally, at $T_0 = T_1$, then the absolute value of the specific work would always be less than the work required by an isentropic compressor [2] and this would be the "minimum absolute work." The "minimum work" required for the isothermal compression is actually an algebraic (and thermodynamic) maximum. In the case of air pressurization it is given by the expression:

$$w_{max} = RT_0 \ln \frac{P_0}{P_1} = -293 \, kJ/kg. \quad (4.3)$$

The last result may be interpreted that *at least* -293 kJ of work must be performed for the pressurization of 1 kg of air from 1 atm to 30 atm; or that *at least* 293 kJ/kg of work must be consumed for the desired action. This becomes the *benchmark of the desired action* that engineers strive to achieve. Although the calculated specific exergy value, -293 kJ/kg, is an algebraic maximum, in engineering practice it is referred to as "the minimum work." It is apparent that an exergy analysis of the gas pressurization process indicates the optimum alternative process: isothermal compression. The specific work consumed during the isothermal compression process of any gas is significantly lower than the work of any other pressurization process including the idealized isentropic compression. It must be noted that, during the isothermal compression, heat must be transferred from the air (gas) to the environment and, because of this heat transfer, the entropy of the air decreases by $R \ln(P_0/P_1)$. The entropy of the environment, which receives this heat, increases (by the same amount if the heat transfer is reversible).

It is apparent that the correct use of thermodynamic theory, and in particular of the exergy method, assists in the determination of the numerical value of the "minimum work," which is the benchmark of a process and provides an indication for the type of processes to be used for the consumption of the "minimum work," in the case of gas compression the isothermal process. With this in mind engineers may strive to achieve compression processes as close to isothermal as it is practically feasible.

It may be easily proven, using the first law of thermodynamics, that heat must be transferred from the gas to the environment during the isothermal compression and that the specific heat to be transferred is equal to the specific work:

$$w_{max} = q_{out} = RT_0 \ln \frac{P_0}{P_1} < 0 \quad (4.4)$$

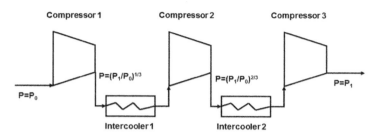

Figure 4.2 Three-stage gas compression with intercoolers.

This kind of heat transfer is difficult to achieve in practice, if a significant quantity of the gas is to be compressed at a reasonably fast rate. For this reason, engineers have devised multistage isentropic processes that take place at fast rates, and use intercoolers that cool the gas between the compression stages. The intercoolers are heat exchangers that receive the hotter gas at the end of a partial compression stage and cool it to a lower temperature, which is close to the environmental temperature. The cooled gas is then fed to the next compressor stage for further pressurization. This compression-intercooling process is repeated until the gas is pressurized to the desired pressure.[2] The arrangement of the equipment in a three-stage compression process with two intercoolers is shown in Figure 4.2. For the optimum operation of the entire compression process with intercoolers, the pressure ratios in each compressor are equal to $(P_1/P_0)^{(1/n)}$, where n is the number of compressors, and the number of intercoolers is $n - 1$.

The continuous isothermal compression of a gas may be achieved, in principle, by an infinite number of compressors and intercoolers. Because of the higher cost of large numbers of compressors and intercoolers, in practice, the number of intercoolers is limited to three or less. Table 4.2 shows the absolute amount of specific work required for the compression of air from 1 to 30 atm using: a. a single isentropic compressor; and b. compressors with one, two, and three intercoolers. For the calculations it was stipulated that the temperature of the gas at the exit of each intercooler is equal to the environmental temperature, $T_0 = 300$ K. The last column depicts the additional work of the corresponding method in comparison to the benchmark.

It is of interest to note in Table 4.2 that the use of one intercooler is "more efficient" and requires less specific work for the compression than the use of an idealized, 100% isentropic compressor. It is clear from these and similar results that the benchmark for gas compression is the isothermal process, which is indicated by exergy analysis, and not the isentropic compression.

---

[2] Apart from lower compression work, another advantage of intercooling, is that the lower temperatures allow the use of less expensive materials for the compressors.

Table 4.2 Work, in kJ/kg, for the compression process of air from 1 to 30 atm.

	Specific work (kJ/kg)	Additional specific work (%)
Isothermal compression (benchmark)	293	0
Isentropic compression, $\eta = 100\%$	491	68
Isentropic compression, $\eta = 80\%$	614	110
Isentropic, $\eta = 80\%$, one intercooler	468	60
Isentropic, $\eta = 80\%$, two intercoolers	433	48
Isentropic, $\eta = 80\%$, three intercoolers	415	42
Isentropic, $\eta = 80\%$, infinite intercoolers	366	25

Table 4.3 State variables for a typical refrigeration cycle.

	T (K)	P (MPa)	h (kJ/kg)	s (kJ/kg K)	e (kJ/kg)
1	268.9	0.25	396.08	1.7296	23.2
2	321.92	0.9	431.63	1.7577	50.3
3	308.0	0.90	248.78	1.1663	44.9
4	268.9	0.25	248.78	1.181727	40.2
4'	268.9	0.25	194.27	0.97898	46.5
4"≡1	268.9	0.25	396.08	1.7296	23.2

4.3 Refrigeration and Liquefaction

The vapor refrigeration cycle is the most commonly used of the refrigeration cycles and is shown in Figures 1.11 and 1.12. This cycle and its variants are widely used in refrigeration, freezing, and air-conditioning systems. The desired action of the refrigeration process is to remove heat from a colder space (at T_L in Figure 1.12) and this is achieved during the evaporation process 4-1. Mechanical work is supplied to the cycle at the vapor compressor for the process 1-2. Table 4.3 shows the conditions as well as the enthalpy and exergy of a refrigeration cycle with R-134a as the working fluid. The conditions of the cycle are upper pressure 0.9 MPa (9 bar) and lower pressure 0.25 MPa (2.5 bar), which corresponds to the evaporation temperature of $-4.3°C$. The efficiency of the compressor is 75%, the temperature of the environment is 27°C (300 K) and the pressure 0.1 MPa.

The coefficient of performance (COP) of this cycle is: $COP = (396.08 - 248.78)/(431.63 - 396.08) = 4.14$. If the compressor were isentropic, then the COP would have been 5.53.

Let us assume that this refrigerator is used to freeze 1 kg of water from the ambient conditions to a temperature $-2°C$ (271.15 K) at 0.1 MPa. The exergy of the ice at $-2°C$ is 38.9 kJ/kg (signifying that, because ice is at low temperature, it may be used to produce mechanical work, as shown in Figure 2.12). For the production of 1 kg of ice the refrigeration cycle removes 450.9 kJ of heat at the expense of 108.9 kJ of mechanical work in the form of electricity. The exergetic efficiency of ice production via this refrigeration cycle is $\eta_{II} = 35.7\%$. If the compressor were isentropic, the exergetic efficiency of ice production would still have been relatively low at $\eta_{II} = 47.7\%$.

An exergetic analysis of the four processes indicates where work is lost:

1. The compression process, 1-2, increases the exergy of the refrigerant by $50.3 - 23.2 = 27.1$ kJ/kg and consumes work 35.6 kJ/kg. The specific exergy destruction is 8.5 kJ/kg.
2. The condensation process, 2-3, removes enthalpy from the refrigerant and dissipates heat in the atmospheric air. The specific exergy destruction in this process is 5.4 kJ/kg.
3. The isenthalpic expansion process, 3-4, produces a two-phase mixture of vapor and liquid refrigerant. The exergy of the mixture is 40.2 kJ/kg and the specific exergy destruction in this process is 4.7 kJ/kg.
4. The evaporation process, 4-1, produces vapor at 0.25 MPa (state 1), with exergy 23.2 kJ/kg. This process also produces $(396.08 - 248.78)/450.9 = 0.327$ kg of ice with exergy 12.7 kJ. The specific exergy destruction in this process is 4.3 kJ/kg of the refrigerant.

It is observed that most of the exergy destruction in vapor refrigeration cycles occurs in the compression process. The process may be improved by the choice of more efficient compressors, with an isothermal compressor being ideal. For large refrigerator systems a series of compressors with intercoolers reduces the required power and improves the exergetic efficiency of the cycle.

Example 4.1: The refrigeration cycle of Table 4.3 is used for the cooling of pasteurized milk, from 80°C to 4°C. The volumetric flow rate of milk is 1.40 L/s, its specific heat capacity is $c = 4.14$ kJ/kg K and its density is $\rho = 985$ kg/m³. Calculate the mass flow rate of the refrigerant in the refrigeration cycle and the power supplied. What is the benchmark for the power consumption for this cooling process?

Solution: The rate of heat to be removed from the milk is: $\dot{Q} = \dot{m}c(T_2 - T_1) = \rho \dot{V} c(T_2 - T_1) = 985 * 0.0014 * 4.14 * 76 = 433.9$ kW (1.40 L/s $= 0.0014$ m³/s). Since 1 kg of the refrigerant removes 147.3 kJ of heat, the mass flow rate of the refrigerant is: 2.946 kg/s and the power to drive the compressor is: 104.7 kW.

For the calculation of the benchmark power, one notes that the specific exergy difference of the refrigeration process as calculated from Eq. (2.21) with $T_0 = 300$ K is 13.52 kJ/kg of milk. Therefore, the benchmark minimum power is: $13.52 * 985 * 0.0014 = 18.64$ kW.

4.3.1 Liquefaction of Gases

Gas liquefaction processes consume a great deal of energy and exergy. The liquefaction process has been used since the end of the 19th century for the production, transportation, and storage of several industrial chemicals, such as oxygen, nitrogen, helium, carbon dioxide, and hydrogen. In the second decade of the 21st century Liquefied Natural Gas (LNG) is being produced in increasingly larger quantities to be transported in transoceanic vessels. The liquefaction process of all gases, includes compression at supercritical pressures (a very much energy consuming process) followed by an isenthalpic expansion (similar to the expansion of the refrigeration cycle) before the liquid phase is produced. Figure 4.3 shows the equipment and states for the liquefaction of methane, the main constituent of LNG. Similar equipment and processes are used for the liquefaction of other gases. At first, the gas is compressed and cooled in a series of compressors with intercoolers to a supercritical pressure; and to temperature close to that of the environment (state 5 in Figure 4.3). Secondly, the high-pressure gas is further cooled in a counter-flow heat exchanger by the colder vapor effluent of the separator, at state 7". During this process the vapor effluent is heated to state 8 and the pressurized gas is cooled to state 6. Thirdly, the gas expands at constant enthalpy, typically to 1 atm, in an expansion valve. The expansion produces a two-phase mixture of saturated liquid (denoted as state 7') and saturated vapor (denoted as state 7"). The two phases in state 7 are separated by gravity. The liquid part is removed at state 7' to be transported. The vapor part – at state 7" – is returned to the heat

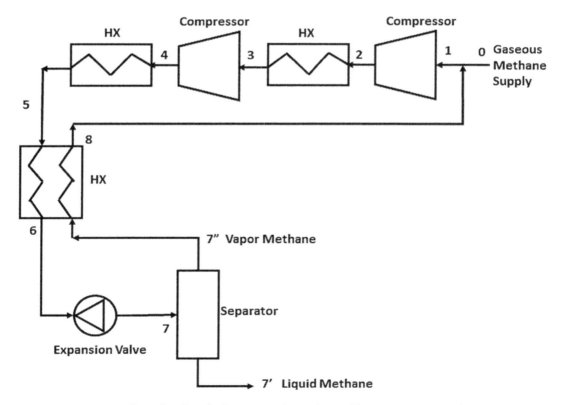

Figure 4.3 Cascading liquefaction process for methane with two stage compression.

exchanger and the compressors to repeat the cycle. This liquefaction process, often referred to as the *cascading refrigeration* process, is repeated and continuously produces LNG for transportation. The *yield* of the process is defined as the dimensionless ratio of the mass of liquid produced per unit mass of gaseous input.

Exergetic calculations were performed for the production of liquid methane at 0.1 MPa (where the saturation temperature is 111.5 K), from a stock of gaseous methane at the environmental conditions, $T_0 = 300$ K and $P_0 = 0.1$ MPa. Two compressors, each with efficiency 84%, followed by intercoolers, as shown in Figure 4.3, compress the methane gas to 40 MPa. The properties of the several states of the cascading refrigeration cycle are calculated using the REFPROP® software [3] and the results are listed in Table 4.4. The pressurized methane gas exits the counter-flow heat exchanger at 40 MPa and 212.6 K, with specific enthalpy 398.1 kJ/kg. The expansion process produces saturated, two-phase mixture with dryness fraction $x_7 = 0.780$, which implies that the liquid yield is 0.220 kg of liquid per kg of gaseous methane input.

The specific work supplied to the cycle in the two compressors is $(1,687.0 - 905.6 + 1,715.8 - 907.3) = 1,589.9$ kJ/kg of gaseous methane. Since the yield of the cycle is 0.220 kg of liquid methane, the work consumed for the liquefaction process is 7,229 kJ/kg of the liquid methane produced. With the product exergy being 1,095 kJ/kg, the exergy destruction (work lost) in the entire liquefaction process is 6,134 kJ/kg of liquid methane and the exergetic efficiency of the process is 15.2%. The exergy destruction in the specific processes of this liquefaction cycle is as follows (all values are per kg of the yield, the produced liquid methane):

1. First compression, process 1-2: $w_{lost} = 335.9$ kJ/kg.
2. Cooling in the first intercooler, process 2-3: $w_{lost} = 1,121.8$ kJ/kg.
3. Second compression, process 3-4: $w_{lost} = 298.9$ kJ/kg.

Table 4.4 Properties of methane at the several stages of the *cascading refrigeration* cycle. The subscript *s* designates the isentropic state. The superscripts ′ and ″ denote saturated liquid and saturated vapor respectively. $\eta_C = 84\%$, $T_0 = 300$ K, and $P_0 = 0.1$ MPa.

	T (K)	P (MPa)	h (kJ/kg)	s (kJ/kg K)	$e - e_0$ (kJ/kg)
0	300.0	0.1	914.1	6.6951	0.0
1	296.2	0.1	905.6	6.6666	0.1
2s	548.6	2.0	1,563.3	6.6951	649.2
2	587.6	2.0	1,687.0	6.9128	707.6
3	305.0	2.0	907.3	5.1361	460.9
4s	572.4	40.0	1,586.4	5.1361	1,140.0
4	609.0	40.0	1,715.8	5.3552	1,203.6
5	305.0	40.0	703.4	3.0622	879.2
6	212.6	40.0	398.1	1.8504	937.4
7	111.5	0.1	398.1	3.5703	421.5
7′	111.5	0.1	−0.6	−0.0050	1,095.3
7″	111.5	0.1	510.6	4.5787	231.4
8	295.0	0.1	902.95	6.6577	0.1

4. Cooling in the second intercooler, process 4-5: $w_{lost} = 1{,}475.4$ kJ/kg.
5. Heat exchanger, processes 5-6 and 7"-8: $w_{lost} = 555.9$ kJ/kg.
6. Expansion, process 6-7: $w_{lost} = 2{,}346.2$ kJ/kg.

It is observed that most of the exergy destruction occurs in the isenthalpic expansion process. An improvement to the cycle would be to expand the gas from state 6 to state 7 in a turbine, thus making the expansion process almost isentropic. The use of a turbine improves the exergetic efficiency of the cycle in two ways: At first, it produces useful work that may be used in the compression processes; secondly, it increases the liquid yield because the enthalpy at state 7 would be less than the value listed in Table 4.4. The partial use of a turbine with gas liquefaction processes is used in the so-called *Claude liquefaction*. Liquefaction with a throttling device – the expansion valve – is sometimes called *Linde liquefaction*. High exergy destruction also occurs in the intercoolers. This suggests the utilization of the higher exergy of the hotter gas for a useful process, such as the production of work via a bottoming vapor cycle.

Since all of the work supplied to the cycle is for the compression of the gas, significant work and exergy savings will be realized by an isothermal compression of methane, leading directly from state 1 to state 5. For an isothermal compression process in the cascading refrigeration, the properties of the several states are listed in Table 4.5.

In this case of isothermal compression, the yield of the entire cascading refrigeration process is 0.238 kg of liquid methane per kg of gaseous methane; the work input is 3,688 kJ/kg of liquid methane; and the lost work is 618 kJ/kg of gaseous methane or 2,593 kJ/kg of liquid methane. The exergetic efficiency of the cyclic process is 29.7%. This is the *benchmark* for the liquefaction of methane at the chosen 40 MPa pressure.

The exergy analysis of cascading refrigeration processes suggests that the exergetic efficiency of gas liquefaction improves with the addition of more compressor stages and intercoolers. One may also improve the exergetic efficiency by optimizing the compression pressure. The following example illustrates this type of optimization.

Table 4.5 Properties of methane at the several stages of the *cascading refrigeration* cycle with isothermal compression from state 1 to state 5.

	T (K)	P (MPa)	h (kJ/kg)	s (kJ/kgK)	$e - e_0$ (kJ/kg)
0	300.0	0.1	914.1	6.6951	0.0
1	296.2	0.1	905.6	6.6666	0.1
5	296.2	40	686.75	3.0071	879.1
6	211.19	40	388.7	1.8291	934.4
7	111.5	0.1	388.7	3.487779	437.0
7'	111.5	0.1	−0.6	−0.0050	1,095.3
7"	111.5	0.1	510.6	4.5787	231.4
8	295	0.1	902.95	6.6577	0.1

Example 4.2: Calculate the work per kg of liquid methane produced, the yield per cycle, and the exergy destruction per cycle, when three compression stages are used with intercoolers and the higher pressure of the cycle is in the range 10–60 MPa. Methane is supplied at the environmental conditions, $T_0 = 300$ K and $P_0 = 0.1$ MPa. Each intercooler reduces the methane temperature to 305 K and the compressors have 84% isentropic efficiency.

Solution: The calculations were performed using the REFPRO® software [3] and the results are listed in Table 4.6.

It is observed that the addition of the third compression stage with the intercooler at 40 MPa decreases the needed work input from 7,229 to 6,108 kJ/kg of liquid methane and improves the exergetic efficiency to 17.9%. It is also observed that, with this configuration of three stage compression, there is an optimum pressure for the liquefaction of methane between 40 MPa and 30 MPa. Figure 4.4, which shows graphically the results of the calculations indicates that this optimum pressure is approximately 34 MPa.

Table 4.6 Liquid yield, supplied work, work lost, and exergetic efficiency for the liquefaction of methane in the range of pressures 10–60 MPa.

P (MPa)	Yield (dimensionless)	w (kJ/kg liquid)	w_{lost} (kJ/kg gas)	w_{lost} (kJ/kg gas)	η_{II} (%)
10	0.092	11,105	917	10,009	9.9
20	0.175	6,839	1,008	5,744	16.0
30	0.216	6,099	1,080	5,003	18.0
40	0.230	6,108	1,152	5,013	17.9
50	0.231	6,384	1,224	5,288	17.2
60	0.227	6,804	1,293	5,709	16.1

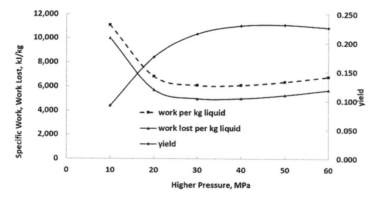

Figure 4.4 Liquid yield, exergy destruction, and work supplied for the production of liquid methane with three compression stages.

Example 4.3: A liquefaction plant, built with the specifications of the one in Figure 4.4, operates at the optimum pressure of 34 MPa and produces 72 tons of LNG per hour. Assuming that the natural gas has the properties of methane, calculate the power consumption of the plant and the rate of exergy destruction.

Solution: From Figure 4.4, the specific work for the liquefaction of natural gas in this unit is $w = 6{,}050$ kJ/kg of liquid produced. The mass flow rate of the LNG is $72{,}000/3{,}600 = 20$ kg/s and the power consumed is $6{,}050 * 20$ kW $= 121$ MW.[3] The rate of exergy destruction is $121 - 1.095 * 20 = 99.1$ MW.

4.4 Drying

Drying is the process of water removal from products ranging from foods, to chemicals, to forestry/wood products. There are two main methods of drying: drying by air; and freeze drying. In the first process a stream of air removes the moisture from the surface of the product and this causes an increase of the specific humidity of the air. In freeze drying the product is first frozen at a temperature below the triple point of water (0.01°C) and the water content is removed by sublimation under partial vacuum. Diffusion of water from the interior to the surface of the product facilitates water transport and its removal in both methods. Because mass diffusion is a slow process, the drying of most products takes very long time. Figure 4.5 is a schematic diagram of the drying process of a generic product. Heat is supplied to the drying equipment and increases the air temperature. Part of the heat is dissipated through the surface of the equipment to the environment. Depending on the process and the product, the moist air output may carry small water droplets in addition to the higher humidity.

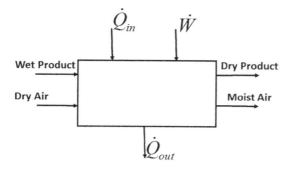

Figure 4.5 Energy exchanges during the drying process with air.

---

[3] Because LNG facilities require a great deal of power they usually have on-site, dedicated plants to produce the needed power, typically one or more gas turbines.

The following are the governing equations for the drying process of Figure 4.5:

1. The mass balances of the three species:

$$\begin{aligned}&\text{For air:} \quad \dot{m}_{ai} = \dot{m}_{ae} = \dot{m}_a \\ &\text{For the product:} \quad \dot{m}_{Pi} = \dot{m}_{Pe} = \dot{m}_P \\ &\text{For water:} \quad \dot{m}_{ai}\omega_i + \dot{m}_{wi} = \dot{m}_{ae}\omega_e + \dot{m}_{we}\end{aligned} \quad (4.5)$$

where ω is the absolute humidity of the air, measured in kg water vapor per kg of dry air; the subscripts a, P, w, i, and e refer to air, product, water, inlet, and outlet respectively; and the mass flow rate of water in the last term in the third expression refers to any liquid water that may be carried out.

2. Energy balance:

$$(\dot{Q}_{in} - \dot{Q}_{out}) - \dot{W} = \dot{m}_a(h_{ae} - h_{ai}) + \dot{m}_P(h_{Pe} - h_{Pi}) + \dot{m}_{we}h_{we} - \dot{m}_{wi}h_{wi}. \quad (4.6)$$

The air enthalpy in Eq. (4.6) is the enthalpy of humid air, which includes the enthalpy of the water vapor carried by the air stream. This property is obtained from air and steam tables or from psychometric charts and is a function of the temperature and the relative humidity (or, equivalently, of the absolute humidity). Because water is removed by diffusion, all the water content cannot be removed from the product within a finite time, even if the air surrounding it were dry.

When drying with ambient (humid) air, the minimum moisture level a product may attain is the *equilibrium moisture content*, ω_{eq}, the moisture content where the partial pressure of water in the product is equal to the partial pressure of water in the ambient air. The moisture content of the product, ω, during the drying process is modeled by the empirical equation:

$$\frac{\omega - \omega_{eq}}{\omega_i - \omega_{eq}} = \exp\left(-\frac{t}{\tau}\right), \quad (4.7)$$

where ω_i is the initial moisture content; and τ is the characteristic time of the drying process. The latter depends on the type of product, the incoming air temperature, and the air velocity in the dryer. An experimental study for the drying of cocoa beans in air streams at temperatures in the range 55–81°C, determined that τ is in the range 3,280–6,260 s [4].

4.4.1 Exergy Analysis of Drying

A moment's reflection proves that, if a wet product is left for a long time in the environment (under cover for protection from rain), the product will attain equilibrium with the environment and the moisture content of the product will become equal to ω_{eq}. A consequence of this observation is that the drying process requires zero work. With heat delivered by the environment it only requires sufficient time for the equilibrium to be reached. Examples of this type of drying are: sundried tomatoes, sundried figs, and sundried tobacco leaves. One may conclude, therefore, that the benchmark exergy supply (minimum work) for the drying of all products is zero. During the drying process, there is exergy destruction as water diffuses through the product and finally evaporates and diffuses in the air. This exergy may only be recovered as work by semipermeable membranes, a technology that has not been developed yet.

Heat and work are supplied to an actual drying process, as shown in Figure 4.5, because of the need to accelerate the process and dry the product faster. In a typical drying process, the air and the product are heated and are mechanically mixed to accelerate the removal of moisture. The higher exergies of the outlet air and product are not recovered as useful work and are, eventually, dissipated in the environment as the product cools and the air is mixed with the ambient air. Therefore, the following sources of exergy destruction are encountered in drying processes:

1. The exiting warmer air with higher humidity.
2. The exiting warmer product.
3. Direct heat loss from the drying equipment to the environment.
4. Air friction and the mechanical energy dissipation of the air/vapor mixture at the exhaust.
5. In the case of freeze drying, the power dissipated in the vacuum pumps and the operation of the refrigeration cycle.

Example 4.4: A stream of ambient air at $T_0 = 20°C$ and 50% relative humidity with volumetric rate 0.015 m³/s is first heated to 60°C and then fed to a dryer of coffee beans that operates in a batch process. The power supplied to the blower is 75 W and the heat is supplied by a methane burner. It takes four hours to dry 25 kg of coffee beans. Determine the exergy destruction per unit mass of the product during this batch process.

Solution: None of the exergy supplied to the air and the product of this process is recovered and, therefore, all the exergy supplied to the dryer system is dissipated.

The exergy dissipation for the operation of the blower during the four-hour process is: $75 * 4 * 60 * 60 = 1,080,000$ J $= 1.08$ MJ.

From the psychrometric chart, the enthalpy of the humid air at 20°C and 50% relative humidity is $h_i = 36.2$ kJ/kg, and when this air is heated to 60°C, its enthalpy increases to $h_e = 76.5$ kJ/kg (kJ/kg of dry air). At the inlet (ambient) conditions the density of the air is 1.19 kg/m³ and, hence, the rate of heat needed for the air temperature increase is: $0.015 * 1.19 * (76.5 - 36.2) = 0.719$ kW. The total heat supplied to the dryer during the four hours of operation is: 10.36 MJ. Since the heat of combustion of methane is 890 MJ/kmol, this quantity of heat is supplied by $10.36/890 = 0.01164$ kmols of methane with exergy 9.52 MJ (values from Table 2.2).

The total destruction of exergy is $9.52 + 1.08 = 10.60$ MJ, or 0.424 MJ/kg of the product (the produced coffee is let to cool and all its exergy is dissipated).

It is apparent in this example that most of the exergy destruction is due to heating the air, which is exhausted in the environment. The exergy analysis suggests the total or partial utilization of this exergy. This may be achieved with the use of a counter-flow HX (often called regenerator) that transfers part of the enthalpy of the exhausted air to the inlet air.

Let us consider that such a regenerator, with effectiveness $\varepsilon = 80\%$ is added to the dryer system and that as a result of this addition the power of the blower needs to be increased to 125 W. The exergy dissipation due to the power in the blower is 1.80 MJ. With the HX, the inlet temperature of the air to the heater is: $20 + 0.8 * (60 - 20) = 52°C$, and the heater only has to supply 0.150 kW of heat to increase the air temperature from 52°C to 60°C. The heat supplied during the four hours of operation is 2.16 MJ and this corresponds to 1.98 MJ of exergy, which is finally dissipated. The total exergy destruction in this case is $1.80 + 1.98 = 3.78$ MJ or 0.152 MJ/kg of the product. Therefore, the addition of the regenerator decreases the exergy required for the drying of this product by a factor of $10.60/3.78 = 2.80$.

4.5 Petroleum Refining

Petroleum (crude oil) is an inhomogeneous mixture of a large number of chemicals, primarily hydrocarbons that cover a wide range of molecular masses. The constituent compounds of petroleum span all three states of matter – gas, liquid, and solid – under ambient temperature and pressure. The petroleum composition varies with its production location and the timing of the extraction. Oil fields produce mixtures of lighter, more volatile hydrocarbons in the first stages of the extraction process and mixtures of heavier hydrocarbons in the later stages. When secondary and tertiary recovery methods are implemented in the oil fields the produced mixture is laden with heavier hydrocarbons. The principal chemical elements of petroleum and their composition by weight are shown in Table 4.7 [5]:

Engines that consume petroleum products by design do not use crude oil in the form it is extracted. Gasoline and diesel are primarily used in internal combustion engines; kerosene in aviation (jet) engines; heavier liquids are used for lubricants; and synthetic gas (a mixture of butane, propane, and other gases) are used for cooking. In order to generate these consumer products, the crude oil undergoes *refinement*, which entails the physical process of distillation and the chemical processes of cracking and production of marketable chemicals.

The distillation process is schematically depicted in Figure 4.6. The chemical constituents of the crude oil are heated to approximately 400°C in a furnace (evaporator) where 75–80% of the species are in a vapor form. At this state the mixture is fed to the

Table 4.7 Elemental composition of petroleum by weight.

Carbon	83–85%
Hydrogen	10–14%
Nitrogen	0.1–2%
Oxygen	0.5–1.5%
Sulfur	0.05–6%
Metals	0–1%

4.5 Petroleum Refining

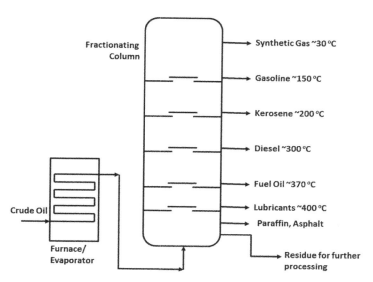

Figure 4.6 The distillation process with a single fractionating column.

fractionating column, where a temperature gradient is maintained using a number of heat exchangers. The lower part of the fractionating column is at approximately 400°C and the top is close to the ambient temperature. The liquid part of the high temperature mixture and any dissolved solids separate by gravity and are removed at the bottom of the column to be used as solids (paraffin waxes, asphalt) or as residue for further processing (cracking). The liquid hydrocarbons that constitute the several lubricants are removed at approximately 400°C; the heavy fuel oil at about 370°C; diesel at 300°C; kerosene at 200°C; and gasoline at about 150°C.[4]

The gaseous components of petroleum that make up the *volatile matter* are removed at the top of the column. They are used as fuel in the refinery or they are further processed to produce hydrogen and additional gasoline components, as for example in the following reaction:

$$8CH_4 \xrightarrow{heat} C_8H_{18} + 7H_2. \tag{4.8}$$

The molecular structures of the heavy hydrocarbons in the residue stream are broken at very high temperatures with the partial addition of hydrogen in the *cracking* process, as for example in the following reaction of eicosipentane, $C_{25}H_{52}$, which produces two of the gasoline constituents:

$$C_{25}H_{52} + 2H_2 \xrightarrow{heat} 2C_8H_{18} + C_9H_{20}. \tag{4.9}$$

Let us consider a simplified example of a distillation process in which a mixture of four petroleum constituents – hexane (C_6H_{14}), octane (C_8H_{18}), decane ($C_{10}H_{22}$), and

---

[4] Actual distillation columns have 20–40 product removal stages and trays.

dodecane ($C_{12}H_{26}$) – is supplied to the fractionating column at 300 K and 0.1 MPa. The mixture consists of equal fractions, 25% each by weight. The boiling points of the four constituents are respectively 341.5 K, 398.3 K, 446.7 K, and 489.0 K. In the typical distillation processes the three lighter components of the mixture are first vaporized and then condensed at separate heights of the distillation column. Propane, which abounds in the gaseous products of the refineries, is the most common fuel that provides the heat for the furnace.

In order to separate the four components of the mixture the entire mixture is heated to 452 K – approximately 5 K above the boiling point of decane – at atmospheric pressure (0.1 MPa) and fed to the distillation column. At this state the hexane, octane, and decane are in the vapor form and dodecane is the liquid residue removed at the bottom of the column. The vapor mixture flows in the column where decane condenses first and is removed slightly subcooled at 442 K; octane is removed at 393 K; and hexane is removed at 336 K. To ensure complete condensation the removal temperatures are approximately 5 K below the boiling points of the corresponding chemicals. Table 4.8 shows the relevant properties of the four constituents of the mixture: a. at the entrance of the furnace; b. at the exit of the furnace, which is the same as the state at the entrance of the fractionating column; c. at the exit of the distillation column; and d. at a few other states that are relevant to the calculations. All the property values are obtained from REFPROP [3].

The heat needed in the furnace for the evaporation of 1 kg of the three-component mixture is:

$$q = 0.25 \sum_{i=1}^{4} [h_i(452) - h_i(300)], \tag{4.10}$$

Table 4.8 Pertinent properties of hexane, octane, decane, and dodecane. The pressure in all the states is 0.1 MPa.

	T (K)	h (kJ/kg)	s (kJ/kg K)	e (kJ/kg)
Hexane	300	−98.5	−0.3072	0.0
	346	342.9	1.0041	48.1
	452	568.6	1.5705	103.8
Octane	300	−240.3	−0.6895	0.0
	346	−133.5	−0.3586	7.5
	403	311.4	0.7816	110.3
	452	421.8	1.0399	143.2
Decane	300	−367.9	−0.9896	0.0
	346	−262.6	−0.6634	7.4
	403	−120.2	−0.2828	35.7
	452	287.6	0.6436	165.6
Dodecane	300	−488.4	−1.2470	0.0
	346	−382.7	−0.9195	7.5
	403	−240.61	−0.5398	35.6
	452	−108.0	−0.2295	75.1

which amounts to 591.3 kJ/kg of the original mixture. Using propane combustion as the heat source, 0.2664 kmol of propane with exergy 561.5 kJ are burned for the fractionation of 1 kg of the mixture.

Using the values in Table 4.8, we calculate the exergy associated with this fractionation process as: $0.25 * (103.8 + 143.2 + 165.6 + 75.1) = 121.9$ kJ/kg of the mixture. This signifies that the separation of the mixture into its four components by evaporation may be accomplished with significantly lesser exergy expenditure. One also realizes that hexane and octane do not need to be heated up to the higher temperature of 452 K in order to be separated. Hexane evaporates at 341.5 K and only needs to be heated up to this temperature to be separated from the mixture. Similarly, octane only needs to be heated up to 398.3 K. In addition, it is not necessary to burn propane for the evaporation since the combustion process is associated with high exergy destruction. Another method of heat addition, such as a heat pump, may be used to minimize exergy destruction.

Such exergetic considerations point to a multistage distillation process, which entails more than one stage. A general multistage distillation process with n stages and $n+1$ components is schematically depicted in Figure 4.7. The mixture to be distilled is heated in the first heater (H-1) and is partly evaporated. The vapor of the first fraction is separated by gravity in the first separator and this fraction is removed as vapor to be condensed. The residue of the first stage is heated in the second heater (H-2) to a higher temperature where the second vapor fraction is separated and removed. The process is repeated to the last, nth stage, where the nth vapor fraction is removed at the top of the last separator to be condensed. The liquid effluent of the last stage is the final residue that contains all the remaining components in the liquid and solid phase to be further processed with chemical methods (cracking).

Let us now apply this multi-stage process to the mixture of hexane, octane, decane, and dodecane that was previously considered. Three distillation stages will separate the

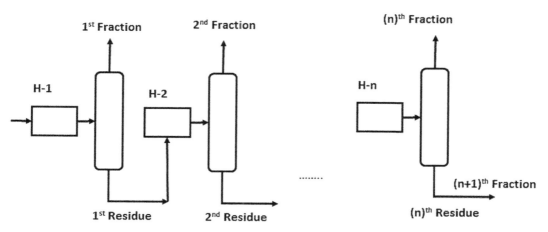

Figure 4.7 A multistage distillation process uses less exergy for the separation of the petroleum fractions.

mixture into its four constituents. In order to achieve complete evaporation, again, each heater raises the temperature of the mixture to a temperature approximately 5 K above the corresponding boiling point. Thus, the first heater exhausts the mixture at 346 K, where hexane is in the vapor state and is separated. The second heater increases the temperature of the remaining components to 403 K, where the octane vapor is removed and separated. The third heater increases the temperature of the remaining fractions of decane and dodecane to 452 K, where decane is separated as vapor and dodecane is the liquid residue. The specific heat input in the three stages is:

$$q = 0.25 \sum_{i=1}^{4} [h_i(346) - h_i(300)] + 0.25 \sum_{i=1}^{3} [h_i(403) - h_i(346)]$$
$$+ 0.25 \sum_{i=1}^{2} [h_i(452) - h_i(403)]. \qquad (4.11)$$

This amounts to 507.3 kJ/kg of the original mixture, a 14.2% reduction compared to the single-stage distillation process. The exergy imparted to the hydrocarbon fractions associated with the three-stage distillation is 99.8 kJ/kg (18.1% reduction). The calculations with multistage distillation processes are in qualitative agreement with the results of a detailed study that compared single and two-stage distillation units for crude oil [6]. It must also be noted that, when a multistage distillation process is designed, sequential columns may operate at different pressures (as well as different temperatures).

The exergy analysis also indicates that improvements can be made by using a heating method, other than fossil fuel combustion. If ideal heat pumps were to be used as heaters in the three distillation stages to raise the temperature of the mixture, the coefficient of performance of the first heat pump would have been 346/(346 − 300) = 7.52; of the second heat pump 3.91; and of the third heat pump 2.97. The three ideal heat pumps would have needed 117.4 kJ of mechanical work per kg of the original mixture. It must be noted that this value of the mechanical work is different than the 99.8 kJ/kg of the total exergy of the discharged hydrocarbons because the heat to the three heaters is supplied at 5 K higher than the corresponding evaporation temperatures.

A further improvement of the exergetic efficiency of distillation – one that is used in single-stage refineries – is the use of HXs to utilize the heat released by the condensing and cooled fractions at the higher temperatures to preheat the original mixture. For example, the heat released during the condensation of the diesel fraction at 300°C and the subsequent cooling of this fraction may be used to preheat the original mixture at temperatures well below 300°C. Such exergetic considerations point to the following design improvements of the distillation processes:

1. Use multi-stage rather than single stage distillation.
2. Use heat pumps rather than fossil fuel combustion in the heaters.
3. Use heat exchangers to transfer part of the energy and exergy of the discharged hydrocarbon fractions to partially preheat the original mixture.

4.6 Water Desalination

Freshwater (potable water) availability is essential for the sustenance of life and the survival of humans and animals. The high demand of an ever-growing human population has caused *water stress* in several regions of the earth, where demand for freshwater exceeds the available supply for at least part of the annual water cycle. Water stress damages the quantity and quality of freshwater resources and induces economic hardship in the affected regions.

Water stress and water scarcity are rather ironic because more than 70% of the earth's surface is covered by water. However, approximately 97.5% of the available water is saline (in the sea and saline groundwater) and unsuitable for drinking and agriculture. Of the freshwater on the planet, most of it is in the form of ice and snow in remote glaciers, and in groundwater. Less than 0.01% of the total water on earth is available as freshwater in lakes, rivers, and swamps with the lakes and the swamps accounting for more than 98% of the available surface freshwater. In particular, the Great Lakes in North America, Lake Baikal, and the Great African Lakes account for 72% of the global surface freshwater [7]. Because of the uneven freshwater distribution and the significant population increase, several regions of the globe – the Saharan region, the West Andean region, the Middle East, and Central Asia – suffer from chronic water stress. Artificial desalination is the most common method to mitigate the water stress in such regions.

Among the several methods used for water desalination are: electro-dialysis; vapor compression; forward and reverse osmosis; flashing; membrane distillation; and boiling. In the 21st century solar energy has been increasingly used (as heat and as PV-generated power) in water desalination plants. All the methods for the production of freshwater encompass significant energy and exergy expenditures that range from 3–6 kWh/m^3 for reverse osmosis, to 30 kWh/m^3 (total electric equivalent) for boiling [8].

Given the significance of freshwater supply for the human society and the increasing use of desalination systems for the production of freshwater, it is important to determine the "minimum work" required for the conversion of seawater to freshwater. The exergy method will be used to make this determination.

Seawater is a mixture of several salts (primarily NaCl) and water. The seawater contains approximately 35 kg/m^3 of salts (35,000 ppm); its molality is approximately 0.6 m^5 and, because of this, it is considered a dilute solution. At the environmental pressure and temperature, the process of mixing of salts with water is irreversible and causes an entropy increase. For the dilute salt-water solution, the entropy increase is [9, 10]:

$$\Delta S_{mix} = -\tilde{R}\left(n_w + \sum_s n_s\right)\left[x_w \ln(x_w) + \sum_s x_s \ln(x_s)\right], \quad (4.12)$$

where $\tilde{R} = 8.314$ kJ/kmol K is the ideal gas constant; x denotes the mole fraction of the several constituents; n denotes the number of mols; and the subscripts w and s pertain to

---

[5] m is the symbol for unit molality; m is the symbol for the length unit, the meter.

the water and the salts respectively. The summation in Eq. (4.12) is over all the constituents (salts) other than water. The internal energy and enthalpy of the dilute salty solution are equal to the sum of the enthalpies of its constituent elements and, therefore, there are no internal energy and enthalpy changes during the process of isothermal mixing of salts and water to produce seawater.

Desalination is the inverse process of mixing salts and water, a process during which the sum of the entropies of the separated constituents decreases by an amount equal to the entropy of mixing. Of all the available desalination methods isothermal desalination, which is achieved by electro-dialysis and reverse osmosis, require minimum work input. The "minimum work" for the production of a given volume of freshwater (e.g. 1 m^3) is given by Eq. (2.7) in terms of the exergy change during the isothermal desalination process:

$$W = \left(n_w + \sum_s n_s\right)(\tilde{e}_1 - \tilde{e}_2)$$

$$= \left(n_w + \sum_s n_s\right)(\Delta \tilde{u}_{mix} + P_0 \Delta \tilde{v}_{mix} - T_0 \Delta \tilde{s}_{mix}) = -\left(n_w + \sum_s n_s\right) T_0 \Delta \tilde{s}_{mix} \Rightarrow$$

$$W = T_0 \tilde{R} \left(n_w + \sum_s n_s\right) \left[x_w \ln(x_w) + \sum_s x_s \ln(x_s)\right], \qquad (4.13)$$

where the ~ overhead symbol denotes molar quantities (e.g. kJ/kmol). When an open system is considered for the desalination, a similar equation is obtained for the minimum power:

$$\dot{W} = T_0 \tilde{R} \left(\dot{n}_w + \sum_s \dot{n}_s\right) \left[x_w \ln(x_w) + \sum_s x_s \ln(x_s)\right]. \qquad (4.14)$$

Water desalination plants do not separate entirely the salts from the final product, the freshwater, which may contain a maximum of 1,000 ppm salt (0.1%). Slightly saline water, which contains salts in the range 1,000–3,000 ppm and has several industrial and agricultural uses, may also be the product of desalination. Figure 4.8 shows the minimum work, in kJ and kWh, for the production of one cubic meter of freshwater or slightly saline water using seawater of 35,000 ppm salt content as input. The salt in the seawater is sodium chloride (NaCl). It is observed that the minimum work for the production of 1 m^3 of freshwater with 1,000 ppm salt content from seawater is approximately 2.1 kWh. The production of slightly saline water with 3,000 ppm needs at least 2.0 kWh/m^3 of work.

In actual desalination plants (e.g. reverse osmosis and electro-dialysis plants), part of the processed water is removed in the several stages of desalination. The common practice is to quote the minimum work for desalination based on the partially recovered water product and not on the total amount of the water that has been handled by the plant. The ratio of the freshwater product to the saltwater input is called the *percent recovery factor*. In this case, the minimum work for the production of 1 m^3 of freshwater would be equal to the value from Figure 4.8 multiplied by the percent recovery factor of the desalination plant.

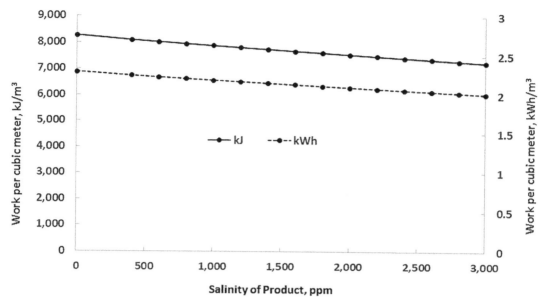

Figure 4.8 Minimum work for the production of 1 m³ of desalinated water as a function of the product salinity. The input is seawater of 35,000 ppm salt content.

4.7 Exergy Use in Buildings

Approximately one third of the global energy consumption is spent in private and public buildings, primarily to fulfill the following desired actions:

1. To provide adequate lighting.
2. To maintain a comfortable interior temperature by heating during the winter months and cooling during the summer months.
3. To provide hot water.
4. To run the electric appliances.

Since the 1970's "energy conservation and efficiency in buildings" has become national priority in most countries, with directives issued by national and local governments that require new buildings to conform to energy efficiency standards and older buildings to be retrofitted and reduce their fossil fuels and electricity consumption. In the USA, the Department of Energy (*US-DOE*) instituted the *National Appliance Energy Conservation Act* (*NAECA*), which specifies minimum efficiency standards for appliances in commercial and residential buildings. A result of the *NAECA* directives is the *ENERGY STAR* system for the efficiency assessment of major household appliances and the mandatory labels for their expected annual energy consumption. In the Peoples' Republic of China, household and industrial energy efficiency are the purview of the *China Energy Conservation and Environmental Protection Group Corporation,* which periodically issues directives on energy conservation and environmental protection. In the European Union, the European Commission issues periodically the *Energy Efficiency Directives,* for the minimum efficiency of household appliances. In Japan, building energy efficiency is part of the mission of the *Ministry of Economy, Trade, and Industry* (*METI*), which has established several innovative energy efficiency policies and programs that were adopted by several nations. Similar agencies in all countries issue reports and directives for energy conservation and efficiency of household appliances; advise on national and local policies; and adopt directives to improve appliance efficiencies. As a result of such efforts, the energy and exergy efficiency of buildings and appliances has significantly increased since the 1970s.

4.7.1 Lighting

The desired action for the lighting of buildings is to make spaces comfortable by providing an adequate amount of light power, in Lumens (lm). This is usually accomplished by converting electric power, to light power with lighting devices. Three types of lighting devices are commonly used in buildings:

1. Incandescent bulbs with conversion factors in the range 4–14 lm/W.
2. Fluorescent lights with conversion factors in the range 40–65 lm/W.
3. Light Emitting Diodes (LED) with conversion factors in the range 60–100 lm/W.

The substitution of incandescent bulbs in buildings with fluorescent lights and LEDs significantly reduces the electricity consumption.

A consequence of the energy conservation principle is that all the electric power used for lighting is dissipated in the buildings and contributes to the temperature increase in their interior. For the calculation of the total exergy savings by light bulb substitution in buildings one must take into account the removal of the dissipated energy by air conditioning during the summers and the additional heat supply during the winters (using a fossil fuel or a heat pump). Since the cooling and heating requirements depend on the geographic location of buildings, savings from lighting device substitutions vary, as the following example illustrates:

Example 4.5: Calculate the annual amount of electricity saved from the substitution of incandescent bulbs with fluorescent bulbs and LEDs in two large buildings: 1. in Houston, Texas, USA where air-conditioning is required 70% of the days in a year and heating for 10%; and 2. in Berlin, Germany, where heating is required for 60% of the days and cooling for 15%. Both buildings use 120 kW of power with incandescent bulbs that have conversion factors 10 lm/W. The conversion factor of the fluorescent lighting is 45 lm/W and that of the LEDs is 100 lm/W. Lighting is needed in the buildings for 12 hours per day during all days of the year. Heating and cooling are accomplished with HVAC systems that have coefficients of performance 3.4 for cooling and 4.4 for heating.

Solution: For a smooth substitution, the same quantity of lumens should be provided to the buildings with all types of lighting. The incandescent bulbs provide annually to the two buildings with: $120{,}000 * 10 * 12 * 60 * 60 * 365 = 1.892 * 10^{13}$ lm * s at the expense of $120 * 12 * 365 = 525.6 * 10^3$ kWh. This energy is dissipated as heat in the buildings. Given the lm to W conversion factors, for the same quantity of $lm * s$, the substitution with fluorescent lighting consumes $116.8 * 10^3$ kWh and that with LEDs $52.6 * 10^3$ kWh. The energy savings from the lighting device substitution alone in the two buildings are: $408.8 * 10^3$ kWh with fluorescent bulbs and $473.0 * 10^3$ kWh with LEDs.

The lesser energy for lighting does not need to be removed by the air-conditioning units in the summer, but needs to be supplied as heat during the winter.

1. The building in Houston requires annually 3,066 hours of cooling and 438 hours of heating. With fluorescent lighting and air-conditioning COP = 3.4, the building will save during the cooling season $3,066 * 120 * (1 - 10/45)/3.4 = 84.165 * 10^3$ kWh and with the LEDs $3,066 * 120 * (1 - 10/100)/3.4 = 97.391 * 10^3$ kWh. For the heating season (COP = 4.4) the building will need to consume an additional $438 * 120 * (1 - 10/45)/4.4 = 9.291 * 10^3$ kWh when fluorescent lighting is used and $438 * 120 * (1 - 10/100)/4.4 = 10.751 * 10^3$ kWh when LEDs are used.

Hence, the total exergy savings for the Houston building are: $408.8 * 10^3 + 84.165 * 10^3 - 9.291 * 10^3 = 483.675 * 10^3$ kWh when fluorescent lighting is used; and $473.0 * 10^3 + 97.391 * 10^3 - 10.751 * 10^3 = 559.64 * 10^3$ kWh when LEDs are used.

2. The building in Berlin requires annually 657 hours of cooling and 2,628 hours of heating. With the fluorescent lighting the building will save during the cooling season $657 * 120 * (1 - 10/45)/3.4 = 18.035 * 10^3$ kWh and with LEDs $657 * 120 * (1 - 10/100)/3.4 = 20.869 * 10^3$ kWh. For the heating season the building will consume an additional $2,628 * 120 * (1 - 10/45)/4.4 = 55.745 * 10^3$ kWh when fluorescent lighting is used and $2,628 * 120 * (1 - 10/100)/4.4 = 64.505 * 10^3$ kWh when LEDs are used.

Hence, the total energy and exergy savings for the building in Berlin are: $408.8 * 10^3 + 18.035 * 10^3 - 55.745 * 10^3 = 371.09 * 10^3$ kWh when fluorescent lighting is used; and $473.0 * 10^3 + 20.869 * 10^3 - 64.505 * 10^3 = 429.364 * 10^3$ kWh when LEDs are used.

This example illustrates that exergy savings in buildings from the reduced power for lighting are not the only exergy savings. Additional exergy savings and additional exergy consumption are associated with the heating and cooling of the buildings. It also demonstrates that the total exergy savings are significantly higher in the parts of the globe where air-conditioning is frequently used. In general, replacing home appliances, such as lamps, television sets, refrigerators, and microwave ovens with more efficient appliances is always a good exergy (and natural resource) saving measure.

4.7.2 Air-Source Heat Pump Cycles for Heating and Cooling

The interior of buildings is maintained within specified ranges of temperature and humidity to ensure the living comfort of the occupants throughout the year. When the environmental temperature, T_0, is higher than that of the building interior, T_{in}, heat enters the building and is typically removed by air-conditioning. When $T_0 < T_{in}$, heat leaves the building and energy must be supplied by a heat pump or a fossil fuel burner. The rate of heat transfer between a building and the environment is:

$$\dot{Q} = UA(T_0 - T_{in}) = \frac{A(T_0 - T_{in})}{R}, \qquad (4.15)$$

where A is the area of the building; U is the overall heat transfer coefficient (similar to that of the heat exchangers); and R ($=1/U$), is the heat transfer resistance, usually referred to as the *R-value* of the building. This equation applies to entire buildings as well as to parts of buildings, such as roofs, windows, walls. The interior temperature of most buildings is usually different during the heating and cooling seasons and will be denoted as T_{in}^H and T_{in}^C respectively. The average environmental temperature is also different during the two seasons.

The rate of exergy supplied to maintain the interior temperature, T_{in}^H, during the heating season, when $T_0^H < T_{in}^H$, is:

$$\dot{E} = \dot{Q}/\left(\frac{T_{in}^H}{T_{in}^H - T_0^H}\right) = -\frac{UA}{T_{in}^H}\left(T_{in}^H - T_0^H\right)^2 = -\frac{A}{RT_{in}^H}\left(T_{in}^H - T_0^H\right)^2. \qquad (4.16)$$

The negative sign signifies that exergy (work) is supplied to the building.

The rate of exergy that needs to be supplied during the cooling season, when $T_0^C > T_{in}^C$, is:

$$\dot{E} = -\dot{Q}/\left(\frac{T_{in}^C}{T_0^C - T_{in}^C}\right) = -\frac{UA}{T_{in}^C}\left(T_{in}^C - T_0^C\right)^2 = -\frac{A}{T_{in}^C R}\left(T_{in}^C - T_0^C\right)^2, \qquad (4.17)$$

Absolute temperatures must be used in Eqs. (4.16) and (4.17) for all numerical calculations.

The negative sign in the rate of heat signifies heat removal and the negative sign in the rate of exergy signifies exergy supply. Equations (4.16) and (4.17) express the "minimum power" (the algebraic maximum power according to thermodynamic convention) that would maintain the building at the desired temperatures during the heating and the cooling seasons. Equations (4.16) and (4.17) also provide the benchmarks for the minimum power required to heat and cool the buildings.

Example 4.6: The area of a large commercial building is 4,050 m² and the average R-value of the entire building is 11.2 m² K/W. The interior of the building is maintained at 22°C during the heating season and at 24°C during the cooling season. 1. Calculate the minimum mechanical work required for the heating and cooling of the building during a winter day (24 hours) when the outside temperature is 0°C and during a summer day when the outside temperature is 36°C. 2. The building uses a methane burner with 88% combustion efficiency for the heating and an air-conditioner with 3.4 COP during the summer. Calculate the exergy destruction during the same two days.

Solution: 1. The minimum mechanical power (rate of exergy supplied) is given by Eqs. (4.16) and (4.17) as 593.3 W for the winter day and 175.3 W for the summer day. During the 24 hours of the two days the exergy needed to keep the building at the specified temperatures is 51.25 MJ and 15.14 MJ respectively. These amounts of exergy are dissipated in the environment with the heat that enters or leaves the building.

2. The rate of heat currently used during the two days is obtained from Eq. (4.15) as: 7,955 W during the winter day and 4,339 W during the summer day. The total heat during the two days is 687 MJ and 375 MJ respectively.

Using $-\Delta H^o$ and $-\Delta G^o$ for methane from Table 2.2: $687/(890*0.88) = 0.8772$ kmol of methane are used during the winter day in the combustion process at the expense of 717.5 MJ of exergy. During the summer day the air-conditioning unit consumes $375/3.4 = 110.3$ MJ of exergy, supplied as electricity. Since all the exergy supplied to the building is dissipated with the heat transfer to the environment, the exergy destruction is 717.5 MJ during the winter day and 110.3 MJ during the summer day.

It is observed in this example that significantly higher exergy destruction occurs during the heating season, when a fossil fuel is burned for the production of low-temperature heat. The exergy destruction would be significantly reduced with a heat pump system that uses electricity. For example, if a heat pump with COP 4.4 were to be used in the building for heating, the exergy destruction would be reduced from 717.5 MJ to 156.1 MJ. Even when one accounts for the exergy destruction in the power plants and the electricity transmission lines (with exergetic efficiencies approximately 35% and 90% respectively), significantly less exergy destruction (495.6 MJ instead of 717.5 MJ) is achieved with the use of a heat pump. In general, if the overall thermal efficiency of the power plant is η_t, and the transmission efficiency for the electric energy is η_{tr}, the substitution of fossil fuel burners with heat pumps destructs lesser exergy overall (implying that primary energy is saved) if the following inequality is satisfied:

$$\beta_{hp}\eta_t\eta_{tr} > 1. \tag{4.18}$$

An additional, practical advantage of using heat pumps is that, because they use a refrigeration cycle for their operation, the same mechanical system may also be used for air-conditioning, when cooling of the buildings is the desired action to be fulfilled.

4.7.3 Ground-Source Heat Pumps and Air-Conditioners

A drawback of conventional (air-source) refrigeration/heat-pump cycles for the heating and cooling of buildings is that they exchange heat with the ambient air. The evaporator of the heat pump cycle removes heat, Q_L, from the atmosphere and the condenser of the air-conditioning cycle dissipates heat, Q_H, also to the atmosphere. However, the atmospheric temperature, where the two heat exchange processes occur, varies and ranges in several locations from $-20°C$ in the winter to more than $40°C$ in the summer. This variability makes the design of the refrigeration cycles difficult to optimize. Another disadvantage is that the COP of a heat pump significantly drops when the environmental temperature is very low and the COP of the air-conditioning drops when the outside air temperature is very high.

Detailed measurements have shown that the underground temperature remains almost constant at depths between 10–300 m in most locations on earth. The surface temperature of the ground (the soil) follows the annual variability of the air temperature, but the variability of the underground temperature is reduced by a factor of 2 at a depth of 1.2 m and by a factor of 4 at 2.4 m. The temperature is almost constant throughout the year at depths below 10 m to approximately 300 m and slightly depends on the latitude of the location: The underground temperature is approximately 6°C (42°F) in Scandinavia throughout the year; 10°C (50°F) in Germany and England; 14–15°C in France;[6] 16°C (59°F) in Southern Italy; 18°C (65°F) in Dallas, Texas; and 20°C (69°F) in San Antonio, Texas.

The Ground Source Heat Pump (GSHP), also called Geothermal Heat Pump, utilizes a refrigeration cycle. The cycle dissipates heat to the underground environment when it is operating as an air-conditioner and receives heat from the underground environment when it is operating as a heat pump. The heat exchange in the subsurface is achieved by a system of tubes or coils [2, 11]. The substitution of conventional air-conditioning and heat pump systems with GSHPs is usually accompanied by a very significant increase in the COP (and SEER) of the equipment. The COP of an optimized GSHP cycle may reach values as high as 8 (27.3 SEER) and that of an air-conditioner 7 (23.9 SEER). Such efficiency improvements entail significantly lower cost for the cooling and heating of buildings and lesser environmental pollution associated with the production of the consumed electric energy. The following example illustrates the exergetic advantages for the use of GSHPs in buildings.

Example 4.7: A building in Shanghai (P.R. China) uses a refrigeration cycle for air-conditioning and heating. R134a is the working fluid in the cycle, which operates with low pressure 2.4 bar; high pressure 12.5 bar; and mass flow rate 0.82 kg/s. The cycle operates for 1,450 hours during the hot season and 640 hours during the cold season. As a result of retrofitting with a GSHP, the new cycle operates with pressures 3 bar and 10 bar. The same mass flow rate is used in the GSHP cycle, the fluid enters the compressor as saturated vapor, and exits the condenser as saturated liquid. The efficiencies of both compressors are 80%. Determine the exergy destruction in the two cycles, the annual energy savings in kWh, and the hours of operation of the GSHP cycle.

Solution: The following table shows the properties at the four states of the conventional and the GSHP cycles. The states correspond to the notation in Figure 1.12. The compressor efficiency was considered in the determination of state 2.

---

[6] This is also the recommended temperature for storing wine, because it mimics the (almost constant) temperature of underground cellars in the Burgundy and Bordeaux regions of France.

	Conventional Cycle				GSHP Cycle			
State	T (K)	P (MPa)	h (kJ/kg)	e (kJ/kg)	T (K)	P (MPa)	h (kJ/kg)	e (kJ/kg)
1	267.8	0.24	395.5	0.0	273.9	0.3	399.0	0.0
2	334.6	1.25	438.5	36.0	322.6	1	430.3	25.9
3	321.0	1.25	268.4	7.5	312.5	1	255.4	4.1
4	267.8	0.24	268.4	0.0	273.8	0.3	255.4	0.0

The electric energy consumed in the conventional cycle is equal to the work input of the cycle, $(=h_2 - h_1)$ 43.0 kJ/kg of refrigerant. For the GSHP cycle the electricity consumed is 31.3 kJ/kg of refrigerant.

When the systems operate in the heat pump mode, the rate of heat added to the building is: $\dot{Q} = \dot{m}(h_2 - h_3)$.

With the mass flow rate of the refrigerant 0.82 kg/s for both cycles, the rate of heat added to the building is 139.5 kW for the conventional cycle and 143.4 kW for the GSHP cycle. The conventional cycle adds to the building 139.5 * 60 * 60 * 640 = 321.4 GJ during the cold season at the expense of 43.0 * 0.82 * 640 = 22,566 kWh of electricity. For the GSHP cycle to add the same quantity of heat, it will operate for 640 * (139.5/143.4) = 622.6 hours (instead of 640 hours) during the cold season. Thus, the GHSP cycle consumes 31.2 * 0.82 * 622.6 = 15,929 kWh during the cold season resulting in 6637 kWh (29.4%) savings in electricity.

When the systems operate in the in the air-conditioning mode the rate of heat removed is: $\dot{Q} = \dot{m}(h_1 - h_4)$.

For the conventional cycle the rate of heat removed is 104.2 kW and for the GSHP cycle 117.8 kW. The conventional cycle removes from the building 104.2 * 60 * 60 * 1,450 = 543.9 GJ during the hot season at the expense of 43 * 0.82 * 1,450 = 51,127 kWh of electricity. For the GSHP cycle to remove the same quantity of heat from the building, it will operate for 1,450 * (104.2/117.8) = 1,282.6 hours during the hot season and will consume 31.2 * 0.82 * 1,282.6 = 32,814 kWh for 18,313 kWh (35.8%) electricity savings.

For the entire year, the savings from the installation of the GSHP system add to: 24,950 kWh (89.8 GJ). This number also represents the annual exergy destruction avoided by the use of the GSHP system.

An additional advantage of the heat pump/air-conditioning systems, especially those using the GSHPs, is that the heat dissipated in their condenser may be used to produce a fraction of the hot water needs of a building. In the southern part of the USA 100% of the hot water needs of households during the months April–October may be produced by the air-conditioning condensers without any additional use (and cost) of energy. When operating in the heat pump modes, these systems also provide hot water to the households during the remainder of the year at lower energy and exergy expense.

4.7.4 Hot Water Supply

Approximately 18% of the energy consumed in buildings in the USA is used for the production of low-exergy hot water, with the average individual using daily 150 L of hot water. The hot water temperature is typically in the range 40–50°C, it is stored in tanks and is primarily used for: baths and showers; shaving; dishwashers; and clothes washers. Part of the enthalpy of the hot water is dissipated by heat transfer from the tank and internal piping to the interior of the house and the environment. This dissipation may be minimized with better pipe and tank insulation.

Let us consider a household with four individuals that consume a total of 600 L of hot water daily. Heat losses in the storage tank are accounted by an additional 20% of the daily energy used, and this is modeled as an increase of the daily hot water need to 720 L. Underground pipes at 18°C, which is also the environmental temperature, supply water to the household.

Water is an incompressible fluid with density approximately 1,000 kg/m³ (1 kg/L). The total exergy in the daily quantity of hot water at temperature 50°C is:

$$E = m\left[c_P(T_1 - T_0) - T_0 c_P \ln \frac{T_1}{T_0}\right] = 4,941 \text{ kJ}. \tag{4.19}$$

The exergy of the hot water is the benchmark for this desired action.

The heat needed for the production of hot water is equal to the enthalpy difference, $Q = mc_P(T_1 - T_0) = 96,399$ kJ. Let us consider four commonly used systems for the production of the 720 L of hot water:

1. An electric heater operates at close to 100% efficiency and uses 96,399 kJ of electricity. The exergetic efficiency of the hot water heater is $4,941/96,399 = 5.1\%$ and the daily destruction of exergy for the hot water production in the household is 91.5 MJ. Additional exergy destruction occurs in the power plant that produces the electricity.
2. A methane (natural gas) burner, which operates with 72% efficiency – due to incomplete combustion and gaseous products exiting the burner at relatively high temperature. For the daily production of 96,399 kJ of heat, the burner consumes $96.399/(890 * 0.72) = 0.1504$ kmol of methane at the expense of 123.1 MJ of exergy. The exergetic efficiency of the burner is 4.0% and the daily exergy destruction is 118.16 MJ.
3. A dedicated heat pump with coefficient of performance 5.2. The heat pump uses $96,399/5.2 = 18,538$ kJ of electricity, it has exergetic efficiency $4,941/18,538 = 26.7\%$, and the daily exergy destruction is 13.6 MJ. As with the electric heater, additional exergy destruction occurs in the electric power plant.
4. A solar collector with 52% overall efficiency. This solar collector receives daily 185,383 kJ of solar energy, which corresponds to 177,970 kJ of exergy. The exergetic efficiency of the solar collector is 2.8% and the daily exergy destruction is 173,029 kJ.

The exergetic advantage of the heat pump for the production of hot water for the household is apparent in these calculations. The solar collector appears to be the most inefficient way to provide hot water to the household. However, the exergy of the solar irradiance, which is used in the collector, would have been anyway destroyed in the environment if it were not intercepted by the collector. Solar exergy is free to be used, while the exergy used by the other three systems is derived from the consumption of primary energy sources. Using the solar exergy in the collector actually reduces the exergy destruction of other primary energy sources and justifies the use of solar collectors for the production of low temperature heat.

This example illustrates the differences between *energy efficiency* and *energy substitution* projects. The most efficient production of hot water is the use of a heat pump, which operates with electricity. The production and transmission of electricity partially entails the use of other primary energy sources – coal, natural gas, and nuclear. Comparing the operation of the first three systems – electric heater, burner, and heat pump – the use of the heat pump minimizes the consumption of these primary energy sources. However, the solar collector does not use primary energy resources, which are "saved" for another use. Instead, the solar collector uses a higher amount of solar exergy that would have been destroyed anyway. Solar and the wind exergy may be considered as "free exergy" (solar and wind energy are also free) whose destruction is of no consequence to the overall conservation of the TPESs on earth. Such energy substitution projects are justified by environmental and sustainability considerations, despite the seemingly higher exergy dissipation.

4.8 Exergy Consumption in Transportation

The global transportation sector uses 36% of the TPESs and is dominated by the road vehicles – primarily passenger cars and freight trucks – that account for 89% of the total transportation energy use. Air transport accounts for 7% of transportation energy, while water and rail transport together account for approximately 4% [12]. Liquid hydrocarbons – gasoline and diesel – supply most of the energy used in the transportation sector. The number and usage of passenger cars and freight trucks has been steadily increasing since the early 20th Century. In 2016 these vehicles accounted for approximately one third of the global CO_2 emissions.

The mechanical power consumed by road vehicles is primarily spent to counter the rolling friction and air drag of the vehicles; the hydrodynamic drag of ships; and the air drag of aircraft. These resistance/friction forces are modeled with coefficients, as for example for a road vehicle [13]:

$$F = F_R + F_D = C_R mg + \frac{1}{2} C_D \rho A V^2, \quad (4.20)$$

where C_R and C_D are the rolling friction and the aerodynamic drag coefficients respectively; m is the mass of the road vehicle; ρ is the air density; A is the frontal area of the vehicle; and V the speed of the vehicle. The rolling friction coefficient depends on the road surface and the tires of the vehicle, while the drag coefficient is a function of the Reynolds number of the moving vehicle. Because these are resistance forces, all the power consumed by vehicles during trips is dissipated in the environment.

$$\dot{W}_{lost} = T_0 \dot{\Theta} = V\left(C_R mg + \frac{1}{2} C_D \rho A V^2\right). \tag{4.21}$$

Other expenditures of engine power are vehicle acceleration, which is eventually dissipated when the vehicle comes to a stop; change in altitude; power for instruments and lights; and air-conditioning. Eq. (4.21) shows that the power dissipation of vehicles is a monotonically increasing function of the speed of the vehicles.

Example 4.8: A large passenger car with frontal area 1.75 m² weights 1,900 kg and cruises on a flat road at 120 km/hr. The rolling friction coefficient of the car is 1.4% and the aerodynamic drag coefficient is 0.42. Determine the power needed for the car to cruise at this speed. If the power is supplied by an engine with 21% thermal efficiency and the car uses octane as fuel, what is the fuel consumption of the car in km/L?

Solution: The cruising speed of the car is 33.3 m/s. The forces that oppose the movement of the car are the rolling resistance force ($= 0.014 * 1,900 * 9.81 = 261$ N) and the aerodynamic drag ($= 0.5 * 0.42 * 1.19 * 1.75 * 33.3^2 = 485$ N). The power needed for the car to cruise at 120 km/hr is $33.3(261 + 485) = 24,841$ W.

At 21% thermal efficiency, the engine of the car uses $24.841/0.21 = 118.3$ kW of heat (0.1183 MW). Since for octane M = 114 kg/kmol and $-\Delta H^o = 5,471$ MJ/kmol, this rate of heat is supplied by $0.1183 * 114/(5,471) = 0.00246$ kg/s of octane fuel.

The car takes 30 s to travel 1 km and during this time period it consumes 0.0739 kg of fuel or 0.1052 L ($\rho_{oct} = 0.703$ kg/L). Hence the fuel consumption of the car is 9.51 km/L.

Given the exergy of octane ($-\Delta G^o = 5,297$ MJ/kmol) it is of interest in this example that the car consumes $0.0739 * 5,297/114 = 3.43$ MJ of exergy per kilometer. In order to counteract the resistance forces for the 30 s, during which it travels for 1 km, the car only needs $24,841 * 30/10^6 = 0.75$ MJ of exergy per km. The difference of 2.68 MJ/km is primarily exergy dissipated in the fuel cycle that powers the car.

The reduction of the cruising speed of a road vehicle is accompanied by the reduction of the dissipated power and decreased fuel consumption, but it also increases the time for the vehicle to reach its destination – an inconvenience for the passengers. Figure 4.9 shows the effect of the cruising speed on the fuel consumption of this vehicle. For the calculations in Figure 4.9 the rolling and aerodynamic drag coefficients are assumed constant at 0.012 and 0.42 respectively.[7] The thermal efficiency of the engine is 21%, an average value for mid-size sedan cars that also includes the transmission losses [2, 14]. It is observed in Figure 4.9 that the fuel consumption increases significantly with the

---

[7] The drag coefficient of any solid object is normally a function of its velocity. In the case of cruising cars, the Reynolds number is very high and the flow around the cars is turbulent, which implies that the drag coefficient is a weak function of the cruising velocity, thus, justifying the assumption of constant C_D.

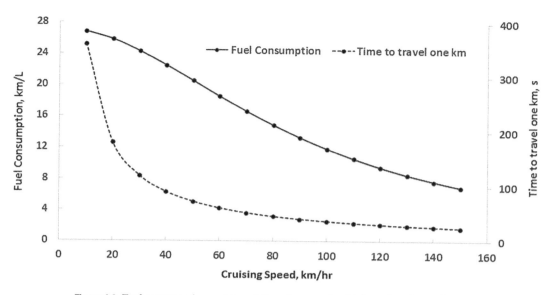

Figure 4.9 Fuel consumption and travel time for a road vehicle as functions of the cruising speed.

cruising speed of the vehicle. Despite the higher fuel consumption, the time to travel a given distance decreases and this may be preferable for the operator of the vehicle, who will choose a higher speed for the convenience of an earlier arrival.

The inverse of the fuel consumption, L/km, is linearly correlated with the exergy spent for the vehicle to travel 1 km. It is apparent from Figure 4.9 that significantly less exergy is spent per km travelled at the low cruising speeds, but this comes with the inconvenience of a late arrival.

The desired action to be fulfilled in transportation is the timely and comfortable conveyance of persons from one location to another; and the transportation of goods and merchandize from the production to the marketplace. For example, desired actions may be: the transportation of 100 tons of potatoes from Idaho to Dallas; or the transportation of 350 passengers from Nice, France to Barcelona, Spain. The methods of transportation are a matter of choice and convenience. The transportation of potatoes is fulfilled when the potatoes are transported by truck, by train, by airplane, or even by a fleet of passenger cars (a very much energy-intensive method). The passenger transportation may be accomplished by airplane, train, bus, a fleet of cars, or by boat. One of the available methods would require the least amount of transportation fuel. Ship and train are typically the least energy-consuming methods for the transportation of large numbers of passengers and commercial goods. Simple calculations show that the transportation of potatoes from Idaho to Dallas by trains, which have rolling friction coefficients in the range 0.15–0.20%, consumes the least amount of fuel and of exergy. The transport of 350 passengers from Nice to Barcelona by ship also consumes the least amount of fuel and exergy.

Of the usual modes of transportation, rail and ships typically consume the least amount of fuel, while airplanes and automobiles/trucks use significantly more fuel (and exergy) to fulfill the desired action. However, airplanes and passenger cars are

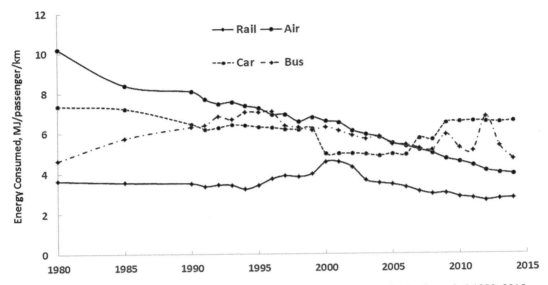

Figure 4.10 Energy consumption per passenger-kilometer in the USA in the period 1980–2015.

more convenient and significantly faster. For this reason airplanes are used for long trips (e.g. Shanghai to New York) and for the transportation of perishable goods (e.g. sushi-grade tuna from the Gulf of Mexico to Tokyo). Figure 4.10 depicts data on the energy consumed per passenger-kilometer in the USA for the most commonly used modes of transportation: airplanes, railways, passenger cars and busses in the period 1980–2015 [15]. Rail transport – not the most popular mode of transportation in the USA – consumes the least amount of energy per passenger-kilometer. It is also apparent that the average energy spent per passenger-kilometer has decreased for airplanes, reflecting the improved efficiency of jet engines and higher capacity factors.[8]

Example 4.9: A new bus line transports on average 25 passengers per trip to an urban center and back, with 10 round trips per day. The one-way trip is 43 km, the average fuel consumption of the cars in this area is 8 km/L, and of the buses 3.2 km/L. Assuming that the fuel of cars and buses is octane, what are the fuel savings per year, the savings in exergy per year, and the annual CO_2 avoidance?

Solution: 25 passengers using their own car with average fuel consumption 8 km/L consume a total $25 * 2 * 43/8 = 268.8$ L of fuel octane per round trip. The bus consumes $2 * 43/3.2 = 26.9$ L. The savings per round trip are 241.9 L of fuel and the savings per year are $10 * 365 * 241.9 = 882{,}935$ L of fuel.

The density of octane is 703 kg/m^3 or 0.703 kg/L. Therefore, $882{,}935 * 0.703 = 620{,}703$ kg of octane is saved with the switch to public transportation. The exergy of this amount of fuel is: $28.84 * 10^6$ MJ, all of which would have been dissipated.

---

[8] The apparent increase of the energy consumed in cars reflects statistical modifications after 2007.

From the equation for the combustion of octane:

$$C_8H_{18} + 12.5O_2 = 8CO_2 + 9H_2O,$$

we deduce that 114 kg of octane produce 352 kg of CO_2. Therefore the annual CO_2 avoidance is: $352 * 620{,}703/114 = 1{,}916{,}557$ kg (~1,917 tons of CO_2).

Increasing the capacity factor of public transportation media saves large quantities of fuel and exergy: trains, trucks, airplanes, and ships that are filled to capacity transport goods and passengers cheaper and with lesser exergy dissipation per passenger-km than partially filled vehicles.

Because of the high energy consumption in the transportation sector and the importance of the sector to the national economies, most governments have introduced regulations for the energy consumption of road vehicles – typically specifications for the mileage of new vehicles. For example, in the USA the *Corporate Average Fuel Economy (CAFE)* standards were introduced in the 1970s and require for new cars and trucks to comply with the mileage set by the *US-Department of Transportation (US-DOT)*. As a result of the national standards – and despite the use of larger and heavier vehicles – the mileage (measured in miles per gallon or kilometers per litter) of the entire automobile fleet in all OECD countries has continuously improved [16]. In particular, the introduction of the hybrid cars, early in the 21st century, has significantly increased the efficiency of personal automobile transportation and has reduced the fuel and overall exergy spent in the transportation sector per passenger-km.

Example 4.10: A saleswoman uses her car to make calls on customers. On the average she travels 700 km per week for 42 weeks per year. She recently decided to trade her car with mileage 9 km/L with a hybrid that averages 19 km/L. Determine how many litters of gasoline she saves annually. Assuming that gasoline is solely composed of octane, determine the annual exergy savings and the annual CO_2 emissions avoidance.

Solution: During an entire year the saleswoman travels $42 * 700 = 29{,}400$ km and consumes 3,267 L of gasoline. When the same distance is travelled by the hybrid, only 1,547 L of gasoline are consumed, bringing the annual gasoline savings to 1,720 L. Since the density of octane is 0.703 kg/L, 1,209 kg of octane are saved with the vehicle substitution corresponding to 56.2 GJ exergy savings per year.

From the equation for the combustion of octane: $C_8H_{18} + 25/2 O_2 = 8CO_2 + 9H_2O$, we deduce that 114 kg of octane produce $8 * 44 = 352$ kg of CO_2. The annual savings of 1,209 kg of octane fuel also alleviate the emission of 3,733 kg of CO_2 in the atmosphere.

4.8.1 Battery-Powered Electric Vehicles

Calculations similar to those of Example 4.8 reveal that IC engines, which operate with the combustion of liquid fossil fuels, consume a great deal more exergy than what is needed to counteract the vehicle resistance. The principal reason for this is the lower first law efficiency of the IC engines. The overall efficiency of road vehicles with IC engines – often referred as *well-to-wheels efficiency*, η_{ww} – is in the range 10–30% [2, 13, 14]. The significant research in this area by most engine manufacturers has improved in the last 40 years the thermodynamic efficiency of the IC engines especially those of passenger cars, while keeping the vehicles affordable for the consumers. Another way to improve the well-to-wheels efficiency of road vehicles is to use electric vehicles. The needed energy for transportation is produced in more efficient electric power plants; it is transmitted to the charging stations; stored in the batteries of cars; and finally used to power the cars.

The use of electricity for passenger transportation is not new. It has been applied for more than a century to rail transport in underground (subways) and above-ground railways (intercity trains) as well as in the trolleys and trams of several cities. Environmental concerns with CO_2 and other combustion product emissions as well as improved battery technology have ushered a new era of electric battery-powered vehicles, such as the one in Figure 4.11. Several governmental incentives – tax credits and rebates – have been adopted to promote the substitution of IC engine passenger cars with electric cars, especially in countries with high renewable energy capacity, such as Norway [2, 13].

The electric car engines are essentially battery-operated motors, with high efficiency, η_M, in the range 90–95%. The average charging–discharging efficiency of the car batteries, η_B, is also high, in the range 80–90%. If electricity is produced by fossil or nuclear fuels, the efficiency, η_t, of the power plant is in the range 30–42%. Electric energy losses during the power transmission to the consumer account for 5–15%, which implies that the electric energy transmission efficiency, η_{tr}, is in the range 85–95%. When all these factors are taken into account, the well-to-wheels efficiency for electric cars is:

$$\eta_{ww} = \eta_t \eta_{tr} \eta_B \eta_M. \tag{4.22}$$

Figure 4.11 An electric car charging in the TCU Campus. Several universities and corporations in Texas provide free electric charging in their parking lots.

With the ranges of efficiency values given above, the well-to-wheels efficiency of electric cars is in the range 19.5–34.1%. This range is comparable to the efficiency of new IC engine automobiles, but lower than the efficiencies of hybrid cars that sometimes exceed 45%. Energy and exergy savings with electric vehicles are chiefly generated from the lower average size and weight of electric cars, especially the very small electric cars used by commuters in some urban areas. The weight benefit is illustrated in the following example:

Example 4.11: An urban commuter drives a mid-size car of 1,400 kg and commutes 24 km/day (round trip) at an average speed 60 km/hr. The rolling friction coefficient of this car is 1.15%; the drag coefficient is 0.4; and the frontal area of the car is 1.7 m². A few stops during the commute increase the total mechanical energy needed for the trip by 80%. The commuter decides to substitute this car with a small, battery driven electric car that weights 430 kg. The rolling friction coefficient of this car is 1.1%; the drag coefficient is 0.35; and the frontal area is 0.90 m². Determine the exergy needed to drive the two cars during the 230 working days of the year.

Solution: Using Eq. (4.21) the power needed by the mid-size car to overcome the friction and aerodynamic drag is: 4.51 kW. Since the round trip takes 1,440 s, the exergy needed for a round trip is 6,494 kJ. With the additional 80% loss for the stops, the total exergy needed is 11,689 kJ. During the entire year this car consumes 2,688 MJ of exergy.

With the smaller electric car, the power needed is 1.64 kW; the mechanical energy needed for a round trip is 2,363 kJ; and with the additional energy for stops, 4,253 kJ. During the entire year, this car consumes 978 MJ exergy (mechanical energy).

The annual savings in exergy from the car substitution amount to 1,710 MJ (64%).

The next example relates the exergy consumption of the two cars to the energy resources used for the operation of the two cars.

Example 4.12: The mid-size car of Example 4.11 has an IC engine with 16% thermal efficiency. The parameters for the operation of the electric car are: $\eta_M = 93\%$; $\eta_B = 88\%$; $\eta_{tr} = 90\%$; and $\eta_t = 39\%$. Considering that the gasoline is octane and all the electric power is produced by methane (natural gas), determine the total annual exergy of fuel savings by the substitution of the mid-size car.

Solution: In order to provide the annually needed 2,688 MJ of energy to the mid-sized car, the 16% efficient IC engine uses 16,800 MJ of heat. With the values in Table 2.2 this heat is generated by the combustion of 16,800/5,471= 3.07 kmols of octane (approximately 500 L). The exergy of this mass of octane is 16,266 MJ.

With the electric car the heat supplied by the methane to the power plant is: 978/(0.93 ∗ 0.88 ∗ 0.90 ∗ 0.39) = 3,404 MJ of heat, and this is generated by 3,404/890 = 3.825 kmols of methane. The exergy of this quantity of methane is 3,129 MJ.

Therefore this vehicle substitution generates 13,137 MJ fuel exergy savings (81%).

Examples 4.11 and 4.12 illustrate that significant fossil fuel exergy is saved by the substitution of commuting IC-powered vehicles with smaller electric vehicles. Most of the savings arise not by the substitution of the car engine, but by the reduction of the size and weight of the cars. This substitution usually goes together with environmental benefits and reduced CO_2 emissions, depending on what is the primary source used for the production of electricity [2, 13]. Another observation in Examples 4.11 and 4.12 is that a great deal more fuel exergy is spent with both electric and IC-powered cars than the mechanical work (exergy) needed to overcome the resistance forces. For the operation of the IC-powered car, an annual exergy of 2,688 MJ is needed to overcome the rolling resistance and the aerodynamic drag, but the fuel exergy consumed is much higher, 16,266 MJ. The corresponding numbers for the electric car are 978 MJ and 3,129 MJ respectively. Most of the excess fuel exergy is destroyed during the combustion process of the fuel (either in the IC engine or the electric power plant) and this highlights the potential advantage of direct energy conversion (DEC) for transportation.

4.8.2 Fuel Cell-Powered Vehicles

Fuel combustion, either in the IC-engines or at the electric power plants, entails very high exergy destruction for all transportation vehicles. All considerations for the reduction of exergy destruction indicate the elimination of the combustion process and the utilization of DEC devices in vehicles, such as fuel cells. Fuel cells directly convert the exergy of a fuel to electric energy that provides power to the motor of the vehicles. In the second decade of the 21st century the use of fuel cells for transportation is primarily limited to busses and cars with hydrogen fuel cells – among them are the *Honda FCX Clarity*, the *Kia Borrego*, the *Hyundai ix35 FCEV*, the *Ford Focus FCV*, and the *Toyota Mirai*. Hydrogen supply facilities are visible in several locations, most notably in Japan, the European Union, and several urban areas of the USA.

One of the problems associated with the extensive use of hydrogen for transportation is that this element is not derived from a primary energy source and must be artificially produced by electrolysis, by steam reforming of natural gas, or by chemical reactions as a by-product of petroleum distillation processes. If hydrogen is to be widely used as transportation fuel, a very large fraction of it will have to be produced by electrolysis. However, during the several energy transformations from primary energy sources (fossil fuels, nuclear, or renewables) to electric energy, to hydrogen, and back to electricity that powers the vehicle motor, there is high exergy destruction, significantly more than in the IC-powered hybrid engines. The well-to-wheels exergetic efficiency of vehicles powered by hydrogen fuel cells is in the range 10–15%, significantly lower than the exergetic efficiency of hybrid IC-powered engines whose exergetic efficiency may reach 45% [2].

It is apparent that, from the exergetic point of view, it does not make good sense to produce hydrogen and use it with IC engines or hydrogen fuel cells to power the vehicles. However, if the fuel cells were to operate with a naturally occurring fuel – natural gas, synthetic gas, or any other of the petroleum distillation products – this fuel would be directly fed to the fuel cell system, where hydrogen would be released in a reforming stage. In this case there would be fewer irreversibilities and the exergetic efficiency of the fuel cell powered vehicles would be higher than 60% [2]. Clearly, the development of fuel cells operating with natural gas, light hydrocarbons, and similar energy sources, will revolutionize the transportation industry. Several automotive corporations are spending substantial resources in research efforts aimed at the development of fuel cells that operate with natural gas (methane) or with one of the products of syngas [17]. An additional advantage to using natural or synthetic gas with fuel cells is that the distribution network of the gas already exists and does not need to be developed. In summary, if the chemical-to-thermal-to-electric processes are avoided for the generation of electric power and the more direct chemical-to-electric process is adopted with natural resources, exergy destruction would be significantly curtailed in the transportation sector and the finite energy resources would last longer.

4.9 Energy Storage

The demand for electricity is subject to diurnal, weekly, and seasonal fluctuations. Electricity consumption is higher during the day; it falls off during the evening and night hours; and it is lower during the weekends. In most OECD countries the widespread use of air-conditioning has significantly increased the use of electricity during the hotter summer season and has triggered a daily demand peak for electricity in the afternoon summer hours [2]. Since electric power is not stored in the electricity grids, the instantaneous demand must be satisfied by the generation units. Power systems that supply electricity to the grids must be flexible enough to accommodate demand fluctuations and this is accomplished by introducing into the mix of electricity generation stations a number of flexible units that may produce electric power at demand, such as gas turbines, diesel generators, and hydroelectric power plants.

The adverse effects of carbon dioxide accumulation in the atmosphere are the driving global forces for the substitution of fossil fuel units with nonpolluting, renewable energy sources in the mix of power generation units. Solar and wind are the most abundant and most widely used renewable energy sources on the global scale. However, solar energy is periodically variable and wind power is intermittent. During a calm summer night, neither the sun nor the wind is capable of providing the energy demand for households and industry. On the other hand, the significantly higher utilization of these renewable energy sources has the adverse effect of surplus electric power production at times when the demand is too low to absorb all the produced power. The surplus power production results in severe supply-demand mismatches that are exhibited in the so called *U-shaped* or *duck demand curve* [2, 18, 19], and impose significant limitations on the higher penetration of renewable energy sources in the electricity markets. Figure 4.12 demonstrates the U-shaped demand curve with reference to the electricity grid (ERCOT) that generates and supplies electric power to 92% of Texas. The current demand of the electric grid during a typical day in the spring is shown as well as the part of the demand that is supplied by the four nuclear power units in the grid. Two other curves (in dotted lines) depict the power that would be produced during that day by solar energy if PV cells generated a total of

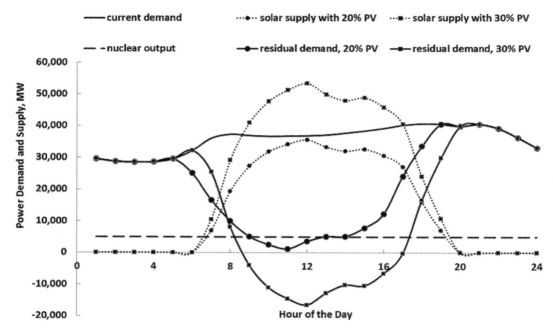

Figure 4.12 Demand-supply mismatches in an electric grid with the higher penetration of renewables for electricity generation.

20% and 30% of the total annual electric energy in the grid. The other two curves depict the residual demand (demand minus solar supply) when the PV cells generate 20 and 30% of the annual total electricity. It is observed in Figure 4.12 the residual demand drops below the level supplied by the nuclear units for two hours at the 20% solar generation. Because the output of the nuclear units cannot be adjusted daily, the excess supply must be stored or dissipated in the electricity grid. The situation is more severe at the 30% solar generation level, when the residual demand drops to negative levels from 10:00 am to 4:00 pm, when the PV cells generate maximum electric power. Again, the excess power produced must be stored or dissipated. The intermittent wind generation causes similar demand-supply mismatches during windy days. These considerations indicate the importance of energy storage, which becomes necessary, with the higher penetration of renewables in the electric power generation sector [20, 21].

Energy storage and reuse entails significant exergy destruction. Nuclear, solar, and wind power plants produce electric energy, which is converted to another energy form to be stored. The stored energy is converted back to electricity, when it is needed. The *round-trip* storage energy efficiency – the ratio of the energy available at the end of the discharge process to the energy consumed by the storage system at the beginning of the storage process – is a figure of merit that best describes the effectiveness of energy storage systems [2, 22]. Round-trip efficiencies of most practical electric energy storage systems are close to 50%, signifying that approximately half of the generated energy is dissipated in the storage-discharge processes.

4.9.1 Pumped Hydro Systems

The *Pumped Hydro Systems (PHS)* elevate water from a lower-level source; and store it in a natural or artificial lake at higher elevation, where the water has high potential energy. When power is needed, the water is returned to its original, lower elevation and produces electric power in hydraulic turbines. Natural or artificial lakes at high elevations may store millions of cubic meters of water for long periods of time and this makes the PHS an ideal medium for daily and seasonal energy storage. The stored exergy at an elevation Δz, higher than the power production plant is:

$$E = mg\Delta z. \qquad (4.23)$$

Let us consider the exergy transformations during the charge and discharge in a PHS, powered by wind or solar energy, which transports and stores water in an artificial lake of average dimensions $200 * 200 * 3$ m^3 at an elevation 750 m higher than the generating station. The water flows through the same route to hydraulic turbines that produce electric power to satisfy the fluctuating demand. The mechanical efficiency of the pump is 78% and of the turbine 80%. The one-way transportation of water in the pipelines entails 4.5% loss of the stored energy because of friction and the minor (fittings and valves) losses.

The water volume when the lake is full is 120,000 m^3; the mass of the water is $120 * 10^6$ kg (120,000 tons); and the potential energy (exergy) stored is $883 * 10^9$ J (245,250 kWh). Taking into account the frictional losses for the transportation of the water, the pump supplies $1.045 * 883 * 10^9 = 923 * 10^9$ J; and the electric energy consumed by the pump is $1,183 * 10^9$ J. Assuming no net amount of water is lost or added at the artificial lake (because of evaporation or rain), during the discharge process the available $883 * 10^9$ J potential energy is reduced to $843 * 10^9$ J by the return transportation losses. Of this energy, the 80% efficient turbine produces $675 * 10^9$ J of electric energy (187,392 kWh).

It is observed in this example that the PHS uses $1,183 * 10^9$ J of electricity generated by the renewable sources and delivers $675 * 10^9$ J of electricity, with $508 * 10^9$ J exergy losses. The exergetic efficiency of the storage system is 57%. Because all the work supplied and produced is electrical and mechanical, the round trip energy efficiency of the PHS is also 57%. The exergy destruction in the system occurs in the pumping process ($260 * 10^9$ J); during the transportation losses ($80 * 10^9$ J); and in the hydraulic turbine ($168 * 10^9$ J). It is apparent that most of the exergy destruction is due to the irreversibilities in the pump and the hydraulic turbine. When the performance of these engines deteriorates with the time of use, the round-trip efficiency of the entire system may drop to less than 50%.

4.9.2 Compressed Air

The *Compressed Air Energy Storage* (CAES) operates in combination with Brayton power cycles. The CAES systems use a water reservoir and a water column as shown in Figure 4.13 to maintain constant pressure, P, in the cavern. When the

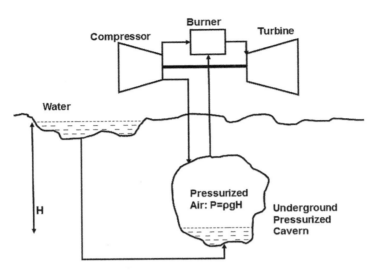

Figure 4.13 Schematic of a CAES system utilizing a water column to maintain constant pressure.

demand for power is low the gas turbine drives the compressor that supplies pressurized air to an underground cavern or an artificial compressed air reservoir. Constant pressure is maintained by water displacement. When the electricity demand is high, the pressurized air is directed to the combustor and then to the turbine, where it produces additional power. A variation of this energy storage method – but of lesser efficiency – is to feed the compressed air directly to the turbine without heating it in the burner.

Let us consider a CAES system in which a compressor with 80% efficiency pressurizes air from $P_0 = 1$ bar, $T_0 = 300$ K to 25 bar and stores the air in an underground cavern with 50,000 m^3 capacity. The air temperature as it exits the compressor is 839 K and the mass of the air in a fully pressurized cavern is approximately 519,700 kg ($PV = mRT$). The mechanical work (exergy) spent for the pressurization of the air is: $519,700 * (992.3 - 426.3) = 294 * 10^6$ kJ. The pressure in the cavern is maintained constant with a water column at 25 bar. Because of air leaks through fissures, 7% of the air mass is lost during the storage period; and because of heat transfer with the water, the air temperature drops by 45–794 K. The exergy of the compressed air is:

$$E = m[h - h_0 - T_0(s - s_0)]. \qquad (4.24)$$

The exergy of the air in the cavern after the losses is $237 * 10^6$ kJ. If the air is used in a turbine with 80% isentropic efficiency, the total energy produced is $190 * 10^6$ kJ (52,778 kWh) and the exergetic efficiency of the storage system is 54.7%.

The two numerical examples in this section and Section 4.9.1 highlight the significant loss of exergy associated with large, utility-level storage systems. An observation on the magnitude of power that may be produced by the PHS and the CAES systems is that, even though both have very high volumetric capacity, the actual power to be produced from the stored energy is not as high as the typical power plants that generate power: The PHS unit of the previous section is capable to generate approximately 45 MW for four hours, while the CAES system may produce approximately 13.3 MW for four hours. Storage systems that would reliably produce hundreds of MW of power must be significantly bigger than the two systems analyzed in Sections 4.9.1 and 4.9.2. Such systems are associated with high capital and operational costs as well as higher irreversibilities from loss of water and loss of the pressurized air.

4.9.3 Thermal Energy Storage

Thermal energy storage (TES) is primarily used for three purposes:

1. Hot water supply in buildings, as described in Section 4.7.4.
2. High temperature storage, typically as latent heat of steam or molten salts, for the production of electric power.
3. Cold TES, typically as chilled water, for the air-conditioning of buildings during periods of high power demand.

TES at high temperature for the production of electricity is a practical method to smoothen the electric power production from solar thermal power plants. A TES system with molten salts is used in the *Gemasolar (Solar Tres)* power plant in Andalucía and has enabled this unit to continuously produce power for 36 consecutive days.

One of the disadvantages of all TES systems is the heat loss to the environment. Given enough time all the stored energy eventually dissipates. For this reason TES systems are designed for short-term energy storage, on the order of several hours, but not for weekly or seasonal storage. TES for 12 hours is sufficient to respond to the diurnal fluctuations of solar energy availability and the demand for power.

A TES vessel with external area A, at an elevated temperature, T, transfers heat rate to the environment:

$$\dot{Q} = UA(T - T_0), \tag{4.25}$$

where U is the overall heat transfer coefficient of the vessel. If the initial temperature of the TES material is denoted by $T(0)$, after a time period t, the average temperature in the vessel becomes:

$$T = T(0) - [T(0) - T_0][1 - \exp(-t/\tau_{st})], \tag{4.26}$$

where τ_{st} is the thermal timescale of the TES material. This expression demonstrates that, if left for a long time, thermal energy storage units would attain the ambient temperature, T_0, and their exergies would vanish.

For a material with density, ρ, and specific heat capacity, c, contained in a vessel of volume, V, and external area, A, the thermal timescale, τ_{st}, is:

$$\tau_{st} = \frac{V\rho c}{AU}. \tag{4.27}$$

The thermal energy (enthalpy) loss is sometimes referred to as *thermal discharge*. During short time periods there is no phase change during the thermal discharge and the enthalpy change of the storage material may be given by the expression:

$$H(0) - H(t) = V\rho c[T(0) - T] = V\rho c[T(0) - T_0][1 - \exp(-t/\tau_{st})]. \quad (4.28)$$

The corresponding exergy destruction during thermal discharge is:

$$E(0) - E = H(0) - H - T_0[S(0) - S] = V\rho c[T(0) - T] - V\rho c T_0 \ln\left(\frac{T(0)}{T}\right). \quad (4.29)$$

Example 4.13: A molten salt mixture of lithium, calcium, and potassium nitrates is considered to be used as TES medium for a solar unit. The combination of sensible and latent heat to be stored and extracted from the mixture of salts is 1,500 kJ/kg. The melting salts are used to produce steam in a Rankine cycle with 38% overall thermal efficiency and generate 10 MW of electric power for 5 hours after dusk. Determine the mass of salts to be used.

Solution: During the 5 hours of the Rankine cycle operation 50 MWh or $180 * 10^6$ kJ of electrical energy is produced. The heat input to the cycle is $473.7 * 10^6$ kJ, which must be stored as enthalpy in the molten salts. Because the total energy stored and recovered in the mixture of salts is 1,500 kJ/kg, the mass of salts required is $315.8 * 10^3$ kg (315.8 tons).

Example 4.14: The solar power plant of the last example uses 400 tons of molten salts with density $\rho = 2,420$ kg/m^3 and specific heat capacity in the liquid phase $c = 2.14$ kJ/kg K. The salts are stored in a cylindrical vessel of 8 m diameter; and their initial storage temperature is $T(0) = 900$ K. The time lapse between the end of the heat storage process and the commencement of heat withdrawal is 2.5 hours and the average heat transfer coefficient U during this period is 10 W/m^2 K. Determine the volume and the area of the molten salts storage vessel; the temperature of the liquid molten salts at the beginning of the heat withdrawal process; the thermal discharge; and the exergy destruction during the 2.5 hours of storage, assuming there is no phase change. $T_0 = 300$ K.

Solution: The volume of the vessel is 400/2.42=165.3 m^3. Since the diameter of the vessel is 8 m, the height of the cylindrical vessel is 3.29 m. Therefore, the total surface area of the vessel is 183.1 m^2; and the timescale of the cooling process is: $\tau_{st} = 467{,}504$ s.

At the end of the heat storage period, the temperature is $T(0) = 900$ K. When the removal of heat for the cycle commences (after 2.5 hours or 9,000 s), the temperature, calculated from Eq. (4.26), is $T = 888.6$ K. The temperature difference of 11.4 K is low enough to justify assumption that the salts remain in the liquid phase.

The heat loss to the environment is calculated from Eq. (4.28): $9.79 * 10^6$ kJ; and the exergy destruction during the 2.5 hour storage is: $6.48 * 10^6$ kJ.

Considering that the Rankine cycle produces $180*10^6$ kJ of electric energy during the 5 hours of after-dusk operation, the exergy destruction due to the storage time is only 3.6% of the total exergy delivered.

Another form of TES, cold storage, or cold TES – essentially storage of chilled water – is increasingly used in large buildings and clusters of buildings to meet the power demand for air-conditioning during the peak afternoon hours. Power distribution corporations provide incentives for such a shift in power consumption patterns by reducing the price of electric energy during the night time and early morning hours. Most of the universities, several large airports, and several hotel clusters in big cities in the southern part of the USA, use chilled water stored in large tanks, to provide air-conditioning during the afternoon hours, when the electricity demand peaks. Chilled water storage may also be used in clusters of grid-independent buildings powered with PVs, to provide air-conditioning during the evenings and night-time and reduce the overall energy storage requirements [23]. The chilled water does not undergo a phase change as it heats up and Eqs. (4.26)–(4.29) apply to this type of exergy storage too.

Example 4.15: Five large buildings have outside area $A = 12,000 \text{ m}^2$ and average heat transfer coefficient, $U = 10.5 \text{ W/m}^2 \text{ K}$. Chilled water at 6°C is used to keep the inside temperature of the buildings at 24°C, during 3 afternoon hours when the outside temperature is 38°C. The chilled water exits the buildings at 19°C. Determine the mass of the chilled water needed and the volume of the storage tank.

Solution: The rate of heat input from the environment to the buildings is: $10.5*12,000*14 = 1,764$ kW. This rate of heat needs to be continuously extracted to maintain constant temperature in the buildings. During the 3 hours of heat extraction, 19,051,200 kJ of heat are removed by the chilled water. Therefore, the needed mass of chilled water is $19,051,200/4.184/(19-6) = 350,257$ kg and the volume of the storage tank is approximately 350 m^3.

Apart from the advantages of power shift and lower cost, an exergetic advantage for the use of chilled water for clusters of buildings is that the refrigeration systems, which chill the water for the clusters, are larger and more efficient than smaller air-conditioning systems that might serve each building separately. As a consequence the COPs (and SEERs) of these systems are higher.

4.9.4 Chemical Storage – Batteries

Most of the energy consumed globally emanates from fossil fuels, where energy and exergy are stored in molecules as chemical energy and exergy. Coal, petroleum, and natural gas consist of chemical substances produced a long time ago from the disintegration of living matter and have been stored in the interior of the earth. As shown in Table 2.2, the materials that are used as fuels have very high specific energy and exergy.

In addition, they are stable, and they are very easy and convenient to use. Several other chemicals – acids, bases, metals, and metallic salts – are also used for the storage of energy and for the direct conversion of their chemical exergy into electric energy. Since the invention of the *Volta Cell,* in the late 18th century, electrochemical cells (batteries) have been widely used to store chemical energy and convert this energy to electricity as direct energy conversion (DEC) devices. The maximum electric energy that may be produced by the corresponding chemical reactions is equal to the change of the Gibbs' free energy, $-\Delta G^o$, and the maximum electromotive force (voltage) delivered by a cell is:

$$-\Delta G^o = \zeta F V_{max} \Rightarrow V_{max} = \frac{-\Delta G^o}{\zeta F}, \tag{4.30}$$

where ζ is the number of electrons deposited in the anode of the cell and F is the *Faraday constant,* 96,500 C/mol (coulombs per mol). Internal and external irreversibilities related to the flow of ions and electrons reduce the voltage developed by 10–25% in actual batteries [2]. Because the voltage developed by a single cell is on the order of 1 Volt, typically a number of cells are connected in series to form the commercial batteries.

The lead cell is the foundation of the automotive and marine commercial batteries. It consists of an anode, made of solid lead, *Pb,* and a cathode made of lead oxide, PbO_2. The electrolyte is a dilute sulfuric acid solution, $H_2SO_4 - H_2O$. The principal chemical reactions in the lead cell are:

$$\begin{array}{l}\textit{At the anode}: Pb + SO_4^{2-} \rightarrow PbSO_4 + 2e^- \\ \textit{At the cathode}: PbO_2 + 4H^+ + 2e^- \rightarrow 2H_2O + Pb^{2+} \\ \textit{In the electrolyte}: 2H_2SO_4 \rightarrow 4H^+ + 2SO_4^{2-} \text{ and } Pb^{2+} + SO_4^{2-} \rightarrow PbSO_4 \\ \textit{Overall reaction}: Pb + PbO_2 + 2H_2SO_4 \rightarrow 2PbSO_4 + 2H_2O\end{array}. \tag{4.31}$$

The Gibbs' free energy of the overall reaction is 393.87 MJ/kmol of lead and the maximum voltage is 2.04 V. Commercial lead cells produce approximately 2.0 V and 6 cells in series constitute the typical automotive and marine batteries that generate approximately 12 V.

While batteries of all kinds are very convenient to use, their charging and discharging involves irreversibilities that reduce the amount of recovered exergy. The battery round-trip efficiency depends on a number of factors, including the rate of charging and discharging. In general, the exergy destruction depends on:

1. The state of charge (SOC) of the battery. When batteries are less than 50% charged, the charging efficiency may exceed 90%; when the SOC is above 80% the charging efficiency may drop to 60%. Permanently charging batteries to lower SOC (e.g. up to 80%) is not an advisable option because it reduces their useful life.
2. The rate of charge and discharge. Faster rates occur with lower efficiency.
3. The age of the battery, with older batteries holding lesser charge; old batteries also charge and discharge at lower efficiencies.

The exergy destruction during a charging–discharging cycle for commonly used batteries is: 5–20% for Lithium-ion batteries; 6–50% for lead-acid batteries; and 10–35% for NiMH batteries. Continuous research and development in the last two decades has significantly reduced the exergy destruction during the charging-discharging processes of commercial batteries [24].

Example 4.16: The owners of a household have decided to install commercial deep-cycle marine batteries to guard against power failures in their area. The batteries have a nominal voltage 12 V, they are rated at 35 Ah (ampere-hours) and they should not be discharged below 20% of their capacity. The charging efficiency of the batteries is 90% when 20% < SOC < 70%; 80% when 70% < SOC < 90%; and 70% when SOC > 90%. The discharging efficiency of the batteries is 85% for the entire range. The owners estimated that the storage system should have enough capacity to supply 6.5 kW for 8 hours. Determine how many batteries the homeowners should install; what is the exergy destruction during a complete charging–discharging cycle; and is what the round-trip efficiency of this energy storage system.

Solution: The energy required for the 8 hours of operation is: W = 52 kWh. This amount of energy must be provided by the batteries when they discharge from 100% to 20% of their capacity. Since the discharge efficiency of the batteries is 85%, the total capacity (stored energy) of the batteries must be $52/(0.8 * 0.85) = 76.5$ kWh. During each charging–discharging cycle the capacity of the batteries drops to 15.3 kWh (20% of capacity) and rises to 76.5 kWh.

From the specifications of the deep-cycle marine batteries each battery has capacity: $12 * 35 = 420$ Wh $= 0.42$ kWh. Therefore, the household will need to install $76.5/0.42 = 182$ deep cycle marine batteries.[9]

Since the batteries are charged from SOC = 20% to SOC = 100%, the energy input during the charging process is: $(76.5 - 15.3)/[(5/8)/0.9 + (2/8)/0.8 + (1/8)/0.7] = 72.6$ kWh.

The exergy destruction during the charging–discharging cycle is: $72.6 - 52 = 20.6$ kWh. The round-trip efficiency of 1 cycle is $52/72.6 = 71.7\%$, and (because electric energy constitutes the input and the output) the exergetic efficiency is 71.7% too.

From the calculations in the last example one concludes that a large number (182 or more in this case) of marine batteries is needed to supply the energy needs of even a moderate household with high HVAC power demand. It is also apparent that a great deal of exergy is dissipated in the charging–discharging process. The use of conventional, lead battery, storage is not the optimum option for larger households. Lithium-based and Redox Flow batteries exhibit higher round trip efficiencies – up to 85% – if charged and discharged during long periods [24]. These batteries are also lighter, they

---

[9] In practical storage systems more batteries will be installed (e.g. 200–220 in this case) for reliability and because it is preferred that batteries do not discharge too close to the recommended SOC minimum of 20%.

store more energy per unit mass (kWh/kg) and may be a better (but not ideal) option for energy storage in households.

4.9.5 Chemical Storage – Hydrogen

Hydrogen is a very convenient fuel because it may substitute all the fossil fuels in power plants; in the automobile IC engines; and even in jet engines. The most efficient use of hydrogen is in the fuel cells that directly convert the chemical energy of the gas to electric energy. An enormous advantage of hydrogen in comparison to other energy storage media is that it is abundant in the seawater. Hydrogen storage has been advocated as an effective, and environmentally friendly energy storage medium [25] with very high specific energy ($-\Delta H^o = 143$ MJ/kmol; $-\Delta G^o = 119$ MJ/kmol). It is chemically stable and, unlike the batteries that suffer from energy drain and partly discharge if left over long periods, hydrogen may be stored for long periods of time to be used for seasonal energy storage (e.g. from spring to summer).

However, hydrogen is not abundantly available in the environment (traces of hydrogen are present in the upper part of the atmosphere) as an energy source. Large quantities of the gas must be artificially produced from hydrocarbon processing or water electrolysis. The exergy efficiency of electrolysis is in the range 70–80% and that of practical fuel cells in the range 50–70%. This implies that the exergetic round trip efficiency of hydrogen storage is in the range 35–56%.

A study on the use of hydrogen storage systems at the household level has shown that hydrogen storage is feasible for the conversion of existing buildings and clusters of buildings to Grid Independent Buildings (GIBs) that derive all their energy from PV cells and avoid the problems of the *U-shaped (duck) demand curve*. The study considered two similar houses: one at a southern location – in Fort Worth, Texas – that uses air-conditioning in the summer; and the second at a northern location – in Duluth, Minnesota – that needs a great deal of heating in the winter [19]. Hydrogen, stored at maximum pressure 50 MPa, is used for the diurnal and seasonal storage of energy. Table 4.9 shows the total annual exergy that was produced by the solar panels and consumed by the two households either directly or after storage. The PV power rating and the maximum storage capacity of the 2 households, in m^3 of hydrogen under 50 MPa, are also included.

Table 4.9 Exergy used and hydrogen storage parameters for the conversion of two buildings to GIBs using hydrogen storage.

Location	Annual Exergy Produced (kWh)	Annual Exergy Consumed (kWh)	PV Rating (kW)	Maximum H$_2$ Capacity (m^3)
Fort Worth, TX	37,584	23,556	18.9	7.0
Duluth, MN	45,957	28,746	29.0	14.2

It is apparent in Table 4.9 that the exergy destruction associated with household energy storage is approximately 60% of the exergy consumed by the 2 households. This occurs despite the fact that a great deal of the PV-produced energy is directly used by the households to satisfy their electric power demand. The amount of energy stored and the exergy destruction may be significantly reduced if thermal storage is used for air-conditioning in addition to the hydrogen energy storage [23].

Another study considered the substitution of coal power plants as well as of all fossil fuel power plants in the ERCOT electricity grid that supplies most of Texas with a combination of wind power and PV cells [21]. The study pertains to utility-level storage and electricity production substitutions that will become necessary for the reduction of the global carbon dioxide emissions in the future. This study concluded that, for the substitution of all the fossil fuel power plants (coal, natural gas, and diesel) by renewables, the needed storage capacity is 12 million m^3 (approximately 1.2 m^3/household). If the entire group of fossil fuel generating units in ERCOT is substituted with renewables, there will be 225 million tons of CO_2 emissions reduction – 0.59% of the total global emissions in 2018 [21].

Problems

(Unless otherwise specified in the problem: $T_0 = 300$ K, $P_0 = 0.1$ MPa.)

1. List four actions in each of the categories: energy efficiency, energy conservation, and energy substitution.

2. Natural gas is compressed and transported in pipelines. Assuming that natural gas may be modeled as the ideal gas methane ($c_P = 2.23$ kJ/kg K, $c_v = 1.71$ kJ/kg K) calculate the specific work when it is compressed from 0.1 MPa 300 K to 12.5 MPa in 1-, 2-, and 3-stage compression processes. The isentropic efficiency of the compressors is 82%. What is the benchmark in this case?

3. A cycle for a freezer operates with R-134a. The upper pressure of the refrigeration cycle is 0.9 MPa and the lowest temperature is $-25°C$. There is a $3°C$ subcooling at the exit of the condenser; the fluid is saturated vapor ($x = 1.0$) at the exit of the evaporator; and the compressor's isentropic efficiency is 70%. Determine the COP for this cycle and the specific exergy destruction in its four processes.

4. What is the minimum specific work to produce liquid ethane at 0.1 MPa?

5. Electric resistance is used to heat 1.2 kg/s of water from $18°C$ to $65°C$. Determine the power used and the exergetic efficiency of this process. Design a system to improve the exergetic efficiency of the process. $T_0 = 18°C = 291$ K.

6. It is desired to desalinate water with 36,000 ppm salt to 400 ppm salt. Calculate the minimum work required for this process in kJ/m^3.

7. The lighting in two university dormitories is to be substituted from fluorescent bulbs to LEDs. Lighting is needed in the buildings for 14 hours per day during all days of the year. The first dormitory is in New Orleans where air-conditioning is needed 50% of the days in a year and heating for 12%. The second dormitory is in Boston where heating is required for 65% of the days and cooling for 12%. Both buildings currently use 95 kW of power for lighting. The conversion factor of the fluorescent lighting is 45 lm/W and that of the LEDs is 100 lm/W. The COP of the equipment in the heat pump mode is 4.5 and the refrigeration mode 3.5. Calculate the annual exergy savings with this substitution in the two buildings.

8. The hot water use in a small house is 750 L/day. Water is supplied from the district pipeline at 18°C and is heated to 65°C. Calculate the daily exergy destruction and exergetic efficiency of this process if the heating is supplied by: a. electrical resistance; b. methane gas in a burner with 85% efficiency; and c. a heat pump with COP = 4.8. $T_0 = 18°C = 291$ K.

9. In order to prepare a bath of 260 L of water at 38°C, a person uses hot water at 52°C and cold water at 18°C. How many liters of hot and cold water does the person use? What are the entropy production and the exergy destruction during mixing if the hot water is produced by methane combustion? Suggest a more efficient way for this person to take a bath. $T_0 = 18°C = 291$ K.

10. Wood chips with initial moisture content 46% are dried in a batch process. 0.05 kg/s of air at 20°C and 80% relative humidity are adiabatically heated by methane combustion to 45°C and supplied to the drier for 2.5 hours. The power of the fan that circulates the air is 45 W. Determine the final moisture of the wood chips and the exergy destruction of this process. For this drying process τ is 5,200 s.

11. An electric clothes' drier operates for ½ hour with 520 W energy input. Calculate the total exergy destruction for this process and suggest a process that largely eliminates the destruction of exergy.

12. A salesman travels on average 820 km/week for 42 weeks/year to call on his customers. He recently decided to trade his large car that consumes 8 km/L with a hybrid that averages 19 km/L. Determine how many litres of gasoline are saved annually with this substitution and (assuming that gasoline is solely composed of octane) determine the annual exergy savings and the annual CO_2 emissions avoidance.

13. A salesman travels on average 820 km/week for 42 weeks/year to call on his customers. He recently decided to trade his large car that consumes 8 km/L with an electric car that travels 5 km/kWh on average. The electric motor of the car has efficiency, $\eta_M = 92\%$; the average charging–discharging efficiency of the car batteries, η_B, is 85%; the yearly average thermal efficiency of the coal power plant that generates the electricity is $\eta_t = 35\%$; and the transmission losses are 12%. Determine the well-to-wheels efficiency of the electric car and the annual exergy savings assuming that gasoline is solely composed of octane. Note that for coal: $\Delta H° \approx \Delta G°$.

14. A PHS is used for utility-level energy storage in a region where a large lake is located 1,200 m above the valley where a nuclear plant has been constructed. The capacity of the lake is 840,000 m^3 of which 75% is to be used for the PHS system. The mechanical efficiency of the pump is 75% and of the turbine 80%. The one-way transportation of water in the pipelines entails 5% loss of the stored energy because of friction and minor losses. Determine: a. how much power the storage system may produce continuously for 5 hours; and b. the round-trip efficiency; the specific exergy destruction.

15. Hydrogen has been proposed as the energy storage medium in a cluster of houses that are to become grid-independent with PV-generated electricity. The electrolysis process has 68% efficiency; the fuel cell is 72% efficient; and there are other 8% energy losses in this energy storage process. Determine the round-trip efficiency of the storage process and the exergy destruction per kWh of electric energy produced by the PV array.

References

[1] E. E. Michaelides, The Concept of Available Work as Applied to the Conservation of Fuel Resources, Proc. 14th Intersociety Engineering Conference, vol. 1, 1762–66 (Boston, MA: August 1979).

[2] E. E. Michaelides, *Energy, the Environment, and Sustainability* (Boca Raton, FL: CRC Press, 2018).

[3] REFPROP – Reference Fluid Thermodynamic and Transport Properties, ver. 9.1 (2012) NIST Standard Reference Data Basis.

[4] M. C. Ndukwu, Effect of Drying Temperature and Drying Air Velocity on the Drying Rate and Drying Constant of Cocoa Bean, *Agricultural Engineering International: The CIGR E-Journal*, **12** (2009), Manuscript 1091.

[5] J. G. Speight, *The Chemistry and Technology of Petroleum*, 5th ed. (Boca Raton, FL: CRC Press, 1999).

[6] H. Al-Muslim, I. Dincer, and S. M. Zubair, Exergy Analysis of Single and Two-Stage Crude Oil Distillation Units, *Journal of Energy Resource Technology*, **125** (2003), 199–207.

[7] P. Gleick, Water Resources, in S. H. Schneider, ed., *Encyclopedia of Climate and Weather*, vol. 2 (Oxford: Oxford University Press, 1996), 817–23.

[8] A. S. Stillwell and M. E. Webber, Predicting the Specific Energy Consumption of Reverse Osmosis Desalination, *Water*, **8** (2016), 601, 1–18.

[9] E. Fermi, Thermodynamics of Dilute Solutions, in *Thermodynamics* (New York: Dover 1936, reprinted 1956).

[10] J. Kestin, *A Course in Thermodynamics*, vol. 1 (Waltham, MA: Blaisdell, 1966).

[11] K. Ochsner, *Geothermal Heat Pumps – A Guide for Planning and Installing* (London: Earthscan, 2008).

[12] International Energy Agency, *Key World Statistics* (IEA: Paris, 2018).

[13] E. E. Michaelides, Thermodynamics and Energy Usage of Electric Vehicles, *Energy Conversion and Management*, **203** (2020), 112246.

[14] R. A. Dunlap, A Simple and Objective Carbon Footprint Analysis for Alternative Transportation Technologies, *Energy and Environment Research*, **3** (2013) 33–9.

[15] US-Department of Transportation Bureau of Transportation Statistics, *National Transportation Statistics* (Washington, DC, 2017).

[16] International Energy Agency, *Energy Efficiency Indicators: Highlights* (IEA: Paris, 2018).

[17] S. Edelstein, Toyota Experimenting with Natural Gas Fuel Cells, *The Drive* (April 27, 2017).

[18] E. Freeman, D. Occello, and F. Barnes, Energy Storage for Electrical Systems in the USA, *AIMS Energy*, **4** (2016), 856–75.

[19] M. D. Leonard, and E. E. Michaelides, Grid-Independent Residential Buildings with Renewable Energy Sources, *Energy*, **148** (2018), 448–60.

[20] M. D. Leonard, E. E. Michaelides, and D. N. Michaelides, Substitution of Coal Power Plants with Renewable Energy Sources – Shift of the Power Demand and Energy Storage, *Energy Conversion and Management*, **164** (2018), 27–35.

[21] M. D. Leonard, E. E. Michaelides, and D. N. Michaelides, Energy Storage Needs for the Substitution of Fossil Fuel Power Plants with Renewables, *Renewable Energy*, **145** (2020), 951–62.

[22] B. Zakeri, and S. Syri, Electrical Energy Storage Systems: A Comparative Life Cycle Cost Analysis, *Renewable and Sustainable Energy Review*, **42** (2015), 569–96.

[23] M. K. DeValeria, E. E. Michaelides, and D. N. Michaelides, Energy and Thermal Storage in Clusters of Grid-Independent Buildings, *Energy*, **190** (2020), 116440.

[24] Batteries, *North America Clean Energy*, **10**, 3 (2017) 40–5.

[25] J. O'M. Bockris, The Origin of Ideas on a Hydrogen Economy and Its Solution to the Decay of the Environment, *Internation Journal of Hydrogen Energy*, **27** (2002), 731–40.

5 Exergy in Biological Systems

Summary

Solar irradiance is the source of exergy for all living organisms. The photosynthesis process in primeval organisms provides the food for all the other living species on earth. It also provides the chemical energy of biomass that is frequently used as fuel. Of particular interest are the energy and exergy conversions in humans, who produce mechanical work using food as energy source. The process of food intake; the metabolic and thermic processes in the human body; the production of adenosine triphosphate – the energy currency of muscles; and the conversion of this energy to mechanical work are analyzed in detail using the principles of classical and nonequilibrium thermodynamics. An interesting conclusion is that humans evolved as good energy storage systems, but as inefficient energy conversion systems, with food-to-work exergetic efficiencies of approximately 10%. The analysis and a number of examples in this chapter elucidate the application of thermodynamics on biological processes including: production and use of biomass; exergy value of nutrients; exergy spent for the performance of vital processes, such as respiration, blood circulation, and maintenance of body temperature; and exergy spent in sports, such as weight-lifting, walking races, the marathon, and bicycling. This chapter also examines the relationship between exergy destruction, the state of health, aging, and life expectancy.

5.1 Photosynthesis

The biological energy systems encompass a wide variety of organisms that extend from land biomass (wood, agricultural crops and residues, grasses, etc.) to aquatic biomass (kelp, algae) to animals and humans. Sunlight is the primary energy source for all biological systems, as it is depicted in the inverted pyramid of Figure 5.1. The sun's energy is captured by several primeval organisms – plants, algae, cyanobacteria – that convert it to chemical energy in a process called *photosynthesis*. These organisms are the primary energy source (food) for all the animals, including humans, and help sustain the life of the higher organisms on earth.

Biomass is a natural method of solar energy storage in the form of the chemical energy in plants. Several intermediate molecules – known as chlorophyll – in the leaves of the plants are the catalysts for the conversion of atmospheric carbon dioxide and

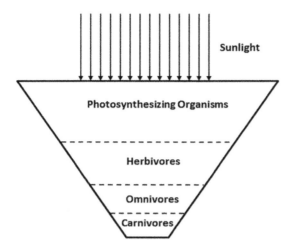

Figure 5.1 Energy sources for living organisms: From sunlight to animal food.

water into the complex and highly energetic molecules of glucose and fructose in photosynthesis. The overall reaction that produces these hydrocarbon molecules may be written in a simplified form as:

$$6CO_2 + 6H_2O \xrightarrow{light} C_6H_{12}O_6 + 6O_2. \qquad (5.1)$$

Approximately 479 MJ of free energy are stored in plant biomass for every kmol of CO_2 fixed during photosynthesis. The glucose formation reaction of Eq. (5.1) is sometimes referred to as the *chloroplast reaction* and has several intermediate stages, in which the light photons activate electrons, ions and intermediate molecules in complex biochemical processes. A schematic diagram for the formation of glucose, which underscores the most important intermediate processes and molecules that catalyze the reaction, is given in Figure 5.2. The glucose itself is produced in a complex cycle of reactions with the commonly used name *the Calvin cycle* (or *Calvin-Benson cycle*), which appears schematically in the right part of Figure 5.2. Inputs to the Calvin cycle are: the nicotinamide adenine dinucleotide phosphate[1] hydrogen (NADPH), and the adenosine triphosphate (ATP), a compound that provides energy in biochemical reactions. The Calvin cycle recycles back $NADP^+$, the reduced form of NADPH, and adenosine diphosphate (ADP) plus phosphoric acid. The ATP molecules are synthesized by the ADP, phosphoric acid, and H^+ ions activated by light (sunlight or artificial light) as it is shown in the bottom-left part of Figure 5.2. The formation of the NADPH follows the processes at the top part of Figure 5.2: At first the reactants (CO_2 and H_2O) and sunlight enter through the leaf membrane. The chlorophyll pigment molecules reach higher excitation states by the energy (photons) of the light and produce highly reactive H^+ ions and free electrons at a higher excitation level. Ferredoxins in the pigments (iron-sulfur proteins with the Fe_2S_2

---

[1] Phosphates are chemical compounds that include the phosphate ion $(PO_4)^{3-}$.

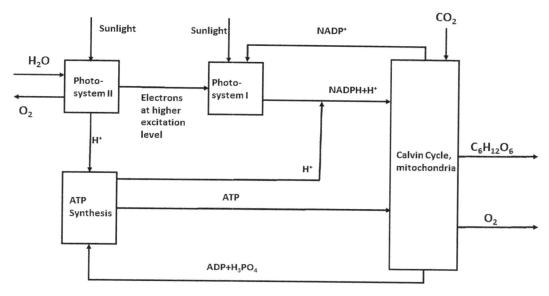

Figure 5.2 The principal stages of glucose production via photosynthesis.

group attached) function as the carriers of the excited electrons and facilitate this process. They also function as electron donors to cellular proteins. A fraction of the high-energy electrons and the H$^+$ ions reduce the NADP$^+$ complexes that are recycled from the Calvin cycle and produce NADPH and molecular oxygen. The NADPH and the remaining hydrogen ions participate in the production of glucose, while oxygen is released as the byproduct of the glucose formation reaction through the leaf membrane.

The overall photosynthesis reaction is highly endothermic with approximately $\Delta H^0 = 2{,}781{,}100$ and $\Delta G^0 = 2{,}848{,}000$ kJ/kmol of glucose and with an entropy change $\Delta S^0 = -224.5$ kJ/kmol K. While the overall formation of the glucose reaction, Eq. (5.1), occurs with a reduction of entropy, the set of the intermediate reactions and processes that are shown in Figure 5.2 as well as the utilization of sunlight exergy result in positive entropy production and significant exergy destruction during the photosynthesis process.

5.1.1 Energy and Exergy Analysis of Photosynthesis

Photosynthesis does not utilize the energy and exergy of the entire solar spectrum, but selectively uses photons in the wavelength range $400 < \lambda < 700$ nm. This is the wavelength range where the plant photo-pigments (of which chlorophyll A and chlorophyll B are the most common) typically operate; it is often referred to as the photo-active region (PAR); and comprises approximately 43% of the incident radiation on the terrestrial surface. As a consequence, 57% of the energy and exergy from sunlight is largely dissipated in photosynthesis and adds to the exergy destruction during this process. Additional exergy destruction occurs in the following processes, which take place in the live organisms that produce glucose [1]:

1. Partial reflectance of the PAR irradiance on the surface of the leaves.
2. Chemical reactions in each one of the subsystems shown as rectangles in Figure 5.2, including the electron transport chain (ETC) processes.
3. In the transpiration and photorespiration processes that continuously occur in the plant leaves.
4. In the metabolic processes of the plants.

The last three processes are necessary for the life sustenance of the organisms that may be considered as the "sunlight-to-chemical energy conversion engines." Because without the sustenance of their lives, such organisms could not photosynthesize the high-energy molecules, all the exergy destruction in these life-sustaining processes must be accounted in the overall exergy balance of photosynthesis.

From the thermodynamic point of view, the photosynthesizing organisms and their parts may be considered as thermal engines that receive solar energy and store it as chemical energy in the form of complex organic molecules. Table 5.1 shows the approximate values of the exergy inputs and outputs during photosynthesis, the exergy destruction for the production of 1 kmol of glucose, and the exergetic efficiency of the several processes involved. Some of the values of exergy input and output are obtained from [1], using the method for the exergy calculation of biochemical compounds outlined in [2].

It is apparent that, when all the processes involved in the production of glucose are taken into account, the exergetic efficiency for the conversion of sunlight to the chemical energy of glucose through photosynthesis is a mere 3.9%. Since the exergy of glucose is approximately equal to its enthalpy of oxidation (ΔH^0) and the exergy of the sunlight is approximately equal to the corresponding energy of irradiance, this efficiency value (3.9%) is also very close to the first law efficiency of photosynthesis. A few studies have not accounted for the exergy needed for the sustenance of the

Table 5.1 Exergy destruction in photosynthesis for the formation of 1 kmol glucose. $T_0 = 298$ K, $P_0 = 1$ atm.

	Input (MJ/kmol)	Output (MJ/kmol)	Destruction (MJ/kmol)	Exergetic Efficiency (%)
Solar spectrum to PAR	72,330	31,102		43.0
CO_2 and H_2O	131			0.0
Total	72,461			
PAR exergy	31,102	27,370	3,732	88.0
Non-PAR exergy	41,228	2,061	39,167	5.0
Exergy available to photosystems		29,431		
Photosystems I and II absorption	29,431	22,871	6,561	77.7
Photosystems I and II ETC	25,169	12,712	12,457	50.5
ATP Synthesis	3,788	2,737	1,051	72.3
Calvin Cycle	9,493	7,704	1,789	81.2
Transpiration, photorespiration, metabolism			4,856	
Total	72,461	2,848	69,613	3.9

organisms that participate in photosynthesis – the engines – and have derived higher values for the exergetic efficiency. The value 3.9% for the first-law and exergetic efficiencies agrees with [1] and is also close to the value obtained by another study [3], which used different data for the exergy of the biochemical species that take part in the process.

As noted in Section 2.11, the definition of the efficiency of thermodynamic processes is not derived from physical principles and is a matter of convenient definition by scientists and engineers. A few studies have used different definitions for the photosynthetic efficiency and other figures of merit for the energy conversion processes in biological systems. For example, [4] considers only the part of the light spectrum that enters the membrane of the leaves, thus excluding from the calculations the light in the non-PAR part of the spectrum and the reflected light. This method reduces the energy and exergy inputs for photosynthesis with the result that the calculated efficiencies become significantly higher, approximately 30%. Other studies (primarily by agriculture scientists) have used more stringent methods for the energy and exergy inputs, for example by considering the energy needed for the organism to grow, or the incident sunlight energy during periods when the organisms are dormant (e.g. plants in the winter). In such cases the numerical values of the energy and exergy inputs in the photosynthesis process become higher and the exergetic efficiency values are closer to 1% [5, 6]. A report commissioned by the United Nations places the exergetic efficiency for the conversion of sunlight to the chemical exergy of glucose in the range 3–6% [7] and this is in line with the data of Table 5.1. Example 5.1 demonstrates the calculations following a different accounting method that includes the energy inputs during periods when the photosynthesizing organisms (corn plants) are dormant.

Example 5.1: A small corn farm in Nebraska, where the yearly average insolation is 170 W/m^2, has 12 ha (approximately 30 acres) and produces 2 crops of corn/year. Each hectare yields 10,700 kg of corn and 12,850 kg of dry stover. The energy per unit mass of corn is 12,000 kJ/kg and that of the stover is 17,470 kJ/kg. Neglecting all other energy inputs, calculate the first law efficiency and the exergetic efficiency of sunlight to chemical energy in the corn conversion process.

Solution: With two crops per year the entire farm produces 256,800 kg of corn with energy $3.081 * 10^9$ kJ and 308,400 kg of stover with energy $5.388 * 10^9$ kJ. The total energy production is $8.469 * 10^9$ kJ/year.

The annual solar energy input in the 12 ha (120,000 m^2) farm is: $365 * 24 * 60 * 60 * 170 * 120,000$ J $= 643 * 10^9$ kJ. Note that, by considering the amount of sunlight energy during the entire year, this method includes time periods (winter and early spring) when the plants are dormant.

The first law efficiency of the sunlight to the chemical energy conversion process (via biomass) is 1.31%. Taking the exergy of the corn and stover equal to their chemical energy and given that the exergy of sunlight is 96% of its energy, the exergetic efficiency of the process is approximately 1.37%.

Starting with glucose and fructose molecules, plants form more complex organic compounds, such as sucrose, starch, lignin, and several other carbohydrates with longer molecules and higher chemical exergy content. In general, plants form carbohydrates according to the reaction:

$$xCO_2 + xH_2O \rightarrow carbohydrate \; molecule + xO_2. \qquad (5.2)$$

The formation of the more complex molecules from glucose and fructose entails additional exergy destruction. For this reason, the formation process of longer and heavier carbohydrate molecules from sunlight has lower exergetic and first law efficiencies than those of glucose formation. This very low efficiency of sunlight energy conversion leads to the conclusion that the biological systems have evolved to preserve and sustain life on the planet rather than become efficient energy conversion devices or efficient energy storage systems. Artificial systems and engines (e.g. PV cells) are significantly more efficient in the conversion of sunlight to other forms of energy.

5.2 Land Biomass

Timber, crops, grasses, human waste, and animal waste are the principal types of land biomass used for the production of energy. Of these, forest timber, such as pine trees, take several years to grow – from approximately 25 years in the southern temperate zone to almost 100 years in the northern temperate zone. Fast growing timber (e.g. sycamore and eucalyptus trees) takes fewer years to grow in favorable climates. Agricultural crops yield one or two crops per year, while grass yields several crops. Human and animal waste is continuously produced at various rates: in the USA a person produces on average approximately 800 kg of waste per year, while Europeans produce less than 250 kg of waste annually [8]. Land biomass, primarily from bushes and trees, is the principal fuel for cooking and domestic heating in many developing nations. In the industrialized nations biomass is partly used for heating and cooking and, recently, for the production of other fuels, the so-called *biofuels*. Small quantities of biomass are also used for the production of electricity, typically utilizing combustion and Rankine cycles. It must be noted that biomass is a *neutral fuel* in the production of CO_2: All the carbon in the biomass has been extracted from the CO_2 of the atmosphere and, hence, its combustion does not add to the atmospheric CO_2 concentration. In addition, the atmospheric oxygen consumed in the combustion process has been released during the production of biomass (as in Eq. (5.2)).

Biomass is composed of a myriad of organic chemical compounds. Because the molecules of these compounds are very large, the direct conversion of their exergy to work (e.g. in fuels cells) is not technologically feasible at present. The conversion of land biomass to work occurs via combustion and is subject to the Carnot limitations. Alternatively, the larger molecules of biomass are split into smaller ones via a fermentation process that produces ethanol or methanol, a highly irreversible process. While several farming communities and corporations that supply farming equipment and chemicals advocate the more widespread use of biomass for the production of biofuels

Table 5.2 LHV (specific energy) of several types of biomass, anthracite and octane.

Biomass Type	kJ/kg	Btu/lb
Rice hull, 0% moisture	13,905	5,970
Rice hull, 30% moisture	9,175	3,940
Bagasse, 0% moisture	17,266	7,414
Bagasse, 30% moisture	11,038	4,740
Dry wood (average)	18,500	7,934
Wood Chips	13,600	5,860
Municipal solid waste (MSW)	4,000–8,000	1,715–3,430
Sewage/animal waste	1,164–1,863	500–800
Corn stover (dry)	17,470	7,502
Corn stover (30% moisture)	12,229	5,251
Anthracite	31,952	13,720
Octane	44,430	19,110

and electric energy, the production and utilization of biomass as energy source has several disadvantages among which are [8]:

1. The production of biomass is periodic and the yield is uncertain.
2. Biomass is a diffuse energy source that uses large tracts of land – a scarce resource.
3. The transportation of the raw biomass to a central energy conversion facility requires significant energy input.
4. Further processing is needed for the production of biofuels and electricity and this entails additional exergy destruction.
5. The use of biomass as an energy source directly competes with the supply of food, which is more important and has higher priority for the growing population of the planet.

The energy content of biomass is characterized by its low heating value (LHV).[2] Table 5.2 [8, 9] shows the energy content of the several types of biomass as well as the energy content of the fossil fuels anthracite and octane, which are included for comparison.

While there is no standard convention on the term, oftentimes, the LHV, either in kJ/kg or Btu/lb, is referred to as the *energy density* of the fuels including biomass. Since the units of this variable are energy per unit mass, a term that is more suitable and consistent with standard thermodynamic nomenclature is *specific energy*. It is apparent from a glance at Table 5.2 that the specific energy of biomass is significantly less than that of conventional fossil fuels. Also, that the specific energy very much depends on the moisture content of the biomass.

The specific exergy of biomass is more difficult to calculate, because every type of biomass is composed of several different solid organic components and their

---

[2] The LHV (rather than the HHV) of biomass is used because biomass is burned in simple equipment that do not allow for the condensation of produced water before the combustion products are rejected.

composition significantly varies among the types of biomass. Given the values of the chemical energy of several fuels in Table 2.2, a good approximation is that the specific exergy of biomass is equal to its specific energy, that is: $\Delta H^0 \approx \Delta G^0$. For the vast majority of the chemicals in biomass, the difference $\Delta H^0 - \Delta G^0$ is less than the uncertainty in the determination of the two values. This implies that, when biomass is used for the production of work (e.g. the generation of electricity), the thermal efficiency and the exergetic efficiency of the conversion process are approximately equal.

Example 5.2: It has been proposed that farms of eucalyptus trees are developed for the production of biomass that will be used for the production of electricity. One hectare (1 ha = 2.5 acres) may be planted with approximately 500 eucalyptus trees that grow in 5 years to mature trees of approximately 400 kg with specific energy 13,600 kJ/kg. When the trees mature they are harvested and used as fuel in a Rankine cycle with 37% thermal efficiency. The annually averaged solar irradiance in this location is 210 W/m². Calculate the total electric energy produced by the entire farm, in kJ and kWh, and the exergetic efficiency of the sunlight to electricity conversion via biomass. You may assume that all other energy expenditures are negligible.

Solution: The 500 eucalyptus trees have total mass 200,000 kg. Their combustion in the Rankine cycle produces $2.72 * 10^9$ kJ of heat, which is converted at 37% efficiency to $1.01 * 10^9$ kJ, or $0.28 * 10^6$ kWh of electricity.

During the 5 years of the tree growth period, the solar energy input in 1 ha (10,000 m²) is: $5 * 365 * 24 * 60 * 60 * 210 * 10,000$ J $= 331 * 10^9$ kJ. Therefore the first law efficiency of the sunlight-biomass-electricity conversion is 0.305%. Given that the exergy of the sunlight is approximately 96% of its energy, the exergetic efficiency of this biomass-to-electricity system is slightly higher at 0.316%.

Example 5.3: A farm in Iowa, where the average insolation is 180 W/m², has 10 ha and produces 2 crops of corn per year. Each crop yields 10,500 kg of corn and 12,300 kg of stover per hectare. The specific energy of corn is 12,000 kJ/kg and that of the stover is 17,470 kJ/kg. The entire biomass production is harvested to be used in a power plant with 36% thermal efficiency. Assuming all other energy inputs are negligible, calculate the first-law and the exergetic efficiency of the sunlight to electricity production process.

Solution: The entire farm produces 210,000 kg of corn annually with energy $2.520 * 10^9$ kJ and 246,000 kg of stover with energy $4.298 * 10^9$ kJ, for a total energy $6.818 * 10^9$ kJ/year. When this energy is utilized as heat in the Rankine cycle, it produces $2.454 * 10^9$ kJ $= 0.682 * 10^6$ kWh of electricity.

The annual solar energy input in the 10 ha farm is: $365 * 24 * 60 * 60 * 180 * 10 * 10,000$ J $= 568 * 10^9$ kJ. Therefore, the first law efficiency of the sunlight-to-electricity conversion via corn biomass is 0.432% and the exergetic efficiency is 0.450%.

Examples 5.2 and 5.3 demonstrate that the use of biomass for the production of electric power entails extremely low overall efficiencies, even in the idealized cases when there are no other energy inputs. For real-world conversions one should also take into account the energy inputs for: tilling and seeding; irrigation; production and supply of fertilizers and insecticides; harvesting and drying; and for the transportation of biomass to the electricity generation station. When these energy inputs and irreversibilities of the several agricultural processes are considered, the energy efficiency of land biomass conversion (excluding the energy from sunlight) is even lower, and sometimes attains negative values [8, 9].

A comparison of the electric energy production via biomass in Examples 5.2 and 5.3 and the energy production from photovoltaic (PV) cells shows unequivocally that the installation of PV cells in the regions of the 2 farms would produce 50–80 times more electric energy than biomass. This is due to the following factors:

1. The PV cells are direct energy conversion devices, not subject to Carnot limitations.
2. The conversion of sunlight to biomass is thermodynamically a very inefficient process as the values in Table 5.2 show.
3. The "engines" of the sunlight conversion (the plants) remain dormant, do not grow, and do not convert the sunlight energy during several months of the year. Corn does not grow in the colder autumn and winter months and all the sunlight during these months is wasted. On the contrary, solar cells produce power daily during all seasons of the year.

5.2.1 Biofuels – Corn-to-Ethanol Production

Biofuels are liquid fuels derived from biomass and wastes. Among these, methanol (CH_3OH) also known as *wood spirit*, ethanol, (C_2H_5OH) or *alcohol,* and *biodiesel* (a mix of several fatty esters produced by chemically reacting lipids) are the most commonly used biofuels. A blend of 10% ethanol and 90% gasoline by volume, commercially known as E10, is the *gasohol*, which is used as a transportation fuel in several countries including the USA. The E15 and E20 fuels contain 15% and 20% ethanol respectively. These mixtures are consumed in several markets globally and may be used in spark-ignition engines without significant engine modifications. The much richer in ethanol mixture, E85 that contains 85% alcohol, is produced in Brazil and a few European countries. This mixture may be used only in modified spark-ignition engines, the *flex-fueled engines*. All the produced biofuels are currently consumed in thermal engines, whose efficiencies are subject to the Carnot limitations.

Brazil and the USA generate 72% of the global biofuels' consumption – produced from sugar cane in Brazil and corn in the USA – with the EU producing 12% primarily from corn and rapeseed [10]. The production and final utilization of biofuels in internal combustion (IC) engines entails the following energy transformations:

1. Sunlight to biomass via photosynthesis.
2. Biomass to biofuel, usually by fermentation and distillation (alcohol) or chemical processes (biodiesel).

3. Biofuel to heat via combustion.
4. Heat to motive power in the IC engines.

As it was shown in Section 5.2, the photosynthesis conversion of sunlight to plant biomass is highly irreversible with exergetic efficiency approximately 3.9%. Irreversibilities in the other three conversion processes further reduce the exergetic efficiency of biomass as a transportation fuel.

The production of ethyl alcohol from corn, an agricultural product, is a common process that currently supplies most of the alcohol used in the gasohol fuel mixture. Figure 5.3 shows the several chemical and physical processes as well as the material inputs and outputs that take place in the conversion of corn to ethanol. The sunlight-corn-ethanol production comprises several energy- and exergy-consuming processes with significant other inputs of materials and energy.

Four independent studies [11–14] conducted detailed calculations on the energy balances of the several systems and processes involved in the production of alcohol from corn. The first two [11, 12] include the energy needed for the periodic replacement of the machinery used in the agricultural processes, while the other two [13, 14] do not consider the machinery replacement. The calculations show that 1 hectare of land produces an average of 3,176 L of ethanol with LHV 67 GJ. The energy expenditures for the production of this quantity of the biofuel are summarized in Table 5.3. The total energy expenditures for the agricultural production of corn are included in the pre-processing column; and the total energy inputs for the conversion of corn to ethanol in the processing column. All four studies do not include the energy of the sunlight for the production of corn. The LHV ratio is the ratio of the energy in the 3,176 L of ethanol, which are produced per hectare of land, to the total energy inputs. The LHV (rather than the HHV) is used for the quantification of energy produced, because it is the LHV of ethanol that is converted to motive power in the IC engines. The last column in Table 5.3 is the exergetic efficiency of the corn-to-ethanol production.

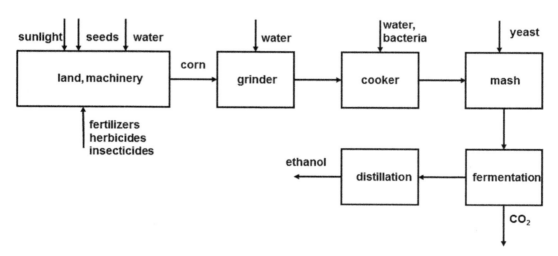

Figure 5.3 Processes, inputs, and outputs for the production of ethanol from corn.

Table 5.3 Total energy inputs, in GJ/ha, and energy output-to-input ratios for the corn-to-ethanol conversion.

Study	Preprocessing Energy (GJ/ha)	Processing Energy (GJ/ha)	Total Energy Usage (GJ/ha)	LHV ratio	Exergetic Efficiency (%)
Pimentel, 2003 [11]	33.8	51.4	85.2	0.79	−21
Patzek, 2004 [12]	29.2	51.4	80.6	0.83	−17
Shapouri et al. 2002 [13]	22.3	46.7	69	0.97	−3
Wang et al., 1997 [14]	19.8	45.9	65.7	1.02	2
Average	26.3	48.9	75.1	0.9	−9.8
Standard deviation	5.5	2.6	8.0	0.10	9.5

It is observed in Table 5.3 that more primary energy is spent for the production of ethanol than the energy content of ethanol itself. In only one of the four studies, which does not take into account the machinery replacement [14], the energy inputs are slightly lower than the energy outputs and this implies that there is a very small energy advantage in the corn-to-ethanol production process. The exergetic efficiencies are negative in 3 of the 4 studies and slightly positive (2%) in the last one, with the average exergetic efficiency being close to −10%.

It must be noted that only the exergy of the input and output energy streams was considered in the calculation of the exergetic efficiency in Table 5.3. A more detailed study [15] includes the energy and exergy expenditures in the agricultural, industrial processing, transportation, and environmental remediation (e.g. the remediation of the water used for fermentation) stages of the corn-to-ethanol production and conversion. This study concluded that the actual exergy input in all the processes is 111.7 GJ/hectare/year and the exergy output is 56.1 GJ/hectare/year. The additional energy and exergy expenditures in this study reduce the exergetic efficiency of the corn-to-ethanol conversion process to approximately −50%.

Exergetic efficiency values, such as those in Table 5.3 and in [15], are strong indications that there is no net energy and exergy to be gained from the production of biofuels using agricultural products and land with inputs similar to those shown in Figure 5.3. On the contrary, the processes for the production of biofuels from agricultural products are net exergy destruction processes. While biofuels may provide some energy in the future, they cannot be viewed as possible replacements for fossil fuels. In addition, their accelerated production and farm commercialization has several problematic global consequences: a. displacing poorer farmers in favor of conglomerates; b. leading to higher food prices;[3] c. increasing the rates of

---

[3] A strong argument against the use of agricultural products for biofuel production is that the growing population of the earth needs these products as food rather than energy sources [8].

malnutrition; d. aggravating biodiversity loss; e. generating GHGs at rates comparable to those of fossil energy production in addition to the GHGs absorbed during the biomass production [16]; and f. large-scale fertilizer production that contribute to the eutrophication of lakes and coastal environments.

5.3 Aquatic Biomass

The production of land biomass from sunlight and plants has very low exergetic efficiency, primarily because the "conversion engines," the plants, do not grow fast and remain dormant during the cold season. This is not a problem with biomass grown in aquatic ponds, because the water temperature may be controlled to be optimum for the growth of the micro-organisms that produce energy. Since the growth of these micro-organisms is not seasonal, aquatic biomass may be continuously produced throughout the year: micro-algae can regenerate and grow to be harvested in 48–72 hours; and cyanobacteria can regenerate in 5–20 hours in sunny regions. Figure 5.4 is a diagram of an algae production pond that may operate continuously with inputs of carbon dioxide from the atmosphere and seawater; and outputs consisting of oxygen and an aqueous solution of algae. The short growth and frequent harvesting times of algae produce more energy per unit area than land biomass. The production of biodiesel from algae in pilot plants has been 10–50 times higher than that of agricultural products [17]. Good aquatic biomass sites in Europe yield 55,000–73,000 kg of biofuel per hectare per year (63–84 kL/ha/yr) [18]. Another study in the USA, which examined the potential for the production of algae under optimum conditions, extended the yield to 182,500 kg/ha/yr (approximately 210 kL/ha/yr) [19].

The amount of available freshwater on the surface of the earth (as surface water in lakes, swamps and rivers) is only 0.01%[4] of the total water mass. Because freshwater is

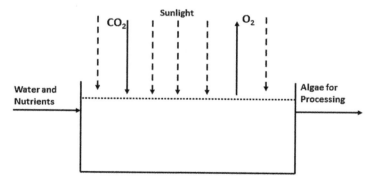

Figure 5.4 An aquatic biomass production pond with algae is an open thermodynamic system operating continuously.

---

[4] The freshwater on earth is approximately 2.7% of the total, but most of it is in ice sheets, the glaciers, and underground.

becoming a scarce resource in several regions, it is not prudent to use large quantities of freshwater for energy plants and algae. On the other hand, the salt-water of the sea and the brackish and marsh-water regions, which comprise 97.3% of the total water mass on the planet, are readily available for the production of biomass and biofuels. Seaweed, kelp, and the several types of algae may produce significant amounts of biomass that can be converted to biofuels. The use of aquatic biomass entails advantages for sustainable energy production because:

1. It does not use land surface, a scarce resource that produces food.
2. No irrigation is necessary.
3. Nutrients, such as K and P, are abundant in the sea and brackish water.
4. If the food of the aquatic biomass is human and animal waste, raising aquatic biomass reduces the volume and pollution effects of urban sewage.

A negative effect of aquatic biomass (in natural settings) is the reduction of dissolved oxygen, which affects the other aquatic life and reduces biodiversity.

Aquatic biomass produces algae and bacteria that convert sunlight first to glucose by photosynthesis and then to lipids (fats). An exergetic analysis of micro-algae biomass production showed that only 4.93% of the exergy inputs – primarily sunlight exergy – in the aquatic ponds is converted to product exergy. Of this, 1.14% is in the exergy of waste products that are not utilized, and the remaining 3.79% corresponds to the exergy of the harvested algae [20]. This value for the overall exergetic efficiency of microalgae production using sunlight as the energy source is approximately equal to the exergetic efficiency of glucose production in plants [1] and within the range 3–6% of the photosynthesis processes in the UN report [7].

The aquatic biomass micro-organisms cannot survive and grow if they are in close proximity. For this reason, the fluid produced from the harvesting of biomass is a dilute mixture of water, waste products, and algae, with the concentrations of the algae in the range 1–2%. Such an aqueous mixture has negative LHV and cannot be used directly for the production of energy. For the harvesting of the aquatic biomass and its conversion to biofuels, energy (and exergy) must be spent in the following processes:

1. De-watering, drying, and salt removal.
2. Cell membrane crashing for the removal of lipids.
3. Physical and chemical processing of lipids to be converted to biofuels.

With the present technology and the large quantities of energy required for the de-watering and drying processes, the cultivation, harvesting, and processing of aquatic biomass for the production of biofuels has negative overall exergetic efficiency. Several studies have concluded that the production of biofuels from aquatic biomass is a costly commercial method for the production of energy at present [21]. Technological advances are necessary for the improvement of the overall exergetic efficiency of biofuels-from-algae production, among which are [9]:

1. Advancements in biomass culturing techniques (new strains of organisms) that would grow faster; would have higher sunlight-to-lipids exergetic efficiencies;

and would be able to grow at higher concentrations in the aquatic environment, thus, lowering the amount of water that needs to be removed.
2. Technological improvements with the processing of the micro-organisms, in particular with the energy required for the drying of the biomass.

5.4 Animal and Human Systems

Of particular interest is the human biological system, which is increasingly receiving attention by scholars. The multitude of chemical reactions that take place in the human body; the physical activities of humans; the entropy production in the human body; and the possible connections between life-long entropy production, health, and longevity have fascinated many researchers including Erwin Schrodinger, who analyzed the biological processes in the human body using " ... a naive physicist's ideas about organisms ... " and elucidated several biological processes using the theory and modeling methods of Physics [22].

Food from plants and other animals is the energy source for animals, including humans as represented in the inverted pyramid of Figure 5.1. A schematic diagram of the vital processes in an animal body, which is modelled as an open thermodynamic system, is shown in Figure 5.5. Chemical energy in the form of food is the primary energy source. The complex food molecules are ingested and processed in the digestive system, where they are converted to simpler molecules, the nutrients, which provide energy and enable the several processes in the body. The nutrients flow in the blood stream and provide all the tissues of the human body with an energy form. The foods-to-

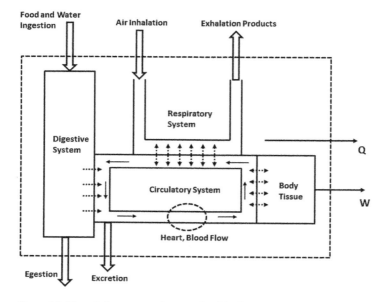

Figure 5.5 The vital processes in an animal body as an open thermodynamic system.

nutrients conversion processes in the animal bodies are analogous to the industrial production of secondary and tertiary energy forms, such as petroleum refinement and electricity. The food residue and wastes of the digestive process are egested in the feces, which also carry a small part of undigested nutrients (5–10%).

The digestive system communicates through permeable cell membranes and transfers the secondary energy forms, the nutrients, to the blood circulation system, which is powered by the heart – a pump made of soft tissue. The circulatory system also communicates with the respiratory system through the lungs, – organs with very large and semipermeable membrane area, through which oxygen diffuses into the blood stream and carbon dioxide diffuses out to be exhaled. Oxygen in the blood stream combines with the nutrients in the blood and the oxidation process produces the body heat, which maintains the temperature of the body. All this heat is transferred to the surroundings. In addition, the oxidation process generates biochemical compounds that provide energy to the tissues of all the organs in the body – heart, liver, breathing muscles, kidneys, muscles, etc. – and enable them to perform the mechanical work, which is necessary for all the processes that define life. In summary, the animal (including human) bodies are open thermodynamic systems that receive chemical energy in the form of food and, through numerous biochemical processes convert this energy to heat and work.

5.4.1 Energy Inputs

Food is the fuel of all animals. The chemical energy of the several types of foodstuff we eat and digest is converted to derivative chemicals that provide the energy of the human body. The basic chemical compounds that make up the human food are classified as carbohydrates, fats, and proteins. The approximate average energy content ($-\Delta H^0$) of the 3 constituents are 18,000 kJ/kg for carbohydrates, 40,000 kJ/kg for fats, and 22,000 kJ/kg for proteins. These organic compounds are mixed with water, fiber (which is not digested by humans), and other nutrients (vitamins, trace metals, etc.) in the several food products. Table 5.4 shows a few common types of food and its nutrient content; the specific chemical energy of the food; and the specific energy of the food[5] constituents in their dry (without water) condition [23, 24]. The high heating value (HHV) of the nutrients is included in Table 5.4 because the human body processes take place at approximately 37°C and any water produced from food oxidation is in the liquid form.

As with fuels listed in Table 2.2, $-\Delta H^0 \approx -\Delta G^0$ for the chemicals in foodstuff and, hence, the HHV values may be approximated with the exergy values of the foodstuff. It must also be noted that, because there are several biochemical molecules classified as carbohydrates, fats, and proteins, the HHV values listed are average values. As a consequence, there is a degree of uncertainty, on the order of 5%, for the actual values of the energy and exergy of foodstuff.

---

[5] What is denoted as 1 Cal (with capital C) in food product labels is actually 1 kcal (4,184 J).

Table 5.4 Nutrient and energy content of common foodstuff.

Food Type	Water (%)	Carbohydrates (%)	Fat (%)	Protein (%)	$-\Delta H^0$, (HHV) (kJ/kg)	dry $-\Delta H^0$ (kJ/kg)	dry ΔH^0 (kcal/kg)
Apple	84.3	11.9	0	0.3	2,500	15,925	3,806
Beef	54.3	0	21.2	23.6	13,000	28,665	6,851
Bread	39	49.7	1.7	7.8	11,200	18,360	4,388
Cheese	37	0	33.5	26	17,000	26,985	6,450
Cod Fish	76.6	0	1.2	21.4	3,100	13,250	3,167
Hamburger	40.9	29.1	14.2	15.8	17,300	29,270	6,996
Milk	87.6	4.7	3.8	3.3	26,000	209,680	50,115
Potatoes	80.5	17.7	0.1	1.4	3,500	17,950	4,290

Example 5.4: A steak and potatoes meal consists of 0.35 kg beef and 0.3 kg of potatoes. Calculate the total carbohydrates, fats and proteins; and the energy value of this meal.

Solution: Using the numbers of Table 5.4, the 0.35 kg of beef has energy value 4,550 kJ and contains 0.0742 kg of fat and 0.0826 kg protein. The potatoes have 1,050 kJ energy value and contain 0.0531 kg of carbohydrates, 0.0003 kg fat and 0.0042 kg protein. The total energy content of this meal is: 5,600 kJ; the total protein intake is 0.0868 kg; the fat intake is 0.0745 kg; and the total carbohydrates intake is 0.0531 kg.

5.4.2 Metabolism of Nutrients

The human body receives energy primarily from the oxidation of glucose and fats. In the absence of these two energy sources, the amino acid constituents of protein may be oxidized too, albeit with a lower energetic efficiency. The oxidation reactions, which are referred to by the general name *oxidative phosphorylation,* produce ATP, a high-energy molecule that causes the movement of the muscles in the body tissues and produces mechanical work.

Glucose is produced in the digestive system primarily from the decomposition of longer carbohydrate molecules. The complete oxidation reaction of glucose, which has molecular weight 180 kg/kmol, is:

$$C_6H_{12}O_6 + 6O_2 \rightarrow 6CO_2 + 6H_2O. \tag{5.3}$$

Glucose molecules are carried in the blood stream, at typical concentrations 0.9 kg/m^3 (0.9 g/L). The actual oxidization of glucose in the human body occurs in the presence of the phosphates and ADP and produces ATP. This oxidative phosphorylation reaction includes the activated phosphate and oxygen. The product is ATP and the byproducts are water and carbon dioxide, as in the following overall reaction:

$$C_6H_{12}O_6 + 6O_2 + 30\text{Phosphate} + 30\text{ADP} \rightarrow 6CO_2 + 6H_2O + 30\text{ATP}. \quad (5.4)$$

There are several intermediate stages of this overall reaction including [25]:

1. Breakdown of glucose to pyruvic acid and NADH, a process called *glycolysis*.
2. Breakdown of the pyruvic acid and production of NADH and $FADH_2$.
3. Respiration process in the mitochondria that entails the oxidation of NADH and $FADH_2$.
4. ATP generation in the mitochondria membranes.

The exergy input (of glucose) in this reaction is approximately 2,955 MJ/kmol [25],[6] and the exergy output is 1,707 MJ/kmol (the exergy of each ATP kmol is 56.9 MJ higher than that of ADP). The difference of 1,248 MJ/kmol is the exergy destruction in the oxidation process that metabolizes glucose in the blood stream to produce ATP. The exergetic efficiency of this metabolic process is 1,707/2,955 or 57.8%.

The overall oxidation of fat molecules also produces carbon dioxide and oxygen. For example, the complete oxidation of palmitic acid, a fat with molecular weight 256 kg/kmol, is:

$$C_{16}H_{32}O_2 + 23O_2 \rightarrow 16CO_2 + 16H_2O. \quad (5.5)$$

The changes of enthalpy and Gibbs' free energy for this reaction are: $\Delta H^0 = -10,017$ MJ/kmol and $\Delta G^0 = -9,994$ MJ/kmol [25].

Fat in the animal and human bodies is a medium for energy storage rather than energy conversion. The body primarily uses carbohydrates for energy production and stores fat to be used when carbohydrates are not available. The several types of fat molecules are metabolized by the mitochondria in the human and animal cells to produce ATP. The overall metabolic reaction of the palmitic acid is:

$$C_{16}H_{32}O_2 + 23O_{23} + 106\text{ADP} + 106\text{Phosphates} \rightarrow 16CO_2 + 16H_2O + 106\text{ATP}. \quad (5.6)$$

The exergy input from the palmitic acid is 9,994 MJ/kmol; the exergy output in the form of the ATP is 6,030 MJ/kmol higher than the ADP exergy [25]; and the exergy destruction in this metabolic process is 3,964 MJ/kmol. Therefore, the exergetic efficiency for the metabolism of palmitic acid is 60.3%.

The calculated exergetic efficiencies of the other types of carbohydrates and fats are very similar to those of glucose and palmitic acid – close to 60% [26].

Proteins are parts of tissue, hormones, and enzymes. In the human body, proteins are created by 20 amino acid molecules, which also contain nitrogen in addition to carbon, oxygen, and hydrogen. The combination of amino acids creates the long macromolecules of proteins, which can be in a rigid state – composed of folded macromolecules – or a flexible state – composed of unfolded, long macromolecules. Humans and animals

---

[6] This value slightly differs (by 2.8%) from the number in [1], which appears in Table 5.1, because it was calculated at the environment of the human cell [24]. The difference is within the uncertainty limits of biochemical processes.

need a continuous supply of amino acids in proteins to form tissue and sustain a healthy structure. Healthy humans of 70 kg weight need 70–100 grams of protein per day to preserve their muscle and tissue mass. If the protein input exceeds this amount it may be metabolized and used as an energy source. Unlike fat, protein is readily used in tissues, enzymes, and hormones and is not stored as an energy source. If the amino acid molecules are metabolized they produce an average of 8 ATP molecules, significantly less than those produced by fats and carbohydrates. For example, albumin ($C_{72}H_{112}N_2O_{22}S$), which is a representative amino acid in protein, is metabolized according to the reaction:

$$C_{72}H_{112}N_2O_{22}S + 77O_2 + 8ADP + 8Phosphates \rightarrow 63CO_2 + 38H_2O + SO_3 \\ + 9CO(NH_2)_2 + 8ATP. \quad (5.7)$$

When albumin and other amino acids are metabolized in the human body, they produce urea [$CO(NH_2)_2$], a compound that removes the nitrogen from the blood stream and contains significant exergy. However, urea is not used by the human body. It is excreted as a waste in the urine and dissipated in the environment. The exergy input in the form of albumin in this metabolic reaction is approximately 33,150 MJ/kmol, while the exergy output of the 8 kmols of ATP – the only useful product of the oxidation for the human and animal bodies – is 455 MJ/kmol. This implies that the metabolism of proteins occurs with 32,695 MJ/kmol exergy destruction and that the exergetic efficiency of the albumin metabolism is a mere 1.4%. In this case, a large part of the exergy input is excreted as urea.

5.4.3 Respiratory Quotient

The oxidation of the digested food – with atmospheric oxygen as the input gas and carbon dioxide as the output gas – is the principal reaction that determines the energy expenditures in humans. An approximation to the total energy spent by the human body may be derived by measuring the oxygen intake and the carbon dioxide output, a process commonly used with athletes by nutritionists and trainers. The *respiratory quotient* (RQ) is defined as the ratio of the CO_2 output to the O_2 input of the body. The RQ depends on the amount of oxygen contained in the chemical constituents of nutrients and is different for the different types of food. For example, the carbohydrates oxidize as:

$$C_x(H_2O)_x + xO_2 \rightarrow xCO_2 + xH_2O, \quad (5.8)$$

and the RQ for carbohydrates is equal to 1. The molecules of fatty acids contain much less oxygen and require more oxygen for complete oxidation. The oxidation of the tri-stearin molecules is given by the reaction:

$$2C_{57}H_{110}O_6 + 163O_2 \rightarrow 114CO_2 + 110H_2O, \quad (5.9)$$

and the RQ of this fatty acid is $114/163 = 0.699$. The oxidation of palmitic acid, as in Eq. (5.5), has $RQ = 16/23 = 0.695$. Both values are approximated as 0.70.

5.4 Animal and Human Systems

The oxidation of protein, which is composed of several amino acids, is more complex because the metabolic oxidation of the amino acids entails the removal of nitrogen, carbon, and oxygen in the form of urea, $CO(NH_2)_2$. The following reaction describes the complete oxidation of albumin:

$$C_{72}H_{112}N_{18}O_{22}S + 77O_2 \rightarrow 63CO_2 + 38H_2O + SO_3 + 9CO(NH_2)_2. \quad (5.10)$$

The RQ for the oxidation of this protein is $63/77 = 0.818$. The values 1.0, 0.7, and 0.8 are accepted as average values for the RQ of carbohydrates, fats, and proteins respectively [27].

Based on the oxygen intake and the carbon dioxide output – variables that are easy to measure in controlled environments – one may determine the amount of energy produced by the oxidation reactions of food as units of energy per unit volume of O_2 consumed and per unit volume of CO_2 produced. The following Table 5.5, with values calculated from data in [27], shows the energy equivalent values of the RQ for carbohydrates and fats. Equivalent values for protein inputs are typically given in graphical form [27].

Example 5.5: The hourly O_2 intake of a person is 30.3 L and the CO_2 output is 26.1 L. Determine the calories this person has consumed during the hour of the measurements.

Solution: The RQ of this person is $26.1/30.3 = 0.86$. From Table 5.5, by interpolation, we obtain that the energy equivalent values for the O_2 intake to be 4.875 kcal/L, and for the CO_2 output 5.669 kcal/L. Using the O_2 value, the energy expenditure of this person is: $4.875 * 30.3 = 147.7$ kcal, and using the CO_2 value, the energy expenditure is $5.669 * 26.1 = 148.0$ kcal.

It is observed in this example that the energy expenditure values obtained from the two methods are almost identical and well within the uncertainty limits of biological energy processes and the uncertainty limits of the gas measurements.

Table 5.5 Energy equivalent of O_2 intake and CO_2 output for nonprotein foods.

RQ	O_2 Intake		Air Intake		CO_2 Output	
	kcal/L	kJ/L	kcal/L	kJ/L	kcal/L	kJ/L
0.71	4.686	19.606	0.984	4.117	6.629	27.736
0.75	4.739	19.828	0.995	4.164	6.319	26.439
0.8	4.801	20.087	1.008	4.218	6.001	25.108
0.85	4.862	20.343	1.021	4.272	5.721	23.937
0.9	4.924	20.602	1.034	4.326	5.378	22.502
0.95	4.985	20.857	1.047	4.380	5.247	21.953
1	5.047	21.117	1.060	4.434	5.047	21.117

5.4.4 Exergy of the ATP to ADP Conversion Process

It quickly becomes apparent in a simple analysis of energy and exergy in biochemical processes that the physical part of the exergy – the exergy associated with higher temperature, potential energy, and kinetic energy – is not as important in the computations as the chemical exergy is. Potential and kinetic energies of humans and animals almost vanish and the thermal exergy associated with the slightly higher than T_0 temperature of the human body contributes very little to the exergy of the biochemical compounds and processes. The chemical energy is the defining component of the exergy of the several molecules that participate in these processes.

Using the data in Table 2.2 for fuels and the state of technology of semipermeable membranes for mechanical power production, it was stipulated in Eq. (2.32) that the exergy change associated with industrial fuel reactions in the terrestrial environment is approximated with the Gibbs' free energy, $-\Delta G^0$. This approximation does not apply to the biochemical reactions, because they occur in living cells and the membranes of the cells are functional semi-permeable membranes, through which the biochemical molecules diffuse. For biological systems the "environment," where the biochemical reactions occur, is not the terrestrial atmosphere but the animal cells that have different constituents and entirely different composition. Consequently, it is the chemical composition of the biological cells – and not that of the atmosphere – that must come in the calculations of the biological exergy. An example of such calculations is the exergy associated with the ATP hydrolysis reaction, which generates the energy for the animal body tissue:

$$\text{ATP} \rightarrow \text{ADP} + \text{Phosphate} \quad \text{or}$$
$$C_{10}H_{15}N_5O_3(OH)(PO_3)_3 \rightarrow C_{10}H_{15}N_5O_3(OH)(PO_3)_2 + PO_3. \quad (5.11)$$

This is one of the essential biochemical reactions that follows the reduction of food nutrients and facilitates the production of mechanical energy in animals and humans. The energetic and exergetic result of the metabolic biochemical processes is the breakdown and oxidation of glucose and fats to produce ATP. The ATP is then hydrolyzed, as in Eq. (5.11), and delivers energy and exergy to the tissues of the human body, including the skeletal muscles.

The molar exergy change associated with this reaction is calculated from the equation:

$$\Delta G^0 = \tilde{E}_{ADP} + \tilde{E}_{Ph} - \tilde{E}_{ATP}. \quad (5.12)$$

A pertinent study offers a comprehensive method for the calculation of the exergy of this reaction by considering the several main processes that occur in the cellular environment, in addition to the processes that would occur in the atmospheric environment [25]. Actually – and because of their higher concentrations in the cells relative to their concentration in the terrestrial environment – the exergy of all biochemical compounds would be higher in the atmosphere than their exergy in the cells. However, this higher exergy is not available to processes inside the intracellular solution, because the biochemical processes take place in the cell "environment" and

Table 5.6 Exergy effects and molar Gibbs' free energy change of bio-chemicals in the ATP-ADP reaction (units in kJ/kmol). Data from [25].

Exergy Effect	ATP	ADP	Phosphate	$-\Delta G^0$ of Effect
Pure Chemicals, excluding adenosine	3,072,400	2,087,400	1,222,100	−237,100
ΔG^0 of formation	−2,672,100	−1,794,500	−1,147,600	270,000
Ionic Interactions	−2,700	−1,900	−3,200	2,400
Intracellular Concentration	−15,600	−22,000	−17,100	23,500
Acid Dissociation	−68,900	−42,000	−32,500	5,600
Mg-ion binding	−16,000	−7,900	−600	−7,500
Totals of Exergy	297,100	219,100	21,100	56,900

not in the terrestrial environment. Table 5.6 includes the principal exergy exchanges and the effects of the processes used for the calculation of the molar exergy associated with the ATP hydrolysis reaction [25].

When the ATP hydrolysis occurs in animal cells all the effects listed in Table 5.6, take place simultaneously and must be included in the calculations. In this case the Gibbs' free energy of the reaction is obtained from the totals in the last line of Table 5.6: $-\Delta G^0 = 297{,}100 - 219{,}100 - 21{,}100 = 56{,}900$ kJ/kmol, or from the totals in the last column that yield the same result, 56,900 kJ/kmol.

5.4.5 Energy and Exergy Expenditures in the Human Body

Animal biological systems receive energy and exergy in the form of food and nutrients and either store them (usually in the form of fat) or use them in several oxidation processes, called the *metabolic* processes or *metabolism* that define and sustain life. The number of cells in the human body is on the order of 10^{14}. Each cell uses part of the energy from the metabolic processes and produces a small rate of heat, often called the *metabolic heat*. The energy conversion processes that describe metabolism are facilitated by the millions of mitochondria cells, which are parts of all body organs and tissue. The energy and exergy produced from metabolism in humans and animals are primarily spent for the following cell functions [24]:

1. The *basal metabolic rate* (BMR) is approximately equal to the energy expenditure of humans at rest and several hours after meals, when digestion has stopped. The BMR is slightly higher than the *minimum metabolic rate*, which typically occurs during sleep, and is the minimum energy expenditure for the maintenance of life. The body does not produce any mechanical work during the BMR processes. However, it produces enough energy to sustain the basic life processes of: a. Heartbeat and blood circulation; b. respiration, including renal function; and c. operation of the vital organs including the brain, the liver, and the kidneys. Approximately 67% of the energy associated with the BMR is spent in the transport of ions against concentration gradients in the human cells and the rest is used for other chemical reactions, including nutrient oxidation [24].

2. The *thermic effect of food* is an increase of the metabolic rate: it represents the energy expenditures associated with the processing of the food in the body and includes the energy spent in digestion, absorption, transport, and storage of food. Thermic energy expenditures follow the digestion of meals, account for 5–30% of the energy value (the caloric value) of food and depend on the type of the digested nutrients. Proteins induce energy expenditures 20–30% of their caloric value; carbohydrates 5–10%; and fat, which may be stored, 0–5% [24]. The thermic effect of food typically reaches a maximum one hour after a meal and decreases to zero approximately four hours after the meal.

3. The energy expenditures for *physical activity* comprise all voluntary activities outside the BMR and include the energy spent by the muscles to keep the body sitting or in an upright position. Although physical activity accounts for 20–40% of the total energy expenditure for an average person, it may be significantly less with sedentary persons. The intensity, duration, and frequency of physical activities as well as the body mass of a person affect this energy expenditure. It has also been observed – and this is considered a benefit of physical exercise – that the increased rate of metabolism, including oxygen consumption, remains elevated for long periods after the physical activity has stopped. A study conducted in metabolic chambers [28], where 10 subjects performed on average 519 kcal of mechanical work during 45 minutes showed that, after the completion of the exercise, the metabolic rates of the subjects remained at higher levels for the next 14 hours resulting in the additional expenditure of 190 kcal on average.

4. *Thermoregulation* refers to the changes in metabolism to maintain the body's core temperature. When the temperature drops and a person does not compensate for it by wearing additional layers of clothes, the metabolism increases and the body produces more heat to maintain constant internal temperature. As shown in Figure 5.6, the thermoregulation effect is very small on average, primarily because most persons adjust their clothing to compensate for the fluctuating temperature of their environment.

Thermoregulation of the human body is inherently connected with the subject of heat transfer and has been studied extensively. A model of the human body, composed of a 15-cylinder assembly, was used to simulate the thermal and respiration exchanges of humans and their environment. The exergy destruction of these interactions was correlated with the thermal comfort of individuals [29]. This study concluded that the exergy destroyed by the body is minimal at low and intermediate relative humidity, even when the temperatures are high. Higher exergy destruction was observed at the higher humidity levels of the environment suggesting that that the body may not be at its thermal comfort condition.

The energy expenditures in humans vary, depending on the physical activity of the person. Figure 5.6 shows the average of these expenditures for humans as well as the observed variability of the energy expenditures [24]. It is observed in Figure 5.6 that most of the energy intake of a person is spent in the BMR for the maintenance of basic life functions. Even athletes spend on average more energy for BMR than for the sports

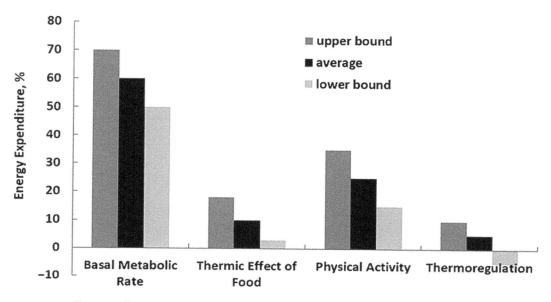

Figure 5.6 Components of energy expenditures in human bodies – average, minimum, and maximum range.

activities. All the energy expenditure processes in human bodies are irreversible and involve considerable exergy destruction.

It must be noted that, even though the human body is an open thermodynamic system, it does not receive energy continuously. The energy intake (feeding) of humans as well as of all the animals occurs intermittently, at the times of the meals. The difference between the intake energy and the energy spent is stored to be used between meals. Any long-term excess of energy intake in comparison to energy expenditure is converted to body fat, which is stored in the body. For several animals the significant increase of body fat becomes necessary to sustain life for long periods when food is scarce (e.g. during hibernation). The animal body has evolved to be one of the most efficient energy storage media. The conversion of foods to fat for storage occurs with significantly higher efficiency than the conversion to ATP, which produces mechanical work and is necessary for all physical activities.

Example 5.6: For the beef and potatoes meal of Example 5.4, calculate the upper and lower limits of the thermic effect. Assume that 8% of the food intake is excreted through the digestive system.

Solution: After accounting for the 8% loss of food in the digestive system the total carbohydrates intake becomes 0.0489 kg; the fat intake is 0.0685 kg; and the total protein intake is 0.0798 kg. With energy values 18,000 kJ/kg for carbohydrates, 40,000 kJ/kg for fat, and 22,000 kJ/kg for proteins the energy intake of these nutrients is 880 kJ, 2,740 kJ, and 1,758 kJ, respectively for a total of 5,378 kJ (1,285 Cal in nutrition units).

Given that carbohydrates induce metabolic energy expenditures 5–10% of their caloric value; fats induce 0–5%; and proteins 20–30% [24], the upper limit of the thermic effect of this food is: $0.1 * 880 + 0.05 * 2{,}740 + 0.3 * 1{,}758 = 752$ kJ and the net energy input to the human body is 4,626 kJ (1,106 Cal in nutrition units).

The lower limit of the thermic effect is: $0.05 * 880 + 0 * 2{,}740 + 0.2 * 1{,}758 = 396$ kJ and the net energy input to the human body is 4,982 kJ (1,191 Cal in nutrition units).

Example 5.7: A person at rest inhales approximately 0.4 L of air with a frequency 14 times per minute. An athlete inhales approximately 1.1 L 18 times per minute. Determine the work performed during an hour by the thoracic muscles and the diaphragm that cause this movement; the associated average power; and the rate of exergy destruction during the breathing process.

Solution: The contraction and relaxation of the muscles produce work at the constant atmospheric pressure. Since the standard atmospheric pressure is 101.3 kPa, during every breath the muscles of the resting person perform $(101.3 * 10^3 * 0.4 * 10^{-3}) = 40.52$ J of work, which correspond to average power $4.052 * 14/60 = 9.45$ W. The muscles of the athlete perform 111.43 J of work and the corresponding average power is $111.3 * 18/60 = 33.43$ W. During an hour (3,600 s) the resting person's muscles have performed 34.0 kJ of mechanical work and the muscles of the athlete 120.3 kJ. Since none of this work is recovered all the work is dissipated and, hence, the rates of exergy destruction are 9.45 W and 33.43 W.

Example 5.8: The human body may be approximated as a cylinder of height 1.75 m and diameter 0.5 m, with the heat rate from the top and the bottom of the cylinder neglected. The interior of the human body is at 37°C and at the skin at approximately 32°C. Calculate the rate of heat transfer and the rate of exergy destruction for humans when they are standing in a wind of temperature 10°C and velocity 10, 20 and 30 km/h. $T_0 = 10°C$.

Solution: The 3 wind velocities are: 2.78, 5.55, and 8.33 m/s and the corresponding Reynolds numbers (calculated with D = 0.5 m as the pertinent length scale) are 95,205, 190,068, and 285,273. For air the Prandtl number is Pr = 0.71. The Churchill-Bernstein [30] correlation for the heat transfer from cylinders is applicable in all cases:

$$Nu = 0.3 + \frac{0.62 Re^{0.5} Pr^{0.33}}{\left[1 + \left(0.4/Pr\right)^{0.67}\right]^{0.25}} \left[1 + \left(\frac{Re}{28{,}200}\right)^{0.625}\right]^{0.8}.$$

Therefore, the Nusselt numbers for the three velocities are: 472.3, 858.2, and 1,233.5.

Since the temperature at the outer surface of the cylinder is the skin temperature (32°C), the heat transfer from the human body occurs with a temperature difference 22°C through the side surface area of the cylinder, which is: $\pi DL = 2.74$ m². The rate of heat transfer is given by the expression:

5.4 Animal and Human Systems

$$\dot{Q} = h_c A(T_{hb} - T_0) = Nu \frac{k_a}{D} A(T_{skin} - T_0).$$

Hence, the rates of heat transfer corresponding to the velocities 10, 20, and 30 km/h are: 1,423, 2,587, and 3,718 W.

The exergy lost when these rates of heat are transferred from the interior of the body to the skin and from the skin to the air are:

$$\dot{W}_{lost} = T_0 \dot{Q} \left(\frac{1}{T_{skin}} - \frac{1}{T_{int}} \right) \quad \text{and} \quad \dot{W}_{lost} = T_0 \dot{Q} \left(\frac{1}{T_0} - \frac{1}{T_{skin}} \right).$$

For the heat transfer inside the body the exergy destruction is: 21.3, 38.8, and 55.6 W. For the heat transfer from the skin to the air the exergy destruction is: 102.6, 186.6, and 268.2 W.

It must be noted that the calculated rates of heat are high and cannot be comfortably sustained by humans for a long time. Under such conditions of cold windy days, humans will add layers of insulation around their skins (clothes) to further reduce the average temperature difference of the heat convection process.

Example 5.9: With the same model of the human body as in Example 5.8 calculate the heat transfer and rate of exergy destruction for a sailor who has fallen into the North Atlantic Ocean, which is at an average temperature of 7°C. The sailor remains stationary, waiting for the rescue boat. Also, because the head is out of the water, the sailor's water-immersed height is 1.50 m.

Solution: The body of the sailor is modeled as a vertical cylinder with $D = 0.5$ m and $L = 1.5$ m. The temperature of the cylinder is 32°C. With the water temperature at 7°C, there is heat transfer by natural (free) convection to the sea water. The heat transfer from the head of the sailor to the air is very small in comparison to the heat transferred to the sea water and may be neglected. The properties of the sea water are: $\rho = 1,025 \text{ kg/m}^3$; $\nu = 1,400 * 10^{-9} \text{ m}^2/\text{s}$; $\alpha = 137 * 10^{-9} \text{ m}^2/\text{s}$; $\beta = 26 * 10^{-6} \text{K}^{-1}$; and $Pr = 10.26$; $k = 0.582$ W/mK.

The Rayleigh number for this human body in the sea water is:

$$Ra = \frac{g\beta(T_s - T_\infty)L^3}{\alpha \nu} = 199 * 10^9.$$

Since $Ra > 10^9$, the thermal plume around the sailor is turbulent and the following heat transfer correlation applies [31]:

$$Nu = \left[0.825 + \frac{0.387 Ra^{1/6}}{\left[1 + \left(0.492/Pr \right)^{9/16} \right]^{8/27}} \right]^2 = 842 \Rightarrow h_{NC} = 327 \text{ W}/m^2 K.$$

The total rate of heat transfer from this immersed body is: $h_{NC} * \pi DL * (T_s - T_\infty) = 19,252$ W and the exergy destruction is given by the equation:

$$\dot{W}_{lost} = T_0 \dot{Q} \left(\frac{1}{T_0} - \frac{1}{T_{skin}} \right) = 1,578 \text{ W}.$$

It is observed in this example that a human immersed in water transfers a great deal more heat to the water than a human in the wind. The heat transfer rate of 19,252 W to the sea cannot be sustained for long. If the sailor is not rescued fast, he/she will die of hypothermia.

5.4.6 Physical Activity – Work Production

Physical and sport activities by the human body are enabled by the functions of the skeletal muscular system. This is an elaborate mechanical system with excellent and fast responding movement control, directed by the human brain via the nervous system. The characteristic response time of the skeletal muscular system is on the order of 1 ms. Muscle movement and control are facilitated by the antagonistic function of muscle pairs: when one muscle contracts, its antagonist paired muscle expands, and vice versa. For example, when the biceps at the front of the arm contract to generate a force and lift a weight, the triceps at the back of the arm expand to add to that force. In addition to the force, the activation of the two sets of muscles provides excellent control for the motion that lifts the weight. ATP in the blood stream plays the central role in the generation of forces by the muscular system. When the nervous system initiates action, a group of biochemical reactions and mechanisms come to play with the end result being the generation of forces and the performance of mechanical work.

The skeletal muscles are composed of long fibers (the muscular cells) that contract and expand. Skeletal muscle fiber contraction and expansion is induced by the translocation of Ca^{2+} ions across the *sarcoplasmic reticulum* membrane. It is this contraction and expansion of the muscular fibers that generates the forces and the performance of mechanical work. The Ca^{2+} ion movement across the sarcoplasmic reticulum membrane occurs against the concentration gradient of the ions and requires energy input, which is supplied by the ATP in the muscle tissue by the following mechanism: Initially, the ATP in the muscles is strongly bound to the *myosin,* a protein that forms the muscle fibers that contract or expand. This is referred to as *energy trapping.* When the muscles receive the command to contract or expand by the nervous system, the bound (trapped) ATP is hydrolyzed and converted to ADP; the energy released by the ATP-to-ADP reaction in the myosin is converted into translational energy that induces the Ca^{2+} ions movement; and the movement of these ions through the sarcoplasmic reticulum causes the contraction and expansion of the muscles and the generation of the forces that perform mechanical work [32].

The blood stream continuously supplies the muscles, with ATP, which is produced by the *aerobic glycolysis* of Eq. (5.4). Normally, aerobic glycolysis occurs when there is sufficient oxygen supply to convert the hydrocarbon to oxygen and carbon dioxide. However, at the initial stages of intense physical activity and during periods of maximum effort, there is not sufficient supply of oxygen in the blood stream. In this case *anaerobic glycolysis* occurs, during which glucose is converted to lactic acid with the production of only two molecules of ATP:

$$C_6H_{12}O_6 + 2\text{Phosphate} + 2\text{ADP} \rightarrow 2CH_3CH(OH)COOH + 2\text{ATP}. \quad (5.13)$$

Because only two molecules of ATP are produced during the anaerobic glycolysis, the exergy output of this reaction is 113.8 MJ/kmol, and the exergetic efficiency of the process is only 3.85%. Anaerobic glycolysis is an effective process for fast energy production by the muscular system during short and very intense efforts. It provides energy for a short period of time – typically from a few seconds to 2 minutes – as well as when maximum effort is needed (e.g. during weightlifting or during a sprint). The glucose for the anaerobic glycolysis comes from stored glucose in the muscles, which is called *glycogen*, and to a lesser extent, from glucose circulating in the blood. Because during anaerobic glycolysis a significant amount of H^+ ions are produced, which lower the pH of the blood stream, the human body quickly switches to the aerobic glycolysis: the rate and volume of respiration increase, and the lungs supply a higher amount of oxygen to the blood stream, which is immediately circulated in quantities sufficient for the aerobic glycolysis to take place as in Eq. (5.4).[7] The aerobic and anaerobic production of ATP during the commencement of a strenuous exercise activity is shown in Figure 5.7. It can be seen that the inefficient anaerobic process dominates during the first few seconds of the activity, its contribution drops exponentially and the more efficient aerobic process takes over and supplies most of the energetic ATP molecules for the remaining duration of the process [24]. A glance at Figure 5.7 proves that the fraction of the ATP produced through anaerobic glycolysis may be approximated by an exponential function:

$$\frac{ATP_{an}}{ATP_{total}} = e^{-t/\tau} \quad \text{and} \quad \frac{ATP_{aer}}{ATP_{total}} = 1 - e^{-t/\tau}. \tag{5.14}$$

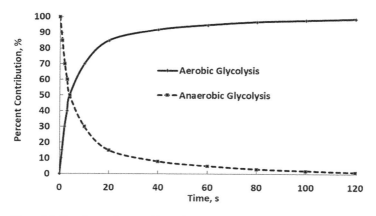

Figure 5.7 Fractions of anaerobic and aerobic production of ATP during the first two minutes of a strenuous physical activity.

---

[7] For earlier terrestrial organisms (before oxygen was at high concentration in the atmosphere) anaerobic glycolysis is considered to have been the primary process of energy production.

In the graphs of Figure 5.7, the time constant, τ, is approximately 8 s. The very low efficiency of anaerobic glycolysis and the fact that the produced lactic acid in muscles gives a sense of fatigue are two of the reasons athletes "warm up" prior to strenuous physical activity. Data from a controlled experiment [33] show that the overall exergetic efficiency of persons subjected to partially anaerobic ATP production, such as weight-lifters, sprinters, and athletes at the brink of exhaustion, is in the range 4–6%.

The preferred primary food source for the ATP production in the human body is carbohydrates, followed by fats. The least preferred energy source is the amino acids in proteins that have lower metabolic efficiency. Significant consumption of amino acids occurs at the end of long endurance events (e.g. the Olympic marathon and the 50 km walking race), when the supply of carbohydrates is almost depleted. A parameter that is often used for the quantification of the physical activity and the production of ATP is the maximal oxygen consumption, (VO_2-max). This is the maximum amount of oxygen the human body can process to produce muscular energy and is often referred to as the maximum aerobic capacity of an individual. At low levels of VO_2-max the human body primarily uses fat from the blood stream and the muscles (triglycerides) with small amounts of energy derived from muscle glucose (the glycogen) and from the blood stream to produce the needed ATP. When the physical activity becomes more intense and the VO_2-max percentage increases, the contribution of fat decreases and the contribution of the glucose from the muscles and the blood stream increases, almost linearly. Figure 5.8 shows the energy expended by an adult of 70 kg from the fat and carbohydrates in the blood and the muscular system as a function of how strenuous the physical activity is, for an activity that lasts at least 30 minutes [34]. The strength of the physical activity is measured as a percentage of the oxygen consumption compared to the maximum oxygen consumption for the person, VO_2-max. It is observed that when the physical activity is not strenuous (e.g. at 25% of VO_2-max, which corresponds to walking) the fat in the blood stream and the muscles contributes most of the power, approximately 68%. The contribution of the fat to the total power produced decreases when the physical activity is strenuous, and becomes approximately 30% at 85% of VO_2-max (for most persons this corresponds to running at 12–14 km/hr pace or playing soccer). At the high levels of physical effort, the glucose in the muscles (the glycogen) and the blood stream provide most of the energy needed, approximately 72%. During strenuous physical activity and short "energy bursts," when more power is needed and not enough oxygen is available in the blood stream and the muscles, the less efficient anaerobic glycolysis of Eq. (5.13) produces the ATP needed for the muscle movement.

It must be emphasized that the ATP is the energy carrier for the muscles and not the cause of the skeletal muscle movement.[8] The cause of movement is the contraction and expansion of the muscle fibers that are enclosed in the sarcoplasmic reticulum. Calcium ions (Ca^{2+}) in this membrane catalyze the conversion of the ATP to ADP; release the chemical energy in the muscular system; and through the muscle fiber contractions and expansions, exert the muscular forces that perform external work. At rest, the

---

[8] Because of this ATP is often referred to as the "molecular currency of cells."

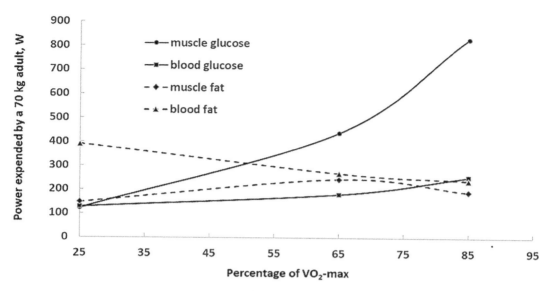

Figure 5.8 Energy expended by a 70 kg adult from fat and carbohydrates in the blood and muscular system during a physical activity of at least 30 minutes. Data from [34]

concentration of Ca^{2+} in the human muscles is approximately 10^{-7} molar. The concentration increases to 10^{-5} molar when the muscle is stimulated by the nerves. This implies that there is a significant Ca^{2+} movement against a concentration gradient, an irreversible process that requires exergy input and is often referred to as the Ca^{2+} *pump of the sarcoplasmic reticulum* [32, 35, 36].

Not all the Gibbs' free energy in the ATP is converted to mechanical work during the contraction and expansion of the muscle fibers. Most of the free exergy in the ATP molecules is consumed by the following processes:

1. ATP and ion transport through the sarcoplasmic reticulum membrane.
2. The formation of intermediate and derivative biochemical compounds.
3. The acceleration of the Ca^{2+} ions and their movement through the membranes that finally dissipates to heat.

The energy and exergy exchanges in these muscular processes are shown in Figure 5.9, which also includes the food and oxygen inputs through the blood stream. A study [32] based on animal measurements estimated that only 20% of the free energy in the ATP of the blood and muscles is converted to actual mechanical work. Studies with human subjects indicate that the muscles' efficiencies for the ATP to mechanical work conversion are higher, in the range 30–46% [37, 38]. The exact value of the muscle efficiency depends on several factors including:

1. The age of the person.
2. The total body mass of the person.

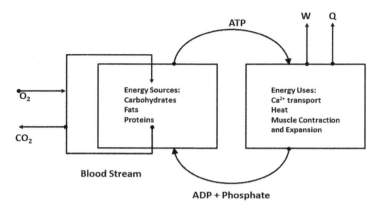

Figure 5.9 Schematic diagram of the energy and exergy exchanges in the muscular processes.

3. The training of the person on the performance of the particular muscular activity. Trained persons and athletes have higher efficiencies.
4. The type of muscular activity.

For the calculation of the overall exergetic efficiency of animals and humans in the production of mechanical work we must take into consideration the efficiency of all the other processes that produce the ATP from food products: the efficiency of food metabolism (approximately 60%); the efficiency of the thermic effect of food (close to 90%); and the loss of nutrients through the digestive system (approximately 7% loss of nutrients, which implies 93% efficiency). With the 30–46% ATP to mechanical work conversion efficiency [37, 38], we arrive at an overall efficiency in the range 15–23% for the conversion of food intake to mechanical work in humans.

This range of efficiency is also reflected in the conversion of power (as watts integrated over a period of time) to total "calories burned" in exercise equipment, which typically calculate the nutrients' conversion efficiency past the digestive track. Commercial exercise equipment calculates the power and work performed on them, in W and J or Cal (kcal), multiply this number by a factor between 4.50–6.25 and display the result either in W or, more commonly, in Cal. The range of these factors corresponds to an overall efficiency in the range 16–22%. Another effect of physical activity, which is sometimes considered in the "calories burned," is that intense physical activity increases the BMR and that a higher metabolic rate continues for several hours after the physical activity has ceased [28].

Example 5.10: One of the most strenuous Olympic events is the 50 km walking race, during which athletes must maintain contact with the ground at all times and the leading leg must be straight and remain straight until the rest of the body passes the leg. This motion causes the athletes' center of gravity to oscillate up and down and the leg muscles of athletes must supply the necessary work. The legs of an athlete are 1 m long and his stride is also 1 m. The athlete weights 65 kg and finishes the race in 4 hours.

Determine: a. the mechanical work done by the athlete during one stride; b. the total work done during the race; and c. the average power of the legs during the race.

Solution: When the athletes' legs are apart the legs and the ground distance covered (1 m) form an equilateral triangle and the height of this triangle is $1 * \sin(60°) = 0.866$ m. Therefore, during every step the center of mass of the athletes' body rises and falls by $1 - 0.866 = 0.134$ m.

a. The mechanical work done at every step is: $65 * 9.81 * 0.134 = 85.45$ J.
b. The stride and the performance of mechanical work are repeated 50,000 times during the race. Hence, the total work performed is: 4,272,255 J.
c. Since this work is performed in $4 * 3,600$ s, the average power is: 297 W.

Example 5.11: The conversion of glucose to lactic acid occurs during the first stages of physical activity and is soon substituted by the more efficient conversion to water and carbon dioxide. A person starts performing a physical activity that requires 0.025 kmols of ATP per second and has a time constant for anaerobic glycolysis 12 s. Determine the total number of kmols of ATP spent in the first two minutes of this activity and the number of kmols produced by anaerobic and aerobic glycolysis.

Solution: The total number of mols for this activity is: $2 * 60 * 0.025 = 3$ kmols.

The amount of ATP produced by anaerobic glycolysis is obtained by integrating Eq. (5.14) for $t = 120$ s, with $\tau = 12$ s: ATPan $= 0.025 * 12 * (e^0 - e^{-10}) = 0.299$ kmol. The remainder, 2.701 kmols is produced by aerobic glycolysis.

Example 5.12: Weightlifting takes place in a very short time and a great deal of the energy consumed by the muscles comes from anaerobic glycolysis. A weightlifter in the "clean and jerk" category raises a 200 kg dead weight from the ground to a height of 1.6 m and the action takes place in 5 seconds. The time constant for the anaerobic glycolysis is 6 s for this athlete and the efficiency of his muscles is 35%. Determine the total number of ATP kmols used for the lift; the numbers of kmols produced by anaerobic and aerobic glycolysis; the corresponding "calories burned;" and the efficiency of the lifting process.

Solution: The mechanical work performed with the lifting of the weights is: $200 * 9.81 * 1.6 = 3.14$ kJ. Since the athlete's muscles' efficiency is 35% and the energy released from 1 kmol of ATP is 56,900 kJ, the muscles use a total of $0.158 * 10^{-3}$ kmol ATP for this clean and jerk lift, or $0.032 * 10^{-3}$ kmol/s.

The amount of ATP produced by anaerobic glycolysis is obtained by integrating Eq. (5.14) for $t = 5$ s, with $\tau = 6$ s to obtain: ATPan $= 0.032 * 10^{-3} * 6 * (1 - e^{-5/6}) = 0.109 * 10^{-3}$ kmol. The remaining $0.049 * 10^{-3}$ kmol are produced by aerobic glycolysis. From the equations for the aerobic (5.4) and anaerobic glycolysis (5.13):

$0.109 * 10^{-3}/2 + 0.049 * 10^{-3}/30 = 0.056 * 10^{-3}$ kmol of glucose were consumed for this lift, corresponding to $0.056 * 10^{-3} * 2{,}955{,}000 \text{ kJ} = 165.48$ kJ of glucose exergy.

Assuming that the metabolic efficiency of the athlete is 60% and the efficiency of the thermic effect of food is 90%, the "calories burned" are equivalent to 306.44 kJ and this makes the efficiency of this clean and jerk lift approximately $3.14/306.6 = 1.03\%$.

Example 5.13: For the 50,000 walking race athlete of Example 5.10 determine the "calories burned" during the race, if the conversion efficiency of his leg muscles is 22%. Several experienced athletes rotate their pelvis while walking and this decreases the amplitude of the vertical oscillation of the center of gravity. For amplitude decreases of 25% and 50%, determine the "calories burned" during the race.

Solution: From Example 5.10 the total work performed by the leg muscles is 4,272.3 kJ. With an efficiency of 22%, the muscles use 19,420 kJ of the ATP free energy. Since the athlete's metabolic efficiency is approximately 60% and the thermic effect efficiency is approximately 90%, the number of "calories burned" by this athlete is: $19{,}420/(0.6 * 0.9) = 35{,}963$ kJ $= 8{,}595$ kcal.

With the circular pelvic movement, the center of gravity of the athlete's body is lifted less than the 0.134 m calculated in Example 5.10. With the 25% decrease, the amplitude of the vertical oscillation is 0.1005 m, the total work done by the leg muscles is 3,204 kJ, and the "calories burned" 6,446 kcal. With the 50% decrease, the amplitude of the vertical oscillation is 0.067 m, the total work done by the leg muscles is 2,136 kJ and the "calories burned" during the 4 hours of the race 4,297 kcal.

Example 5.14: A bicyclist goes around a 36 km loop in the country within two hours. The total weight of bicyclist and bicycle is 82 kg, the rolling resistance is 1.14%; the average aerodynamic drag coefficient is 0.88, the frontal area is A = 0.6 m^2; and the efficiency of the power transmission from pedals to wheels is 86%. Determine: a. the average power and the total work performed by the bicyclist during the entire effort; b. the "calories burned" if the muscle to work conversion efficiency of the bicyclist is 38%, the metabolic efficiency is 60%, and the thermic effect is 90%; and c. the total amount of ATP consumed.

Solution: a. The entire route is a loop, therefore there is zero elevation change. The average speed of the bicyclist is 5 m/s. The total resistance force due to the air drag and the rolling resistance of the bicycle is given by Eq. (4.20):

$$F = F_R + F_D = C_R mg + \frac{1}{2} C_D \rho A V^2 = 0.0114 * 82 * 9.81$$
$$+ 0.5 * 0.88 * 1.19 * 0.6 * 25 = 17.024 \, N,$$

and (based on the average speed of the bicyclist) the average power dissipated is 85.12 W. With the power transmission efficiency of 86%, the bicyclist's muscles must provide average power 99.0 W for 2 hours, during which time the total work performed by the bicyclist's muscles is: 712.64 kJ.

b. The "calories burned" by the cyclist are: $712.6/(0.38 * 0.6 * 0.9 * 4.184) = 830$ kcal.

c. With $\Delta G^0 = -56{,}900$ kJ/kmol (13,600 kcal/kmol) the bicyclist has consumed 0.061 kmol of ATP.

5.4.7 Overall Mass and Energy Balance – Destruction of Exergy

The mass conservation equation for the human body, which is an open thermodynamic system, is:

$$\left(\frac{dm}{dt}\right)_{hb} = \sum_i \dot{m}_i - \sum_e \dot{m}_e, \qquad (5.15)$$

where the subscripts hb, i, and e refer to the human body as a system, the inputs, and the outputs respectively. The mass inlets are primarily the mouth (food and water) and the nose (air to the lungs). Since the feeding of the human body (mass and energy input) is intermittent and the energy is continuously expended even when humans are asleep, the mass and energy of the human body fluctuate throughout any day. Of particular interest is the growth period of humans, when the mass of the body increases to the mass of an adult. During the growth period, significantly more food mass (and the energy associated with it) is ingested by the human body to account for the positive dm/dt as well as for the production of energy.

In the case of adults that maintain constant weight on average, the sum of the time-averaged mass inputs is equal to the sum of the time-averaged mass outlets:

$$\int_0^T \left(\sum_i \dot{m}_i\right) dt = \int_0^T \left(\sum_e \dot{m}_e\right) dt \Rightarrow \sum_i \overline{\dot{m}}_i = \sum_e \overline{\dot{m}}_e, \qquad (5.16)$$

where the time period, T, and the time-averaged period are long enough (e.g. one week) for the calculations not to be influenced by the periodic inputs of food.

Ingested food in the human body is essentially the fuel that provides energy for the several biological processes that sustain life, maintain a constant body temperature, and produce mechanical work. The fuel is oxidized by the air intake in a group of biochemical reactions and primarily produces the following:

1. Waste products, which are egested or excreted.
2. Metabolic biochemical products (ATP) that are used as energy sources in the organs.
3. Body heat, which is dissipated in the environment.
4. Work for the operation of the internal organs.
5. Work for all the physical activities.

The general expression of the first law of thermodynamics for the human body may be written as follows:

$$\left(\frac{dU}{dt}\right)_{hb} = \dot{Q} - \dot{W} + \sum_i \dot{m}_i h_i - \sum_e \dot{m}_e h_e. \quad (5.17)$$

Adults of constant weight are at steady state over long periods of time (e.g. one week or longer) and Eq. (5.17) yields the expression for the mechanical work performed by a human during the time period (0, T):

$$W = Q + \int_0^T \left(\sum_i \dot{m}_i h_i\right) dt - \int_0^T \left(\sum_e \dot{m}_e h_e\right) dt. \quad (5.18)$$

The mass outlets of the human body are wastes. They are not utilized and may be assigned zero energetic and exergetic values.[9] Since the work W is performed during the time period (0, T) one may define the first law efficiency and the exergetic efficiency for humans as follows:

$$\eta = \frac{W}{\int_0^T \sum_i \dot{m} h_i} \quad \text{and} \quad \eta_{II} = \frac{W}{\int_0^T \sum_i \dot{m} e_i}. \quad (5.19)$$

Figure 5.10 summarizes the several processes that occur in a human body for the production of physical activity and mechanical work. It is apparent that between 75–90% of the energy and exergy in the food are wasted before the food is metabolized to be used by the biological system. A fraction (approximately 40%) of the exergy in the digested nutrients is dissipated for the generation of the ATP during metabolism. Most of the exergy in the generated ATP is utilized to maintain the BMR of humans and a small part for thermal regulation. The remainder is used by the skeletal muscles for the production of work, with typical exergetic efficiencies in the range 30–46%. The exact value of the latter depends on the type of physical activity; and on the fitness, age, and body mass of the person performing the activity.

Table 5.7 shows the results of the simulation of the daily exergy inputs and exergy expenditures of 3 human subjects: subject A is a manual worker who is fit and produces 800 Wh (2,880 kJ) of mechanical work[10] daily; subject B uses an exercise bicycle for 1 hour and produces 150 Wh (540 kJ) of mechanical work; subject C is a sedentary person and produces a mere 15 Wh (54 kJ) of work during the day. A high level of physical activity increases the BMR [24, 28] and this is reflected in the BMR values of the Table 5.7. Also, the exergetic efficiency of the nutrients to mechanical work is higher in the 2 active persons (42% and 40%) because they are more trained to produce mechanical work than the sedentary person (38% efficiency) [24, 37]. The food input is

---

[9] A few municipalities use part of the wastes in the sewage systems to produce small amounts of methane, which are used for electricity generation. The energy and exergy of the produced methane are negligible (less by six orders of magnitude) in comparison to the energy and exergy of the foodstuff consumed in the municipalities.

[10] While the horse-power is precisely defined as 736 W, the human-power has not been precisely defined. In most calculations the human power is approximated as 100 W (0.1 kW) and, therefore, the work produced during the 8-hour workday is 800 Wh.

5.4 Animal and Human Systems

Table 5.7 Food input exergy and exergy balances for three humans with different records of physical activity and work production.

	Manual Worker	Active Human	Sedentary Human
Food input (kJ)	24,631	13,488	10,643
Wastes in digestion, $\eta = 93\%$ (kJ)	22,907	12,544	9,898
Thermic effect, $\eta = 95\%$ (kJ)	21,762	11,917	9,404
Metabolism, $\eta = 60\%$ (kJ)	13,057	7,150	5,642
Basal metabolic rate (kJ)	6,200	5,600	5,000
Thermoregulation (kJ)	0	200	500
Nutrients to muscles (kJ)	6,857	1,350	142
Mechanical work (kJ)	2,880	540	54
Exergetic efficiency (%)	11.69	4.00	0.51
Efficiency w/o BMR and thermoregulation	22.15	21.09	20.04
Exergy destruction (kJ)	21,751	12,948	10,589

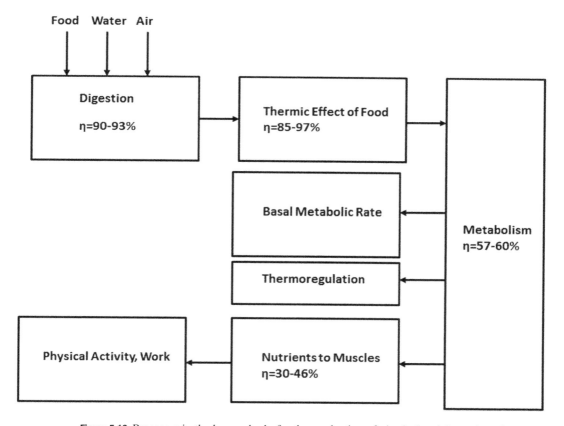

Figure 5.10 Processes in the human body for the production of physical activity and work.

determined from the sum of the daily energy expenditures after considering the efficiencies of the processes.

It is observed in Table 5.7 that almost half of the exergy from food is expended before or at the metabolism stage. This appears in the exergy losses occurred as a result of digestion (in the wastes), the thermic effect of food, and the production of ATP. After the nutrients in the food are metabolized to ATP, the basal metabolic rate consumes a large fraction of the exergy in the generated ATP. The mechanical work produced by the humans is only a small fraction of the total exergy input from foodstuff, but it is not negligible in the cases of the physically active persons. The exergetic efficiency for the production of mechanical work is low – 11.69% for the manual worker, significantly less for the others. The exergetic efficiency of humans at rest, who do not generate any useful mechanical work, vanishes.

It must be noted that the conversion from the chemical energy of nutrients to the mechanical energy of the muscles in the human body is accomplished directly and not through the production of heat. Despite the fact that the human body is a Direct Energy Conversion (DEC) device – and not subject to the Carnot limitations – the exergetic efficiencies achieved by humans are significantly lower than other DEC devices, such as batteries and fuel cells. This is an indication that humans did not evolve to produce manual work, but to think and design engines that have better exergetic efficiencies; are less strenuous to operate; and generate lesser environmental pollution.

A few studies define the exergetic efficiency of humans by excluding the necessary energy for the BMR and thermoregulation, even though the two processes are indispensable for the operation of the "human engine." The penultimate row in Table 5.7 calculates the exergetic efficiency according to this definition. It is observed again that, even if the energy spent for the two vital body functions is not included, the resulting exergetic efficiency is significantly lower than the efficiency of human-made energy conversion engines.

Example 5.15: The blood circulation rate in the human body is approximately 0.1 L/s and the blood density is 1,060 kg/m^3. The heart of a healthy person pumps blood at an average pressure difference 40 mm mercury. The efficiency of the heart muscles to convert ATP to muscular work is approximately 40%. Calculate the power produced by the heart muscles; the ATP mols consumed per day for the pumping of blood; and the corresponding daily energy input to the body for the production of this pumping work.

Solution: Blood is incompressible and, hence, the isentropic specific work needed for its circulation is $w_s = (P_2 - P_1)/\rho$. The pressure difference of 40 mm Hg is 5.33 kPa, and the specific work is calculated: $w_s = 5.03$ J/kg. With a mass flow rate $0.1 * 1,060/1,000 = 0.106$ kg/s, the isentropic power for the circulation of blood is: $5.03 * 0.106 = 5.33$ W. Since the heart is 40% efficient, the heart muscles must produce $5.33/0.4 = 13.3$ W for the blood circulation.

The daily work produced by the heart muscles is: $13.3 * 60 * 60 * 24 = 1,149$ kJ. Since for ATP, $-\Delta G^0 = 56.9$ kJ/mol, this work corresponds to $1,149/56.9 = 20.2$ mols of ATP.

Using the efficiencies for metabolism, thermal effect of food, and digestion from Table 5.7 (60%, 95%, and 93% respectively), the corresponding daily energy input for the heart muscles is: $1,149/(0.60 * 0.95 * 0.93) = 2,168$ kJ (518 kcal).

Since the heart power for the blood circulation only overcomes the frictional losses in the blood vessels (with a small amount lost as waste) all this food exergy is destructed and the corresponding energy is dissipated as heat in the human body.

5.4 Animal and Human Systems

Example 5.16: Air intake measurements have been used to determine the heat loss of a person at rest (zero work production). The person is in a room at 25°C, inhales air at this temperature and exhales at 37°C. The measurements show that the person consumes 450% theoretical air for the oxidation process and that the breathing rate is 6.2 L/min. Determine the rate of heat transferred by this person to the environment as a result of the air intake.

Solution: Since this person consumes 450% theoretical air, the oxidation reaction is:

$$C_6H_{12}O_6 + 4.5*6(O_2 + 3.76N_2) \rightarrow 6CO_2 + 6H_2O + 18O_2 + 101.52N_2,$$

In this case, the mechanical power produced and the rate of energy increase in the body vanish. Hence, the energy balance, Eq. (5.17), may be written as follows:

$$\dot{Q} = \sum_e \dot{n}_e \tilde{h}_e - \sum_i \dot{n}_i \tilde{h}_i.$$

The specific volume of air at 25°C is 24.46 m³/kmol. Therefore the volumetric air intake of 0.0062 m³/min (6.2 L/min) corresponds to $42.2*10^{-7}$ kmol/s. Because we have 450% theoretical air at inhalation, the rate of glucose consumption is $42.2*10^{-7}/(4.5*6) = 1.563*10^{-7}$ kmol/s. From the stoichiometric coefficient of the glucose oxidation reaction, the rates of carbon dioxide, water vapor, oxygen, and nitrogen are: $9.378*10^{-7}$, $9.378*10^{-7}$, $28.137*10^{-7}$ and $158.676*10^{-7}$ kmol/s respectively.

We will take as the reference temperature for the enthalpies, the room temperature of 25°C. Glucose is in the body at 37°C. Hence, the molar enthalpy of glucose is: $-1,268,000 + 12*115 = -1,266,620$ kJ/kmol.

The molar enthalpies of the input oxygen and nitrogen are zero, because they enter the body at the reference temperature.

The molar enthalpy of exiting oxygen is: $0 + 12*29 = 348$ kJ/kmol.
The molar enthalpy of exiting nitrogen is: $0 + 12*29 = 348$ kJ/kmol.
The molar enthalpy of exiting water vapor is: $-241,820 + 12*34 = -241,412$ kJ/kmol.
The molar enthalpy of exiting carbon dioxide is: $-393,510 + 12* = -393,066$ kJ/kmol.
Substitution of these quantities in the last equation yields for the rate of heat:
$-9.378*10^{-7}*393,066 - 9.378*10^{-7}*241,412 + 28.137*10^{-7}*348 + 158.676*10^{-7}*348 + 1.563*10^{-7}*1,266,620$ kW $= -0.3905$ kW.

The negative value of the rate of heat signifies that heat is transferred from the human body to the environment.

Example 5.17: The body of the bicyclist in Example 5.14 has approximately the same exergetic efficiency as that of the manual worker in Table 5.7, 11.5%. For the work effort determined in Example 5.14, and given that 72% of this work was supplied from carbohydrates and 28% from fats, determine the intake amounts of the two nutrients that would replenish the bicyclist with the exergy expended during this effort.

Solution: From Example 5.14, the total work performed by the muscles of the bicyclist is 712.6 kJ. With an exergetic conversion efficiency 11.5%, (which also includes the exergy spent for the metabolism and other effects as in Table 5.7) the exergy supplied by the food to the bicyclist is 6,197 kJ. Of this, 4,462 kJ are supplied from carbohydrates and 1,735 from fats. Using the values of Section 5.4.1 and that for fuels and nutrients $-\Delta H^0 \approx -\Delta G^0$, the bicyclist must consume $4,459/18,000 = 0.248$ kg of carbohydrates and $1,134/40,000 = 0.043$ kg of fats to replenish the nutrients spent during the 2 hours of bicycling.

5.5 Nonequilibrium Thermodynamics of Biological Systems

The growth of biological systems and the maintenance of life processes require the continuous input of material, information, energy, and exergy that are spent for the formation of biological macromolecules, which are necessary for the sustenance of all the biological activities that define life. It is apparent from the last sections that, the preservation of a thermodynamic state for biological systems requires a large number of coupled metabolic reactions and transport processes with complex mechanisms that control the rate and timing of life processes. To the inexperienced the life processes appear to violate the second law of thermodynamics, because the state of the living biological systems appears "highly organized" with lower entropy than the entropy of its constituents. This is not the case because the high degree of organization is achieved at the expense of higher quantities of energy and exergy in the group of coupled biochemical reactions and the other biological processes [39].

Let us consider a system of n coupled chemical reactions, including the reactions for the ionization and transport of electrons and ions, taking place in a biological system at temperature T. There is a total of k species involved in the n coupled reactions. For these coupled processes we may use the theory of nonequilibrium thermodynamics (NET) and Eq. (2.54) to express the rate of exergy destruction – or, equivalently, the rate of entropy production – in the general form as the product of fluxes and their conjugate forces:

$$\dot{W}_{lost} = T\dot{\Theta} = T\sum_{i=1}^{n} J_i F_i \geq 0. \tag{5.20}$$

For any single reaction, r, the flux, J, is the difference between the forward and backward rates of the reaction [39]:

$$J = r_f - r_b, \tag{5.21}$$

and the corresponding conjugate force is the affinity of the reaction [40, 39]:

$$F = \frac{A}{T} = \frac{1}{T}\left[\sum_p v_p \mu_p - \sum_r v_r \mu_r\right] = R\ln\frac{r_f}{r_b}, \tag{5.22}$$

where μ is the chemical potential of the species in the reaction, and ν is the stoichiometric coefficient corresponding to this species. The subscripts p and r denote products and reactants of the reaction. Substituting the backward rate, r_b, from Eq. (5.21) into Eq. (5.22) yields the following expression for the reactant flux J_i:

$$J = r_f \left[1 - \exp\left(-\frac{A}{RT}\right) \right]. \tag{5.23}$$

When the reactions are close to thermodynamic equilibrium the affinities are very low [40], $A \ll RT$, and Eq. (5.23) may be approximated as:

$$J \approx r_f \frac{A}{RT} \Rightarrow \frac{A}{T} \approx R\frac{J}{r_f}. \tag{5.24}$$

With the n coupled reactions in the biological system, the flux of any species, i, is the sum of the fluxes of that species produced or consumed in all the reactions:

$$J_i = \sum_{j=1}^{n} r_{fij} \frac{A_j}{RT}, \tag{5.25}$$

where the summation of the index j is over all the n reactions. A comparison of Eqs. (5.22), (5.25) and (2.52) proves that the phenomenological coefficients, L_{ij}, for the n coupled chemical reactions are:

$$L_{ij} = \frac{r_{fij}}{R}, \tag{5.26}$$

where the coefficient r_{fij} is the forward rate of the jth reaction that results in the production or consumption of the ith species. Since there are k species involved in the system of reactions, the condition of positive entropy generation that stems from Eq. (5-20) is:

$$\dot{W}_{lost} = T\dot{\Theta} = T\sum_{j=1}^{n}\sum_{i=1}^{k} \frac{A_i}{T} \frac{r_{fij}}{R} \frac{A_j}{T} \geq 0. \tag{5.27}$$

This condition ensures that the system of the n reactions with k components proceeds according to the second law of thermodynamics. It also implies that the cross coefficients L_{ij} and r_{fij} are positive definite. In addition, we also have the conditions from the Onsager relationships of Eq. (2.55):

$$L_{ij} = L_{ji} \quad \text{and} \quad r_{fij} = r_{fji}. \tag{5.28}$$

The NET methodology has been applied to the process of oxidative phosphorylation (OP) in the inner membranes of the mitochondria of animal cells for the production of the ATP [41]. Equations (5.4), (5.6), and (5.7) are the overall OP reactions for the oxidation of carbohydrates, fats, and proteins. As it is apparent from Section 5.4.2 there is a large number of secondary reactions in the processes of metabolism that involve groups of electron carriers, several types of biochemical molecules, and side reactions that catalyze and facilitate the complex OP process. The complete modeling of all the chemical reactions, reactant and product transfers, and energetic balances in the entire set of the OP reactions is a difficult task and has not been accomplished yet.

Instead of attempting to model the entire set of reactions in OP, one may take a phenomenological approach and recognize that oxygen is the dominant input in all the OP reactions (glucose, fat, and protein are ingested and exist in the human body) and that ATP is the output species of interest. Therefore, the complex system of the OP processes and biochemical reactions may be phenomenologically described by two fluxes, one for oxygen, J_O, and the second for ATP, J_{ATP} [41]. Their conjugate forces are denoted by the variables F_O and F_{ATP} respectively and the constitutive equations that connect the conjugate fluxes and forces in this phenomenological model are:

$$J_{ATP} = L_{11} F_{ATP} + L_{12} F_O$$
$$J_O = L_{21} F_{ATP} + L_{22} F_O$$
(5.29)

The tensor L_{ij} is symmetric, $L_{12} = L_{21}$ according to the Onsager principle. Hence, the positive entropy production (or positive exergy destruction) as expressed in Eq. (5.20) yields the following inequality:

$$L_{11}(F_{ATP})^2 + 2L_{12} F_{ATP} F_O + L_{22}(F_O)^2 > 0,$$
(5.30)

for all the values of the two forces, F_{ATP} and F_O. This implies that the end coefficients are positive and the discriminant of the polynomial is negative [40]:

$$L_{11} > 0, \quad L_{22} > 0, \quad \text{and} \quad (L_{12})^2 - L_{11}L_{12} < 0.$$
(5.31)

For all the systems of coupled chemical reactions, one may introduce the *degrees of coupling tensor*, which denotes the relationships among the coefficient of any two fluxes. In the general case this coefficient, ψ_{ij}, is defined as follows [41]:

$$\psi_{ij} = \frac{L_{ij}}{\sqrt{L_{ii} L_{jj}}}.$$
(5.32)

For the two fluxes that describe the OP in Eq. (5.29) and because of the Onsager symmetry relations of Eq. (5.28), the degrees of coupling tensor reduces to a single coupling coefficient: $\psi_{12} = \psi_{21} = \psi = L_{12}/(L_{11} L_{22})^{1/2}$. The positive definite condition for the entropy production implies that $\psi \leq 1$.

The ratio J_{ATP}/J_O, is often referred to as the *conventional P/O ratio* in the OP reaction. The ratio of the two generalized forces, F_{ATP}/F_O, is referred to as the ratio of the *phosphate potential to the redox potential* and it is negative.

The conventional P/O ratio, denoted as [P/O], may be obtained from Eq. (5.29) as follows:

$$[P/O] = \frac{J_{ATP}}{J_O} = \frac{L_{11} F_{ATP} + L_{12} F_O}{L_{21} F_{ATP} + L_{22} F_O} = \sqrt{L_{11}/L_{22}} \frac{\psi + \left(\dfrac{F_{ATP}}{F_O}\right)\sqrt{\dfrac{L_{11}}{L_{22}}}}{\psi \left(\dfrac{F_{ATP}}{F_O}\right)\sqrt{\dfrac{L_{11}}{L_{22}}} + 1}.$$
(5.33)

First law and exergetic efficiencies may be defined for the ATP production process in terms of the energy and exergy inputs and outputs. One may define an efficiency for this reaction based on the input and output rates [39, 42]:

$$\eta = \frac{\text{output energy rate}}{\text{input energy rate}} = -\frac{F_{ATP}J_{ATP}}{F_O J_O}. \tag{5.34}$$

The efficiency may be calculated in terms of the conventional P/O ratio, the coupling coefficient and the generalized force ratio as [39, 41, 42]:

$$\eta = -\frac{\psi + x}{\psi + 1/x} \quad \text{with} \quad x = \left(\frac{F_{ATP}}{F_O}\right)\sqrt{\frac{L_{11}}{L_{22}}}. \tag{5.35}$$

with x in the range $(0, -\psi)$. For a given coupling coefficient, ψ, the efficiency of the OP is maximized when:

$$\frac{\partial \eta}{\partial x} = 0 \Rightarrow x = \frac{-1 + \sqrt{1 - \psi^2}}{\psi} = -\frac{\psi}{1 + \sqrt{1 - \psi^2}}. \tag{5.36}$$

Note that the P/O ratio is always negative because the phosphate potential and the redox potential have opposite signs. Under this condition, the maximum efficiency of the PO process is:

$$\eta_{max} = \frac{\psi^2}{\left(1 + \sqrt{1 - \psi^2}\right)^2}. \tag{5.37}$$

It is observed that the maximum efficiency of the OP is only a function of the coupling and that it cannot exceed the coefficient, ψ. Figure 5.11 shows the variation of the OP efficiency with the modified generalized force ratio, x, for several values of the coupling coefficient, ψ. The maxima in these curves correspond to the values of the force ratio x given by Eq. (5.36).

Example 5.18: The efficiency of the OP processes for the production of ATP from glucose, palmitic acid, and albumin was calculated as 57.8%, 60.3%, and 1.4% respectively. Assuming that ATP is produced at maximum efficiency, determine the degree of coupling in the two phenomenological expressions of Eq. (5.29).
Solution: Eq. (5.37) may be solved for the coupling coefficient, ψ, to yield:

$$\psi = \frac{2\sqrt{\eta_{max}}}{1 + \eta_{max}}.$$

Substituting the values for the three efficiencies we obtain: $\psi = 0.964$ for glucose; $\psi = 0.969$ for palmitic acid; and $\psi = 0.233$ for albumin.

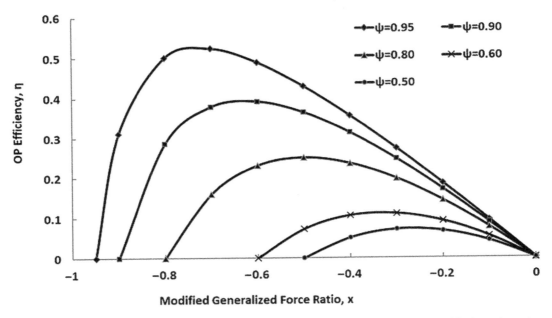

Figure 5.11 Exergetic efficiency of the OP process. The maximum efficiency highly depends on the coupling coefficient, ψ.

5.6 Entropy Production and Exergy Destruction in Humans

The laws of thermodynamics are scientific principles that have explained the operation of thermal engines; have paved the road to numerous improvements in the operation and significant efficiency increases for thermal engines; and have enabled us to precisely define the mechanical work an energy source may generate through the concept of exergy. It was inevitable that the laws of thermodynamics would be applied to predict aspects related to human life and in particular, growth, aging, life expectancy, causes of longevity, and better life quality. Prigogine and Wiame [43] were among the first to attempt the application of the principles of NET to biological and life processes in a short and mostly descriptive paper. They postulated three principles (hypotheses) that may be summarized as follows: a. The biological systems, in general, evolve toward stable steady states; b. the steady states correspond to the internal generation of minimum entropy production; and c. during the process of evolution, the rate of entropy production is minimum.

A shortcoming of this postulate from a theoretical standpoint is that the minimum rate of entropy production has been vaguely described and has never been numerically defined. The second law of thermodynamics allows for any nonequilibrium system to evolve with vanishingly low rate of entropy production. As a consequence, the minimum rate of entropy of a biological system and its environment – two subsystems that together constitute an adiabatic system – may approach zero. Minima and maxima are defined when one or more constraints are satisfied (e.g. for the stability of materials at constant energy, the entropy must be maximum; or at constant entropy the energy must be minimum). A minimum and finite rate of entropy production in biological systems must be related to a constraint. Neither the constraint(s) nor the value of the minimum have been defined by the postulate in [43]. More recent studies that use the NET theory [23, 39, 44, 45] do not define what is the minimum entropy production and do not specify any constraints. Also, more recent work by Prigogine on biochemical evolution [46] points to the fact that structural instabilities during molecular evolution give rise to new processes that tend to increase the number of chemical reactions and chemical species and also increase entropy production. Furthermore, the authors state " ... This is in contrast to the near-equilibrium situations ... in which the entropy production tends to a minimum."

The empirical evidence accumulated since the first publication of the minimum entropy production hypothesis, in 1946, does not support the hypothesis. Several experimental studies indicate low exergetic efficiencies (implying significant entropy production) in biological systems: as it was seen earlier in this chapter, the exergetic efficiency of photosynthesis in plants is in the range 3–7% while the exergetic efficiency of human subjects (Table 5.7) is in the range 0–12% – both values by far lower than the exergetic efficiencies of engines. The low exergetic efficiencies of the biological systems imply that the rate of entropy production associated with them is far from being a minimum.

A group of more recent studies simulated the behavior and evolution of a population composed of two types of exponentially growing *self-replicators*. The simulations revealed that organized and kinetically stable structures evolve from the replicators and subsequently persist because their formation is accompanied by additional work absorption and higher energy dissipation [47]. The several simulations, which are based on the principles of NET, indicate that when a group of particles is driven by an external energy source (e.g. solar energy) and is surrounded by a heat reservoir (e.g. the terrestrial atmosphere), the group will restructure itself in order to dissipate increasingly more energy. This is referred to as the *dissipation-driven adaptation hypothesis of abiogenesis* [47]. According to this simulation-supported hypothesis it may be concluded that – contrary to the evolution of engines – the biological systems have evolved from the primordial soup to waste energy (and exergy). This conclusion is incongruent with the minimum entropy dissipation in [43].

Several other experimental studies have attempted to measure the entropy production in humans. An experimental study [48] that carefully defined the biological systems (several human subjects of all ages and of both sexes) and their environment (a closed chamber at temperatures in the range 24–34°C with controlled air circulation) concluded that the rate of entropy production by the human subjects does not appear to achieve a minimum or to level off during the lives of men and women. Based on measurements, the study promulgates a three-stage hypothesis for the rate of entropy production by humans: During the embryonic and early growth stage, the rate of entropy production per unit surface area of the body increases rapidly; after the age of 2 it decreases, also rapidly; and after the age of 25 the rate of human entropy production per unit area decreases gradually. This study also asserts that similar trends have been observed with the entropy production rates of ecological systems [48].

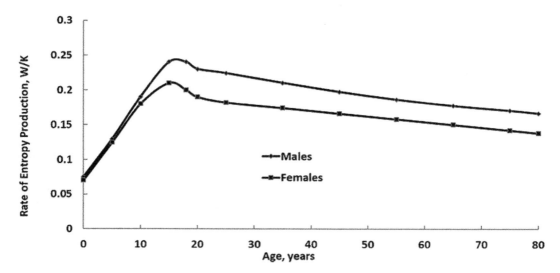

Figure 5.12 Entropy production rates for male and female subjects in Japan. Data from [49]

One observes that the trends in [48] are significantly influenced by the lesser surface area of the human body at the embryonic and early growth stages. In a companion study [49] the same author reported the total rate of entropy production of humans, it appears that the entropy generation in humans increases during the growth period (zero to approximately sixteen years for females and eighteen years for males) and then gradually decreases. The data of the last study are reproduced in Figure 5.12, where the maximum entropy production at the end of the growth period for males and females is easily observed. The observed decline of the rate of entropy production with age is a consequence of reduced physical activity by older persons, rather than an indication that biological processes occur more efficiently in the older subjects of the study.

Since the life expectancy for males and females in Japan is 80 and 82 years respectively, one calculates from the data of Figure 5.12 that the lifetime entropy production of a Japanese male who reaches the age of 80 is approximately 480 MJ/K and that of a female who reaches the age of 82 is approximately 415 MJ/K. The trends in this Figure 5.12 and considerations of the implicit uncertainty of the averaged data also indicate that it is not possible to accurately determine the "aging of a person" based solely on the entropy production. This, because human health and "aging" depend on a multitude of factors that are not accounted by a single variable, the entropy production or, equivalently, the exergy destruction.

5.6.1 Entropy Production, Life, State of Health, and Life Expectancy

Based on the estimated or inferred rate of entropy production for humans, there is a school of thought that humans have the capacity to produce only a certain amount of entropy during their lifetime and would meet death when this threshold is exceeded [23,

5.6 Entropy Production and Exergy Destruction in Humans

45, 50]. One particular hypothesis extends the lifetime entropy production to the aging of humans as well as to commercial corporations and the entire universe [51]. The lifetime entropy production was determined to be approximately 11,500 kJ/kg K for males and 10,120 kJ/kg K for females [45, 50]. Accordingly, a 70 kg male would produce approximately 800 MJ/K of entropy and a 60 kg female would produce approximately 610 MJ/K of entropy during their lifetimes. These numbers correspond to 316 GJ and 176 GJ exergy destruction for males and females ($T_0 = 293$ K) during their lifetime.

A major problem with the hypothesis of an upper lifetime entropy production limit (or exergy destruction limit) for humans, is that it does not account for factors such as heredity, prior health conditions, lifestyles, and access to good health care. When this hypothesis is quantitatively applied to models of human life and aging, it leads to erroneous conclusions, because the models under-predict the vital importance of exercise and fitness, which are considered to be determining factors of longevity by the medical profession. As a case study, let us consider the lifetime entropy production of the three individuals in Table 5.7. Following good dietary principles, the three individuals may have the same weight. With a lifetime limit of 316 GJ exergy destruction, the sedentary person would live for 82 years, the active person for 67 years, and the manual worker for only 40 years. Clearly, these are not realistic predictions for the life expectancy of the three individuals. It must be noted that a few studies [45] somehow mitigate their predictions by highlighting in a qualitative manner the role of proper diet and moderate physical activity level. However, such provisions on the hypothesis of a lifetime entropy production limit (or, equivalently, an exergy destruction limit) only serve to emphasize the importance of the missing/neglected variables in the hypothesis rather than support the hypothesis of an upper limit entropy production for humans.

A second weakness of the hypothesis is the significant discrepancy of the upper limit of entropy production (or exergy destruction) for humans: The experimental data from [48, 49] indicate an upper limit of 480 MJ/K for males and 415 MJ/K for females in Japan, while the calculations in [50, 45] show 67% higher limits for males (802 MJ/K) and 47% higher limits for females (610 MJ/K) in the USA, a discrepancy that cannot be explained by the different surface areas, weights, and life expectancies of the Japanese and American subjects in the two studies. The inconsistency of the hypothesis is exacerbated when one considers that human life expectancy in the USA has increased from 47 years for males and 49 for females in 1900, to 77 years for males and 80 years for females in 2010. Therefore, the ceilings of entropy production in 1900 were 488 MJ/K for American males and 374 MJ/K for American females. There is no reasonable thermodynamic argument that would justify the dramatic increase of the upper ceilings of entropy production (or exergy destruction) in the USA by 64% for males and 63% for females between 1900–2010.

From a careful examination of the empirical data on entropy production and exergy destruction by human subjects it quickly becomes apparent that all attempts to determine or to explain life processes, including human aging and death, through entropy or exergy calculations lead to the conclusion that the life processes are subject to a multitude of other primary variables, which depend on the historical era the lives of humans took place as well as on heredity, the system of public health (if any), and the culture where the individuals have lived. Simply put – and unlike the state of thermal engines – the state of human health, life, aging, and life expectancy are very complex biological processes to be described (even qualitatively) by a single variable, the entropy production or the equivalent exergy destruction.

A more recent analytical study [52] applied the general principles of NET as the foundation for the description of the dynamic behavior of biological systems that are constantly not in equilibrium with their environment. The study recognizes the importance of the "deviations from equilibrium" and uses the concept of internal variables to describe these deviations. It acknowledges that there is a very large number of internal variables in biological systems. Also, that heredity, access to health care, health history, and environmental influences are only a few of the parameters that may be represented as internal variables that affect evolution, health, and human life expectancy. For this reason, this analytical study does not attempt to quantify the internal variables and perform calculations that may lead to quantitative conclusions about aging and life expectancy.

In summary, entropy and exergy estimates cannot be used to derive useful explanations and conclusions for all the processes in biological systems and, more importantly, for any reliable determination about human aging, longevity, state of health, and life expectancy. Several other very important variables – related to health care and heredity – must be considered to clarify the state of health and longevity for humans and even then, any predictions are purely statistical. When applying the principles of physics and chemistry to biological systems one must recall the words of Schrodinger [22] " ... the laws of physics and chemistry are statistical throughout."

Problems

1. A 400 MW thermal power plant with 36% overall thermal efficiency is designed to operate with corn stover, the residue from corn production. The farms in that region yield 2 crops per year and each hectare produces on average 13,000 kg of dry stover with heat of combustion 17,500 kJ/kg. Based on this information alone, determine how many hectares and how many square kilometers need to be harvested for the power plant to continuously operate for a year. What other energy and exergy requirements were neglected in these calculations?

2. It is proposed that the solid wastes of a city be used for the production of electric power in a 250 MW thermal power plant of 36.5% efficiency. Based on the values of Table 5.2 what is the range of waste mass rates that must be supplied to this power plant? What kind of environmental problems for the city do you envision this project will create?

3. The concentration of glucose ($C_6H_{12}O_6$) in individuals after a meal is approximately 110 mg/L. What are the molecular concentration and the mass fraction of glucose in the blood? The density of blood is approximately 1,060 kg/m^3.

4. A student consumes for lunch an apple of 0.150 kg mass; a hamburger of 0.160 kg; and a bread bun of 60 g. Calculate the nutritional value of this meal in kJ and kcal (Cal).

5. Calculate the nutritional value of the three meals you consumed yesterday.

6. Measurements on the metabolic rate of a sleeping person show that her hourly O_2 intake is 28.4 L and the CO_2 output is 26.5 L. Determine the calories this person has consumed during the six hours of the measurements.

7. The heart of a healthy person raises the blood pressure from 78 mm Hg to 140 mm Hg during strenuous exercise. During an exercise session that lasts for 45 minutes the heart of this person pumps 38 L/min of blood. Determine the power of the heart in W and the total work performed during the exercise in kJ. The density of the blood is 1,060 kg/m^3.

8. A person consumes in a meal a 0.4 kg beefsteak, 0.3 kg of potatoes, vegetables with negligible nutritional value, 0.1 kg of bread, and a desert that contains 0.025 kg carbohydrates, 0.012 kg fat, and 0.006 kg protein. Calculate the upper and lower limits of the thermic effect of this meal. Assume that 10% of the food intake is excreted through the digestive system.

9. A marathon athlete, whose stride is 1.2 m and weighs 68 kg, completes the race (42,195 m) in 2 hours and 15 minutes. With every stride the center of gravity of the athlete is raised by 12 cm. Determine: a. the mechanical work done by the athlete during one stride; b. the total work done during the race; c. the average power of the legs during the race; and d. the "calories burned" during the race, if the conversion efficiency of his leg muscles is 36%, the thermic effect 91%, and the metabolic efficiency 55%.

10. A weightlifter lifts 220 kg of dead weight from the ground to a height of 1.75 m in an action that takes place in 5 seconds. The time constant for the anaerobic glycolysis is 6 seconds for this athlete and the efficiency of his muscles is 35%. Determine the total number of ATP kmols used for the lift; the numbers of kmols produced by anaerobic and aerobic glycolysis; the corresponding "calories burned;" and the efficiency of the lifting process (thermic effect 90% and metabolic efficiency 60%).

11. A bicyclist goes around a 50 km loop within 3 hours. The total weight of bicyclist and bicycle is 82 kg; the rolling resistance of the bicycle is 1.08%; the average aerodynamic drag coefficient is 0.95; the frontal area is A = 0.65 m^2; and the efficiency of the power transmission from pedals to wheels is 80%. Determine: a. the average power and the total work performed by the bicyclist during the entire effort; b. the total amount of ATP consumed by the muscles; and c. the "calories burned" if the ATP to work conversion efficiency of the bicyclist is 21%.

12. Based on the data in Figure 5.12 calculate the approximate entropy production of males and females in their 20s and their 50s. Do you think that these results have any significance?

References

[1] C. S. Silva, W.D. Seider, and N. Lior, Exergy Efficiency of Plant Photosynthesis, *Chemical Engineering Science*, **130** (2015), 151–71.

[2] S. Lems, H. J. van der Kooi, and J. de Swaan Arons, Thermodynamic Analysis of the Living Cell: Design of an Exergy-Based Method, *International Journal of Exergy*, **4** (2007), 339–56.

[3] R. Petela, An Approach to the Exergy Analysis of Photosynthesis. *Solar Energy*, **82** (2008), 311–28.

[4] S. Lems, H. J. van der Kooi, and J. de Swaan Arons, Exergy Analyses of the Biochemical Processes of Photosynthesis. *International Journal of Exergy*, **7** (2010), 333–51.

[5] G. Bisio and A. Bisio, Some Thermodynamic Remarks on Photosynthetic Energy Conversion. *Energy Conversion and Management*, **39** (1998), 741–48.

[6] H. A. Reis and A. F. Miguel, Analysis of the Exergy Balance of Green Leaves. *International Journal of Exergy*, **3** (2006), 231–37.

[7] Food and Agriculture Organization of the United Nations, Renewable Biological Systems for Alternative Sustainable Energy Production. In K. Miyamoto ed., *Food and Agriculture Organization of the United Nations*, Bulletin 128 (Rome, Italy: FAO, 1997).

[8] E. E. Michaelides, *Alternative Energy Sources* (New York: Springer, 2012).

[9] E. E. Michaelides, *Energy, the Environment, and Sustainability* (Boca Raton, FL: CRC Press, 2018).

[10] Royal Academy of Engineering, *Sustainability of Liquid Biofuels* (London: RAEng, 2017).

[11] D. Piementel, Ethanol Fuels: Energy Balance, Economics, and Environmental Impacts are Negative. *Natural Resources Research*, **12** (2003), 127–34.

[12] T. Patzek, Thermodynamics of the Corn-Ethanol Biofuel Cycle. *Critical Reviews in Plant Science*. **23** (2004), 519–67.

[13] H. Shapouri, J.A. Duffield, and M. Wang, *The Energy Balance of Corn Ethanol: An Update*, US Dept. of Agriculture, Agricultural Economic Report (2002).

[14] M. Wang, C. Saricks, and M. Wu, *Fuel-Cycle Fossil Energy Use and Greenhouse Gas Emissions of Fuel Ethanol Produced from U.S. Midwest Corn*, Argonne National Laboratory Report (December 17, 1997).

[15] K. J. Ptasinsky, *Efficiency of Biomass Energy: Exergy Approach to Biofuels, Power, and Biorefineries* (Hoboken, NJ: Willey, 2016).

[16] A. Hornborg, The Magic of Money and the Illusion of Biofuels: Toward an Interdisciplinary Understanding of Technology. *The European Physical Journal Plus*, **132** (2017), 82.

[17] J. W. Tester, E. M. Drake, M. J. Driscoll, M. W. Golay, and W. A. Peters, *Sustainable Energy* (Cambridge, MA: MIT Press, 2005).

[18] *Biofuel Fact Sheet*, European Biotechnology Energy Platform, www.etipbioenergy.eu/images/EIBI-7-aquatic%20biomass.pdf (last accessed in September 2018).

[19] J. Sheehan, T. Dunahay, J. Benemann, and P. Roessler, *A Look Back at the US Department of Energy's Aquatic Species Program - Biodiesel from Algae*, Report TP-580-24190 (CO: National Energy Laboratory, 1998).

[20] E. Sorgüven and M. Özilgen, Thermodynamic Efficiency of Synthesis, Storage and Breakdown of the High-Energy Metabolites by Photosynthetic Microalgae. *Energy*, **58** (2013), 679–87.

[21] B. Bharathiraja, M. Chakravarthy, R. M Kumar, D. Yogendran, D. Yuvaraj, J. Jayamuthunagai, R. P. Kumar, and S. Palani, Aquatic Biomass (Algae) as a Future Feed Stock for Bio-Refineries: A Review on Cultivation, Processing and Products. *Renewable and Sustainable Energy Reviews*, **47** (2016), 634–53.

[22] E. Schrodinger, *What Is Life? The Physical Aspect of the Living Cell* (Cambridge: Cambridge University Press, 1944).

[23] K. Annamalai, I. K. Puri, and M. A. Jog, *Advanced Thermodynamics Engineering*, 2nd ed. (Boca Raton, FL: CRC Press, 2011).

[24] S. S. Gropper and I. L. Smith, *Advanced Nutrition and Human Metabolism*, 6th ed. (Belmont, CA: Wadsworth, 2013).

[25] S. Lems, H. J. van der Kooi, and J. de Swaan Arons, The Second-Law Implications of Biochemical Energy Conversion: Exergy Analysis of Glucose and Fatty-Acid Breakdown in the Living Cell. *International Journal of Exergy*, **6** (2009), 228–48.

[26] C. E. K. Mady and S. Oliveira Junior, Human Body Exergy Metabolism. *International Journal of Thermodynamics*, **16** (2013), 73–80.

[27] W. D. McArdie, F. I. Katch, and V. L. Katch, *Exercise Physiology: Nutrition, Energy, and Human Performance*, 7th ed. (Philadelphia, PA: Kluwer, 2010).

[28] A. M. Knab, R. A. Shanely, K. D. Corbin, F. Jin, W. Sha, and D. C. Nieman, A 45-Minute Vigorous Exercise Bout Increases Metabolic Rate for 14 Hours. *Medicine & Science in Sports & Exercise*, **43** (2011), 1643–48.

[29] C. E. K. Mady, M. S. Ferreira, J. I. Yanagihara, and S. de Oliveira, Human Body Exergy Analysis and the Assessment of Thermal Comfort Conditions. *International Journal of Heat and Mass Transfer*, **77** (2014), 577–84.

[30] S. W. Churchill and M. Bernstein, A Correlating Equation for Forced Convection from Gases and Liquids to a Circular Cylinder in Crossflow. *Journal of Heat Transfer*, **99** (1977), 300–6.

[31] S. W. Churchill and H. Chu, Correlating Equations for Laminar and Turbulent Free Convection from a Vertical Plate. *International Journal of Heat and Mass Transfer*, **18** (1975), 1323–29.

[32] T. Kodama, Thermodynamic Analysis of Muscle ATPase Mechanisms. *Physiological Reviews*, **65** (1985), 468–551.

[33] G. Marques-Spanghero, C. Albuquerque, T. Lazzaretti-Fernandes, A. J. Hernandez, and C. E. Keutenedjian-Mady, Exergy Analysis of the Musculoskeletal System Efficiency During Aerobic and Anaerobic Activities. *Entropy 2018*, **20,** 119 (2018), 1–16.

[34] L. L. Spriett, and M. J. Watt, Regulatory Mechanism for the Interaction between Carbohydrate and Lipid Oxidation during Exercise. *Acta Physiologica Scandinavia*, **178** (2003), 205–43.

[35] T. Kodama, N. Kurebayashi, H. Harafuji, and Y. Ogawa, Calorimetric Studies of the Mechanism of the Ca*+-ATPase Reaction of Sarcoplasmic Reticulum. *Journal of Biochemistry*, **96** (1984), 887–94.

[36] T. Kodama, N. Kurebayashi, and Y. Ogawa, Heat Production and Proton Release during the ATP-Driven Ca Uptake by Fragmented Sarcoplasmic Reticulum from Bullfrog and Rabbit Skeletal Muscle. *Journal of Biochemistry*, **88** (1980), 1259–65.

[37] C. J. Barclay, R. C. Woledge, and N. A. Curtin, Inferring Cross-Bridge Properties from Skeletal Muscle Energetics. *Progress in Biophysics & Molecular Biology*, **102** (2010), 53–71.

[38] G. Offer and K. W. Ranatunga, A Cross-Bridge Cycle with Two Tension-Generating Steps Simulates Skeletal Muscle Mechanics. *Biophysical Journal*, **105** (2013), 928–40.

[39] Y. Demirel and S. I. Sandler, Thermodynamics and Bioenergetics. *Biophysical Chemistry*, **97** (2002), 87–111.

[40] J. Kestin, *A Course in Thermodynamics*, vol. 2 (Washington, DC: Hemisphere, 1979).

[41] J. W. Stucki, The Optimal Efficiency and the Economic Degrees of Coupling of Oxidative Phosphorylation. *European Journal of Biochemistry*, **109** (1980), 269–83.

[42] O. Kedem and S. R. Caplan, Degree of Coupling and Its Relation to Efficiency in Energy Conversion. *Transaction of the Faraday Society*, **61** (1965), 1897–911.

[43] I. Prigogine and J. Wiame, Biologie et Thermodynamique des Phenomenes Irreversibles. *Experimenta*, **2** (1946), 451–53.

[44] Y. Demirel, Exergy Use in Bioenergetics. *International Journal of Exergy*, **1** (2004), 128–47.

[45] C. A. Silva and K. Annamalai, Entropy Generation and Human Aging: Lifespan Entropy and Effect of Physical Activity Level. *Entropy*, **10** (2008), 100–23.

[46] D. Kondepudi and I. Prigogine, *Modern Thermodynamics: From Heat Engines to Dissipative Structures* (New York: Wiley 1999).

[47] N. Perunov, R. Marsland, and J. England, Statistical Physics of Adaptation. *Physical Review X*, **6** (2) (2016), 021036.

[48] I. Aoki, Entropy Principle for Human Development, Growth and Aging. *Journal of Theoretical Biology*, **150** (1991), 215–23.

[49] I. Aoki, Entropy Production in Human Life Span: A Thermodynamical Measure for Aging. *Age*, **1** (1994), 29–31.

[50] C. A. Silva and K. Annamalai, Entropy Generation and Human Aging: Lifespan Entropy and Effect of Diet Composition and Caloric Restriction Diets. *Journal of Thermodynamics*, (2009), 1–10.

[51] D. Hershey, *Entropy Theory of Aging Systems: Humans, Corporations and the Universe* (London: Imperial College Press, 2010).

[52] A. A. Zotin and V. N. Pokrovskii, The Growth and Development of Living Organisms from the Thermodynamic Point of View. *Biological Physics*, **512** (2018), 359–66.

6 Ecosystems, the Environment, and Sustainability

Summary

Energy usage by an exponentially increasing human population has created environmental problems that are stressing several ecosystems on earth. The concept of eco-exergy (which is not equivalent to mechanical work) has been used to explain the relationship between energy use and the formation of complex organisms in ecosystems. The harmonious coexistence in ecosystems has inspired the notion of industrial ecology as a paradigm for the improvement of exergetic efficiencies and complete utilization of resources. The exergy-environment nexus and the implications of exergy analyses on sustainable development are critically examined in this chapter. Environmental exergonomics, exergoenvironmental analysis that includes eco-indicators, life-cycle exergy analysis, and sustainability indices are theoretical tools that use exergy and other thermodynamic variables to define the state of the environment and to recommend industrial practices that would alleviate the detrimental effects of energy use and would promote global environmental stewardship and sustainability.

6.1 Environmental Effects of Energy Usage

The exponentially increasing global population, energy use, and industrial output have created major changes in the natural environment and ecosystems that have alarmed the earth's inhabitants. Chief among the environmental threats are industrial pollutants in the biosphere and the continuous increase of the atmospheric concentration of *greenhouse gases* (GHGs) including carbon dioxide (CO_2), methane (CH_4), and the oxides of nitrogen (NOX). The GHGs play the role of a "blanket" over the planet; they have caused the warming in the biosphere; and – according to all indications – they will bring about Global Climate Change (GCC). Other environmental impacts of energy use are: acid precipitation (acid rain) primarily generated from SO_2 emissions and produced by the combustion of sulfur contained in coal and petroleum products [1]; the ozone depletion in the stratosphere (the ozone hole), primarily due to chlorofluorocarbon (CFC) refrigerant emissions [2]; lead contamination, primarily due to the combustion of tetraethyl lead, an antiknock additive in leaded gasoline; nuclear waste, produced by the approximately 450 nuclear reactors in continuous operation worldwide; and the consumption of vast amounts of freshwater for the cooling of power plants [3].

Most national governments have adopted regulations that restrict the emissions of pollutants in the environment. As a result, several environmental threats of the late 20th century have been neutralized or reversed by public policy – either national policy or international treaties and protocols. Concerted public policy triggered the engineering advances and adoption of systems (sulfur scrubbers, fluidized bed reactors, unleaded gasoline, etc.) that have significantly reduced the atmospheric emissions of lead and sulfur dioxide and have alleviated the problems of acid rain and lead contamination [3, 4]. Public policy following the *Montreal Protocol* in 1987 resulted in the global substitution of the CFC refrigerants with hydrofluorocarbons (HFC) that do not harm the ozone layer in the stratosphere, a policy that has put a check on the ozone hole problem [3, 5].[1] Problems related to nuclear waste are addressed by governments at national and regional levels. Solutions for the long-term storage of radionuclides have been formulated and several countries have started implementing them. However, and despite the efforts of the environmental community, there is not yet a comprehensive and binding agreement for the mitigation of the global CO_2 emissions, the principal environmental threat of the 21st century. The main reason for the inability of the international community to mitigate the GHG emissions is that the fossil fuels, which produce most of the emitted CO_2, generate 79.7% of the global energy supply [6]. Unlike the environmental problems of acid rain, lead, and ozone depletion that were successfully tackled in the recent past, the elimination or significant reduction of fossil fuels from the global energy mix is very costly and will require intense international negotiations, price controls, sacrifices and, very likely, life-style changes [3]. As the most significant variable that characterizes the production and use of all energy forms, the exergy concept is intricately connected to the environmental and ecological effects of energy use and plays an important role in attaining the goals of sustainability.

6.2 Ecology and Ecosystems

Ecology is the study of the relationships of living organisms with one another and the relationship of organisms with their habitat – the natural environment. Ecology incorporates parts of the biological sciences, physics, physiology, geology, and chemistry. Central in the study of the subject is the *ecosystem* – a rather loose concept that refers to a subdivision of a landscape or of a relatively homogeneous geographic region. The ecosystem is made up of the interdependent organisms; their habitats; and all the physicochemical and ecological processes. The living organisms that constitute an ecosystem maintain a state far from equilibrium with their environment by continuously producing entropy and dissipating exergy derived from the sun and their nutrients.

---

[1] Ironically, the HFCs are potent GHGs and contribute to GCC. The Kigali agreement of 2016 [3] mandates their substitution with hydro-fluoro-olefins (HFOs) that do not harm the ozone layer and are less potent GHGs.

The study of ecosystems includes: living organisms, species, and populations; soil and water; weather, climate, and other physical factors; and physicochemical processes, such as nutrient cycles; energy flow; the carbon cycle in the biosphere; and water flow, freezing, and thawing. Ecosystems are very complex systems and, if they are considered as thermodynamic systems, their analysis would be complex and would include a large number of variables.

Ecosystems depend to a large extent on their natural environment. All environmental changes have ecological consequences. The increase of atmospheric CO_2 concentration and the expected global and regional climate changes have ecological effects, such as altered patterns of crop production, the migration of animals, and the extinction of species. The discharge of environmental pollutants has a multitude of complex ecological consequences. For example, discharges of pollutants in the subsurface do not only affect the subsurface organisms, but also may affect humans and animals when the pollutants leach into aquifers and are transported to streams and lakes, from which potable water is derived.

Ecosystems possess energy and exergy – the *eco-exergy*, – not only because they are composed of chemical substances, which are not in equilibrium with their environment, but also because their chemical components are part of organizational structures within the living organisms. In an attempt to model the degrees of organization in the creatures that comprise ecosystems several studies have defined the eco-exergy by taking into account factors for the higher organization of the molecules and add to the actual exergy of higher organisms. One of the studies [7] added to the chemical exergy of ecosystem components a factor that accounts for the complexity of the organisms, where the chemical components reside:

$$\tilde{e}_{ec} = (\mu_1 - \mu_1^{eq}) \sum_{i=1}^{N} c_i - \tilde{R}T \sum_{i=2}^{N} c_i \ln(P_{i,a}), \tag{6.1}$$

where c_i is the concentration of the biological component i in the ecosystem; μ_1 is the chemical potential of dead organic matter; μ_1^{eq} is the chemical potential of this matter when it is in equilibrium with the environment; and $P_{i,a}$ is a probability associated with the formation of the component i. $P_{i,a}$ is, in general, estimated using information embedded in the genes and the number of permutations associated with the formation of amino-acids and genes. The latter is not precisely available and is usually approximated. The difference $(\mu_1 - \mu_1^{eq})$ is known for organic matter, which is a mixture of carbohydrates, fats, and proteins [7]. This study also recognized that all ecosystems have been formed/created from inorganic matter with the addition of solar exergy and defined the concept of *emergy* – a measure of the total solar exergy consumed for the formation of organisms – as an approximate measure of the complexity of ecosystems.

A useful expression for the probability $P_{i,a}$ was estimated from the number of amino acids in an organism [8]:

$$P_{i,a} = a^{-Ng_i}, \tag{6.2}$$

where $a = 20$ is the number of possible amino acids; $N = 700$ is the approximate number of amino acids in one gene; and g_i is the number of the coding genes of an organism.

Table 6.1 Amino acid and gene parameters for several organisms; and estimates of their specific eco-exergy.

Organism	g_i	β_i	e_{ec} (MJ/kg)
Detritus	0	1	18.7
Fungus	3000	9.5	178
Plants-trees	10,000–30,000	29.6–86.8	554–1,623
Jellyfish	10,000	29.6	554
Reptiles	130,000	370	6,919
Mammals	140,000	402	7,517
Humans	250,000	716	13,389

With this closure equation for the probability of formation, a simple expression for the estimation of the specific eco-exergy was derived that is based on weighting factors [8]:

$$e_{ec} \approx \sum_{i=1}^{N} \beta_i c_i, \qquad (6.3)$$

where β_i is the weighting factor calculated from the amount of information[2] the component i carries and is estimated from the information accumulated in the genes of the organism, where the component i is found. Because of this, β_i has significantly higher values in organisms that are higher in the evolutionary scale, as it is shown in Table 6.1, which includes examples of eco-energy estimates for several species [8]:

A glance at Table 6.1 and Eqs. (6.1) and (6.2) proves that for the higher organisms – trees, mammals, humans, etc. – the eco-exergy is by far higher than the chemical exergy of the organisms, which is calculated from the first term of Eq. (6.1). The eco-exergy is not simply the maximum work that can be obtained from the chemical constituents of an ecosystem, but also includes the relative importance of the species where these constituents exist. It must be noted that the numerical value of eco-exergy is an estimate; it is dominated by the information stored in the genes of organisms; and is a measure of the relative importance of higher and lower organisms in the ecosystem.

This dominant part of eco-exergy is sometimes called the *info-exergy* and has been extended to encompass the complexity of ecosystem structures, including humans [9]. Info-exergy has been used to distinguish and differentiate species that inhabit the same ecosystem. In this case the complexity of species within an ecosystem structure is evaluated using the information accumulated by the species and the inter-relationships between the members of the species. The info-exergy for humans includes the accumulated information in the culture, customs, traditions, written and verbal histories, technology, etc. [9]. As a result, the value of info-exergy is much higher than the chemical exergy of the components that comprise the species; it is based on subjective

---

[2] The notion that information is equivalent to work (and exergy) may be traced back to *Maxwell's demon*, which uses information to violate the second law of thermodynamics and creates two systems with different temperatures that may be utilized to produce work.

stipulations; and may only be considered as an estimate. Most important, the info-exergy and eco-exergy are not related to mechanical work that may be extracted from an ecosystem.

An approach that is similar, but not equivalent, to exergy analysis for ecosystems is that of entropy flux calculation using an entropy input-output analysis [10]. This approach also serves in the relative characterization of ecosystems and their current state of evolution. Data from large ecosystems in the Amazon suggest that younger ecosystems produce lower rates of entropy. The rate of entropy production accelerates and reaches a maximum (climax), when the ecosystems mature. Afterward the rate of entropy production decreases, when the ecosystems are aging (retrogression) [11].

While the relative characterization and modeling of ecological processes and ecosystems using the concepts of exergy and entropy are of interest (and oftentimes provide useful information) one has to be cognizant of the following limitations:

1. Exergy and entropy are variables rigorously defined in thermodynamics for systems that are well-defined. Ecosystem boundaries are ambiguous and, often, indefinite. This ambiguity propagates as higher uncertainty in the calculated values of ecosystem entropy and exergy.
2. The calculation of eco-exergy is based not only on the state of a system and the state of the environment, but also on the modeled pathway for the formation of the organisms [7, 8]. The exergy associated with the latter is much higher than the actual exergy (the chemical exergy content) of organisms and significantly differs from the definition of exergy – the maximum work that may be obtained from a thermodynamic system.
3. The calculations of the permutations for the formation of the amino-acids and for the weighting factors in Eqs. (6.1) and (6.3) are approximate and subjective. The values of the derived eco-exergy are estimates with significant uncertainty.
4. Unlike classical thermodynamic systems, ecosystems continuously adapt and evolve and their organisms and colonies change in time.

These limitations imply that for a more complete and quantitative description of ecosystems several other variables must be used together with eco-exergy.

Another approach to define human effects on ecosystems is the *Ecological Footprint* (EF) – an indicator that accounts for the human demand on global biological resources [12]. The EF compares the level of consumption with the available areas of productive land and sea (the area resources) and helps draw conclusions about the sustainable nature of operations and industries in regions and nations. As with the eco-exergy and the info-exergy, the EF is also an estimate, based on subjective measures and has only been used in a relative manner. While these methods may produce a few useful qualitative results, they all suffer from the drawback that they attempt to condense the effects of several processes, inter-relationships, and evolutionary developments into a single variable. With the quantification methods of 150 original papers reviewed in [12], the authors concluded that " ... none of the major methods identified can address all relevant issues and questions at once." The environment and the ecosystems are too complex to be described by only one variable.

6.2.1 Industrial Ecology

Industrial ecology pertains to the use of materials and energy in industrial systems. The name is derived from the idea that principles of ecosystems, which coexisted for millennia, may be used to assist with the design of sustainable industrial systems, especially with the use of wastes and material recycling [13]. Industrial systems are considered parts of their surroundings, exist and thrive in concert with the surroundings and not in isolation from them. Figure 6.1 shows schematically the industrial ecosystem that delivers a final product (e.g. a vehicle). The production entails several stages – mining and refining of materials, conversion of the materials to usable forms, fabrication of parts, fabrication of the final product, and use of this product. Each of the processes requires the supply of other materials (solid line) and energy (dashed line). Each stage of the manufacturing generates waste energy and waste products that are captured to be used either in the same stage or in a different stage. Thus, wastes from one process become part of the raw materials and energy for another process. The net result is the minimization of material waste, minimization of energy resource utilization, and a lesser impact on the environment and the ecosystems.

By extending the use of wastes to processes for the production of a whole group of finished products, the notion of industrial ecology takes a global view of industrial facilities and processes and models the entire global industry as a network of facilities and processes that extract resources from the natural environment; utilize energy and exergy derived from parts of these resources; and transform the remaining resources into products that are sold in the global markets to meet the needs of the consumers.

An integral part of industrial ecology pertains to the impact of industrial activities on the environment, the depletion of natural resources, and the problems created from waste disposal. Principles and methods of the natural sciences, engineering, economics, and sociology have been adapted in the analyses of this subject that seeks to quantify the

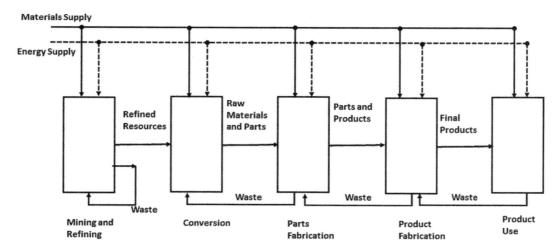

Figure 6.1 The concept of industrial ecology utilizes the wastes from one process as raw materials and energy for other processes.

materials transport, the energy flows, and the industrial processes, which are necessary to satisfy the demands of modern society. At the same time industrial ecology strives to minimize the impacts of human activities on the environment and to design sustainable industrial practices.

Exergy analyses are tools of industrial ecology, because they indicate where irreversibilities occur and natural resources are wasted or diluted and become unusable. The improvement of the exergetic efficiency in industrial processes preserves energy resources and, in most cases, reduces environmental pollution and makes the processes more benign to the ecosystems. A particular theme of industrial ecology is the utilization of wastes – both materials and energy – from one system or process as inputs (resources) for other systems and processes. The co-generation of electric power and heat (Section 3.4); the production of additional power in a bottoming cycle using the waste heat of a gas turbine (Section 3.3.2); and the hot water supply using the condenser of a heat pump (Section 4.7.3) are examples where exergy analyses and industrial ecology can be applied in practice to improve the performance of energy generation systems and preserve natural resources.

In addition to thermodynamic objectives that are addressed by exergy, industrial ecology entails several other objectives, which are represented by other factors and variables. An example of the objectives at different levels of an industrial ecology analysis is given in Figure 6.2. It is apparent that this subject is multifaceted; its operation entails several levels of complexity; and it may not be characterized simply by a single thermodynamic variable, the exergy [14].

Related to the subject of industrial ecology is the representation of the impacts of industrial wastes on ecosystems and the environment solely by means of the exergy of the wasted materials and energy forms [15]. A moment's reflection, however, proves that "environmental impact" is a multifaceted concept with origins in geology, hydrology, chemistry, environmental science, environmental health, and toxicology – not only

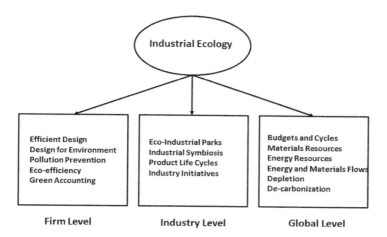

Figure 6.2 Objectives of industrial ecology at the firm level, the industry level and the global level.

thermodynamics. The ecological and environmental impacts of 1 kg of radioactive waste are by far higher than the impacts of 1 kg of petroleum from a refinery, regardless of their exergy. Chemical solvents and heavy metal wastes have significantly higher environmental impacts than paper product wastes, which have higher specific exergy. The toxicity of the waste, its effects on human health and ecosystems, and the potential of the waste to cause significant environmental and ecological disruption (e.g. lake and sea eutrophication by nitrates and phosphates that may kill all the fish species) are primary variables for the environmental impact to be measured or calculated [16]. The exergy of a substance (its potential to produce mechanical work) may be one of the variables to assess the environmental and ecological impact, but it cannot be the only variable, or even the most important variable.

6.3 The Natural Environment

The *natural environment* is everything that surrounds us and where our economic activity occurs. The upper part of the lithosphere (solid earth's crust), the lower atmosphere (air) and the near-surface hydrosphere (sea, lakes, rivers, etc.) are the three components of our environment. Human activity, weather, and climatic events interact with the environment and affect the ecosystems. For example, a hurricane formed in the atmosphere transports large quantities of water, which originate in the hydrosphere. When the hurricane washes over land, it dumps rainwater that floods vast areas; erodes the soil; and caries parts of the lithosphere (as dissolved solids and solid particles) into the rivers, the lakes and the sea, where the solids form sediments. Such weather phenomenal and interactions generate *environmental changes,* which in their vast majority are undesirable and detrimental to ecosystems and human economic activity. Human activity also causes environmental changes: the incessant combustion of fossil fuels since the beginning of the industrial revolution has increased the concentration of CO_2 in the atmosphere from 280 ppm to more than 400 ppm in 2020, an effect that portends GCC, with dire consequences for many ecosystems.

It must be noted that the natural environment – the subject of this section, which is often abbreviated as *environment* – is semantically different than the *model environment* or *reference environment*, of Sections 2.2 and 2.13. The latter is a model of the natural environment that is precisely constructed by thermodynamicists to define exergy and to calculate the numerical values of exergy. The former – the natural environment – is the actual environment, where living organisms thrive and all human economic activities take place. The natural environment is by far more complex than its models: It is composed of a myriad of substances – including human-made pollutants – that are difficult to precisely define and quantify; it is not in thermodynamic equilibrium; it is continuously evolving; and it affects all living organisms and human economic activities. Only a very small subset of the chemical substances in the natural environment comes into the definition of the model environment used in exergy calculations.

6.3.1 The Exergy – Environment Nexus

Since the global supply of energy is the origin of most current environmental problems and since exergy is the indicator of the "quality of energy" and a better representation of energy use, it is logical that exergy would be connected to the state of the natural environment, environmental degradation, and possible mitigating actions. Early studies on the relation between exergy and environment focused on environmental cost functions and penalties – monetary values associated with the environmental and social costs of pollutants. The assignation of environmental costs, which are not real costs incurred by the polluting entity, assists in the calculation of numerical values for possible pollution penalties and defines break-even points for energy conversion operations [17]. An analytical framework for evaluating environmental interactions using the exergy concept and an anthropocentric sensitivity analysis was introduced in [18]. The application of this framework to energy conversion processes entails subjective criteria and numerical values for the different types of pollution.

A figure of merit, the *environmental exergonomics criterion*, Z_{env}, was defined in [19] in an attempt to assess the environmental impacts of the construction and operation of renewable energy systems, which have multiple effects on the environment and the ecosystems. The criterion is chosen to account for the loss of biodiversity and the ability of the eco-system to function. This criterion and its inverse, the *environmental/sustainability efficiency*, η_{ENV} are defined by the expressions:

$$Z_{env} = \frac{1}{\eta_{II}} + \frac{1}{\eta_K} + \frac{1}{\eta_E} \quad \text{and} \quad \eta_{ENV} = \frac{1}{Z_{env}} = \frac{1}{\frac{1}{\eta_{II}} + \frac{1}{\eta_K} + \frac{1}{\eta_E}}, \quad (6.4)$$

where η_{II} is an exergetic efficiency of the system; η_K is the ratio of delivered exergy to invested exergy during the life cycle of the system; and η_E is an efficiency based on the eco-exergy flow through the system during the lifetime of the renewable energy system. While using this expression one must be aware that the specific eco-exergy values, which define η_E, are obtained from [20], they are based on Eqs. (6.1) and (6.3) and include parts related to the organizational complexity of the ecosystems. The exergetic efficiency, η_{II}, is defined by the actual exergy flows in the energy conversion system using the methods of Chapter 3.

A different type of analysis for the exergy-environment nexus, called *exergoenvironmental* analysis – delineated as conventional [21], and advanced [22] – attempts to optimize energy conversion systems based on cost as well as environmental objectives and includes the following steps:

1. An exergetic analysis of the energy conversion system.
2. A life-cycle assessment (LCA) of the components of the system and the input streams.
3. Determination of the environmental impact of the system as cost and assignment of this cost to the exergy stream of the system.

In this method the environmental impact rate, $\dot{B}_j$, of a material effluent, j, is considered to be the product of the exergy rate, $\dot{E}_j$, of the effluent and the specific environmental impact, b_j:

$$\dot{B}_j = \dot{E}_j b_j = \dot{m}_j e_j b_j. \tag{6.5}$$

The coefficients, b_j, are calculated using *eco-indicators*, which assess the relative environmental impact of the system effluents [23]. The numerical values of the eco-indicators are based on subjective criteria, which give higher weight to environmental factors that affect human health. Such criteria vary with time and with the environmental jurisdiction (region or country).

While using subjective or relative criteria, an important contribution of the methods that attempt to measure environmental impact using exergy is that they provide a measure of the wasted exergy in the system and in the environment [17, 18, 21, 22, 24]. This information may be used by the engineers for two purposes: a. To reduce exergy waste and the environmental impacts of energy conversion; and b. the adoption of trade-offs in the design of energy conversion systems that would mitigate their environmental impact.

It must be noted that, while the exergy calculations are objective, the assignment of environmental cost to the LCA of systems and processes introduces subjectivity in the calculations that leads to significant uncertainty of the outcome. The emission of several effluents that are considered as pollutants does not have a conventional and universal cost. In addition, several substances that are considered as pollutants in one country or region are not considered as such in other countries. As a result, the environmental impact of the same effluent (e.g. carbon dioxide, or salty water) is valued differently by scientists in different regions and even by scientists in the same region. The different values of environmental costs have a significant impact on the analysis of the LCA costs of systems and processes.

Assuming that the environmental cost is the environmental remediation cost, a second limitation in the assignment of environmental costs is the well-known dictum for the environment *how clean is clean?* How much remediation effort must be spent to "clean" the effect of an effluent? Let us consider the example of a nuclear power plant whose cooling system receives water from a river. Since part of the cooling water evaporates in the cooling towers, the balance of the water is returned to the river with higher salt concentration. Is the cost of remediating the water returned to the river (e.g. by removing the salts) an environmental cost that should be assigned to the operation of the nuclear power plant? What happens if the river enters the sea (as all rivers finally do) 5 km downstream this power plant, where the salts level is much higher? Again, a level of subjectivity enters the calculations of environmental cost that would affect all the calculations. Clearly, a great deal of standardization, international consensus, and wide appreciation of what constitutes an environmental cost must be achieved before the outcomes of LCAs with such costs become widely validated, accepted, and applied to the economy. Until then, one must be cognizant that such outcomes are subjected to a higher degree of uncertainty, a subject discussed in more detail in Section 7.5.

6.4 Exergy, the Natural Environment, and Ecosystems

The production of primary energy sources and their conversion to useful energy forms are causes of several environmental effects, such as water and air pollution, large-scale environmental accidents, maritime pollution, radioactivity and radioactive waste, solid waste, acid deposition, waste heat, atmospheric ozone depletion, and the most important environmental threat of the 21st century, GCC. Because exergy is a better indicator of the energy used for a given task, several groups of scholars have endeavored to connect exergy to ecosystems – through the concept of eco-exergy – as well as to the impacts of energy use on the natural environment. As with all scientific endeavors, these connections have intrinsic value and lead to reasonable and useful conclusions for environmental mitigation and ecosystem preservation. A moment's reflection on the environment and ecosystem impacts leads to the following observations:

1. All ecosystems are affected by their environment and changes in their environment. Actually, ecosystems adapt to all environmental changes.
2. When considering environmental pollution, it is the type and mass of chemical pollutants – not their exergy, regardless of the way it is measured – that affect the environment. One kmol of methane is 21 times more potent GHG than 1 kmol of carbon dioxide and 1 kmol of refrigerant-134a is 9,470 times more potent GHG, regardless of their exergies.
3. Similarly, for ecosystems and human health it is the type and mass of the chemical substances – and not their specific exergy – that cause impacts. While both sulfur dioxide (SO_2) and hydrogen cyanide (HCN) are considered human health hazards, the latter is by far more hazardous (actually, lethal) to humans, even at low concentrations. A few studies have asserted that, since the exergy content of a pollutant is a measure of the work the pollutant may perform, exergy is also a measure of the damage the pollutant may cause in the biosphere [25, 26]. A moment's reflection, however, proves that poisons with low exergy cause enormously greater environmental and ecological damage than nonpoisonous pollutants with high specific exergy, such as methane and ethane.
4. Damage to the environment and the ecosystems is not absolute, but relative and qualitative. Environmental and ecological damage is weighted a great deal more heavily if it affects (or is perceived as affecting) human health. While arsenic is in relatively high concentrations in seaweeds that thrive in seawater, the same concentrations of arsenic in potable water constitute an unmitigated threat to human health.
5. Unlike thermodynamic systems that have well-defined boundaries, cut-off criteria for the establishment of boundaries in ecosystems and the environment are not defined.
6. Despite several decades of research, there is not a set of objective, incontrovertible, and relatively accurate metrics that would measure the state of the environment and of the ecosystems; or the effects of anthropogenic activities in the degradation and remediation of ecosystems and the environment. Instead,

there are several versions and sets of qualitative indicators and other metrics, which are not universally accepted; have changed with time; and vary with the perceived environmental or ecological threat. One may count nine different indices pertaining to environmental policies for countries and another five indices pertaining to industrial environmental policies [27]. The numerical values of the indicators that constitute the indices are subjected to debate and different interpretations. This has introduced high uncertainty, which is apparent in the several articles of a scientific journal dedicated to ecological and environmental indicators that was first published in 2001 [28].

7. The results of studies using environmental and ecological indicators are subjected to significantly higher uncertainty than thermodynamic calculations of exergy and are considered qualitative.

Based on these observations, one draws the following conclusions:

1. Procedures to adequately and objectively describe environmental and ecological impacts should take into account the type of pollution; how the pollution affects the ecosystems; and must also supply the method to numerically quantify the impacts.

2. Ecosystems and the natural environment are very complex systems. A single variable (e.g. exergy, entropy, or enthalpy) is inadequate to define the state of the environment; the state of an ecosystem; environmental degradation; and environmental mitigation. When exergy is used and Eq. (6.5) is applied for the quantification of environmental impacts, the numerical values of the "damage factors" pertaining to the several classes of pollutants – the eco-indicators, b_j – span nine orders of magnitude [23]. This signifies – actually it proves – that it is the type of pollutant and not its specific exergy that affects the environment and the ecosystems. Similarly, for ecosystems using Eq. (6.3): the weighting factors, β_i, span three orders of magnitude to differentiate components, which have the same chemical exergy, but reside in different organisms.

3. The total exergy of pollutants is proportional to the mass of the pollutants ($E = me$) and, if there is a motive, total exergy may be used as a proxy variable for the mass. While this approach is counterintuitive and does not relate cause and effect, it would generate rational conclusions. For example, in Eq. (6.5) it is the mass flow rates of the effluents (and their chemical composition, which is characterized by b_j) that cause the pollution. Since $\dot{E} = \dot{m}e$, one may use the rate of exergy $\dot{E}$ as a proxy for the mass flow rate $\dot{m}$, in which case the specific exergy, e, becomes proportionality constant. One must note, however, that while this method may appear rational, it does not relate cause and effect, because the laws of thermodynamics do not describe the severity of pollution.

4. Any quantification of the effects of anthropogenic activities on ecosystems and the environment necessitates the adoption of measures, such as the eco-indicators [23] or the weighting factors, β_i, of Eq. (6.3) [8]. The origin of these measures is largely subjective and anthropocentric, assigning higher importance on human health effects.

5. Because the pollutant carrying capacity of the entire planet is unknown, the calculations of environmental deterioration are relative. When the state of the environment is close to capacity the numerical values of the "damage factors" would become very high. For example the release of 1 ton of CO_2 in 1920 did not represent an environmental threat. The release of the same ton of CO_2 in 2020 is an environmental threat and a damage factor is assigned to it. If the atmospheric concentration of CO_2 continues to increase, the damage factor of one ton of CO_2 released in 2120 would be significantly higher.

Regardless of these limitations that pertain to the quantification of environmental and ecological effects, the exergy method is a good *qualitative indicator* for these effects. Striving for the higher exergetic efficiency of energy conversion systems always results in lower lost work and lower natural resource consumption. This implies lesser environmental damage from the extraction of the resource and, in the case of fossil fuels, lesser CO_2, CO, NOX, and SO_2 emissions.[3] The higher exergetic efficiency also generates lesser waste heat and, in the case of water-cooled systems, lesser freshwater consumption. The following describe the emissions reduction and water consumption effects of four energy conversion systems that were quantitatively analyzed using the exergy method in preceding chapters.

1. For the 400 MW coal power plant examined in Section 3.2 and depicted in Figure 3.5: If the combustion efficiency increased from 92% to 96% the exergetic efficiency of the plant would increase from 36.9% to 38.5%; the mass rate of carbon supplied for the delivery of the heat input would be 31.64 kg/s instead of 33.02 kg/s, resulting in fuel savings of 1.38 kg/s. This corresponds to savings of 43,520 tons of coal/year and approximately 160,000 tons/year reduction of CO_2 emissions, with corresponding NOX, CO, and SO_2 emissions. Similarly, if the turbine efficiency improved to 90%, we would have a commensurate improvement of the exergetic efficiency as well as pollutant emissions and freshwater use reductions.
2. In the case of the addition of the regenerator in the Brayton cycle of Section 3.3.1: The methane fuel consumption of the cycle dropped from 2.999 kg/s to 2.179 kg/s for 0.82 kg/s savings. This corresponds to savings of 25,860 tons of methane per year and annual avoidance of 71,100 tons of CO_2.
3. For the fuel cell of Example 3.12: If its efficiency were to improve from 58% to 75%, the amount of hydrogen needed for a trip would be 6.2 kg instead of 8.0 kg and the volume of the hydrogen fuel tank would be reduced from 310 L to 240 L.
4. The substitution of the conventional heat pump system with a GSHP system in the building of Example 4.7, resulted in annual electric energy savings of 24,950 kWh (89.8 GJ). The electricity savings in the building correspond to fuel savings at the power plants that supply the region. If this amount of electricity is produced in a coal power plant with 36% efficiency, the electric energy savings would

---

[3] An exception to this rule is when exergy-consuming components are added to existing systems for the reduction of pollutant emissions, as in the case of SO_2 scrubbers that are now used in coal power plants.

correspond to 27.9 tons annual CO_2 avoidance, with the analogous drop in NOX, CO, and SO_2 emissions as well as ash removal. If the 24,910 kWh electricity savings were produced in a nuclear power plant with 32% thermal efficiency, the annual nuclear fuel savings would be 0.48 kg of natural uranium. Given that most uranium becomes nuclear waste, approximately 0.48 kg of environmentally harmful nuclear waste would be avoided. Regarding freshwater use reduction, the cooling towers of the power plants would use less freshwater −72.4 tons lesser freshwater in the coal power plant and 86.8 tons in the nuclear power plant.

Exergetic efficiency improvements in energy conversion systems always engender primary resource savings, from which emissions reductions and other environmental impacts' mitigations may be calculated. While the emissions reductions are accurate and precise, their environmental and ecological impacts are largely qualitative. It must be noted that exergetic efficiency improvements will not completely eliminate the environmental and ecological effects of energy conversion and will not reduce to zero any rational measure of environmental impact: Referring to fossil fuel power plants, even if it were possible to increase their exergetic efficiency to 100%, there would still be environmental impact, because fossil fuels would still produce CO_2 (albeit in lesser quantities), which is released in the atmospheric environment.

6.4.1 Life Cycle Analysis

A useful tool for the assessment of ecological and environmental impact of engineering systems, including energy systems, is the *life cycle analysis (LCA)*. LCA is a method used to evaluate the environmental impact of a product through its life cycle. It encompasses the extraction and processing of the raw materials, the manufacturing of the parts and the final product, the distribution, the usage, any recycling of materials after decommissioning, and the final disposal of the nonrecycled materials. The LCA assists with the holistic environmental and ecological assessment of systems, products, and services. In the case of engineering systems the LCA assesses the entire range of ecological and environmental effects of a system by quantifying:

1. The effects of all the energy and raw materials that have been used for the construction of the system, including their transportation.
2. The effects of all the inputs and outputs of energy and material flows through the system during the entire lifetime of the system.
3. The effects of all energy and materials used for the decommissioning of the system at the end of its useful life, including transportation, storage, and custody of the produced waste.

This quantitative information provides a holistic assessment of how a system affects the environment and ecology via all the energy and materials utilized during its entire life cycle. The LCA assessment may be used to provide a rational background for informed decisions on the environmental and ecological impacts of systems, processes, and products.

A special case of the LCA method for energy conversion systems is the *life cycle exergy analysis* (LCEA) that applies the exergy concept to electricity generation systems [22, 24] and accounts for the total exergy consumption associated with all the inputs and outputs of the systems during three stages: construction, operation (energy generation), and decommissioning (clean up). The LCEA offers an all-inclusive assessment of the exergy resources spent during the entire life of systems, not just during the time of their operation. Because exergy is an indicator of the impact of energy use on the natural resources, the LCEA is more useful than the exergetic efficiency, η_{II}, of systems, which only assesses the systems' impacts on the resources during the operation stage. Figure 6.3a and b depict schematically the exergy resource consumption and production for two power generation units – the first using a fossil fuel and the second a renewable resource – during the stages of construction, operation, and decommissioning. Both units require exergy (and other resources) for their construction and decommissioning stages. The fossil fuel unit consumes the exergy of the fossil fuel – a natural resource, which is depleted – during the operation stage. As a result, this unit is encumbered with a significant exergy deficit during its life cycle. The renewable energy unit in Figure 6.3b generates significant net exergy during the stage of operation, which surpasses the exergy consumed during the construction and decommissioning

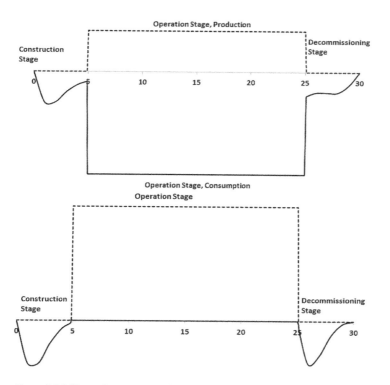

Figure 6.3 Life cycle exergy production and consumption of (a) a fossil fuel power generation system; (b) a renewable power generation system.

stages. As a result, the latter unit produces exergy during its life cycle. On the whole renewable energy power generation units are net exergy producers.[4]

The LCEA offers an accurate measure of the ability of energy conversion systems to positively contribute to the problem of natural resources depletion.

The concept of LCEA is similar to a figure of merit that has been extensively used with PV cells, the *Energy Payback Time (EPBT)* – the time it takes for PV cells to produce all the energy that was consumed during their life cycle. The significant improvements of PV technology have lowered the EPBT of PV cells from approximately 40 years in the 1970s to less than 2 years in 2019, with cadmium-tellurium, and polymeric cells having EPBTs less than 1 year [3]. The conversion of the EPBT of PV cells to an LCEA figure of merit is a trivial task.

6.5 Sustainable Development

Although several synonymous concepts were proposed since the beginnings of the industrial revolution, the term *sustainable development* was coined by the UN *World Commission on Environment and Development* (WCED) that was established in 1963 and was discharged in 1987. The Commission – also known as the *Brundtland Commission* – established a set of guidelines for nations to achieve higher standards of living (development), while taking under consideration the interrelationships between people, resources and the natural environment [29]. The WCED recognizes that the environment does not exist as an isolated entity separate from human actions, ambitions, and needs; that economic development is what humans do attempting to improve their conditions within their habitat; and that the environmental, social, and economic factors, which affect all nations and human communities, are highly interconnected and inseparable. The Commission defined sustainable development (*sustainability*) as the ability of a society to *ensure that it meets the needs of the present, without compromising the ability of future generations to meet their own needs*. In this context, sustainability implies limitations imposed by the current state of technology, recoverable resources, social organization, and the ability of the natural environment to absorb the impacts of human activities. After the publication of the Commission's report, *Our Common Future* [29], the sustainability concept has been adapted and used within several disciplines (e.g. biology, architecture, economics, environmental science, engineering, sociology) in a variety of contexts and connotations, often encompassing the notions of: political empowerment; economic equity; effective citizen participation in the national decision making process; and greater democracy and equity in international decision-making fora.

Sustainability extends well beyond the principles of thermodynamics and encompasses multi-layered issues and diverse subjects. Motivated by the dictum

---

[4] The (low) amounts of exergy needed for the operation and maintenance of the plants are included as a deficit in the exergy produced. In the LCEA analyses the spent exergy of the renewable source (e.g. solar or wind exergy) is not considered as exergy consumption because it is perennially available, it is not associated with resource depletion, and it would have been dissipated (wasted) if not utilized in the power plant.

"you cannot manage what you cannot measure," several groups have defined indicators (metrics) that measure the progress toward sustainable development, globally as well as within nations, regions, and localities. Early work in the USA identified 40 indicators to quantify sustainable development [30]. These indicators were divided into three groups – economic, environmental, and social – that correspond to the three pillars of the sustainability concept. The proposed USA metrics are listed in Table 6.2 [30].

Every indicator in Table 6.2 may be defined by a single variable (e.g. the USA population and life expectancy at birth), or by composite indices (e.g. the global climate response index, which is defined in terms of five different indicators).

Based on such sets of metrics, sustainability initiatives define a single sustainability index, SI, as a linear combination of all the indicators, I_j, with each indicator weighted by a factor, w_i.

$$SI = \sum_j I_j w_j \quad \text{with} \quad \sum_j w_j = 1. \tag{6.6}$$

Table 6.2 Economic, environmental, and social indicators in the USA.

Economic	Environmental	Social
Capital assets	Surface water quality	USA population
Labor productivity	Acres of major terrestrial ecosystems	Children living in single-parent families
Federal debt to GDP ratio	Contaminants in biota	Teacher training level
Energy consumption (usage) per unit GDP	Status of stratospheric ozone	Births to single mothers
Materials consumption per capita and per unit GDP	Quantity of spent nuclear fuel	Contributions of time and money to charities
Inflation	Greenhouse climate response index	Educational attainment by level
Investment in R&D as percentage of GDP	Ratio of renewable water supply to withdrawals	Participation in arts and recreation
Domestic product	Fisheries utilization	People with 40% or greater poverty
Income distribution	Invasive alien species	Crime rate
Consumption expenditures per capita	Conversion of cropland to other uses	Life expectancy at birth
Unemployment	Soil erosion rates	Educational achievement rates
Home ownership rate	Timber growth to removals balance	
Percentage of households in problem housing	Greenhouse gas emissions	
	Identification and management of superfund sites	
	Metropolitan air quality non-attainment	
	Outdoor recreational activities	

The relative value of the factor, w_j, assigns the relative emphasis and significance of the initiative I_j. It is apparent that the numerical values of the weighting factors, which are largely subjective and nation-specific, play a very important role in the determination of SI.

A multitude of studies have proposed lists of indicators similar to those in Table 6.2, which measure sustainability within defined geographical areas – countries, regions, and localities. In the early part of the 21st century (2004) the *Compendium of Sustainable Development Indicator Initiatives* listed more than 600 efforts to establish sustainability indices with global, national, regional, and local applications [31]. Twelve of these initiatives were reviewed in [32] to highlight their motivation processes, technical methods, and similarities. The review concluded that there are no indicator sets, which are backed by compelling theory or are universally accepted in the twelve initiatives. It also determined that, in the application of sustainability initiatives, there are no rigorous data collection methods and analyses. This is very likely due to the ambiguity of the sustainability concept, the plurality of purposes in characterizing and measuring sustainable development, and, often, confusing terminology.

6.5.1 Exergy and Sustainability

Sustainability is a multifaceted concept with its overarching principle rooted in intergenerational ethics, namely that a given generation should pursue a higher standard of living without jeopardizing the environment and without depleting the resources succeeding generations will need. In the energy context, sustainability implies that regional and global economic development should be pursued without causing permanent damage to the ecology and the environment and without excessive reduction of the primary energy sources that will be essential resources for future generations.

It is important to note that the concept of sustainability and its principles are rooted in ethics and not in physical laws and scientific principles. Human existence and economic activity will continue (perhaps in different than the present norms) without the application of sustainability principles. The rampant industrial and economic developments of the 19th, 20th, and early 21st centuries, which has been accompanied by the sizeable depletion of the global fossil fuel resources and significant environmental impacts, are examples that "unsustainable development" does not violate any physical laws and may persist for long periods. However, it is becoming increasingly clear that resource depletion and the environmental deterioration generated by fossil fuel combustion will significantly distress future generations. If the "unsustainable development" of the past continues for long, the irreparable environmental damage caused by the GHG emissions, the GCC, and the depletion of energy sources will hamper the ability of future generations to meet their own needs.

The use of energy and the depletion of energy resources are integral components of the sustainability principles and – in various forms – have become part of all the sustainability indices, SI. The concept of exergy characterizes the use of energy. Exergy destruction offers a very good indicator to quantitatively assess energy resource depletion [3, 33]. Of all the measurable variables, exergy is the one that most accurately and quantitatively represents the use of energy and the primary energy resource depletion. For this reason, exergy consumption in regions and nations and annual rates of exergy destruction should be two of the indicators in all reliable sustainability indices. The appraisal of the four cases 1–4 in Section 6.4 – actually, of most examples in this book – proves that exergetic efficiency improvements always result in primary energy resource savings and lesser resource depletion. As with the environmental impact, this is a qualitative rather than deterministic relationship for global sustainability and is schematically presented in Figure 6.4. Exergetic efficiency improvements are always concomitant with lower environmental impact and improved energy resource sustainability. One must notice, however, that even at 100% exergetic efficiency, the environmental impacts are not nullified.

The *extended-exergy (EE) concept* has been proposed to advance the state-of-the-art using exergy as a measure of – among other concepts – capital, labor, state of the environment, and environmental remediation [34]. While EE has the advantage of measuring everything in exergy units (e.g. kJ or kWh) it suffers from inconsistencies with widely accepted meanings and terminology, chief among which are the discrepancies in the accounting of capital and national labor force. The related concept of emergy, which has been applied to ecosystems [7], may also serve as one of the metrics, but it does not qualify as a holistic and all-encompassing indicator of sustainability, which by definition depends on several objectives and factors [35].

A glance at Table 6.2 and similar lists of sustainability indicators proves that sustainability is a multifaceted concept. There are several goals for the sustainable global development, which are not always congruent with each other [36]. The quantitative assessment of sustainable development and the progress towards these goals requires the application of several metrics. As the gross domestic product (GDP) alone cannot fully characterize the economic development of communities within a nation, exergy and its derivative concepts (EE or emergy) alone cannot characterize sustainability. These concepts cannot serve as *global* or *the only* variables and indicators to assess sustainability, in general, or even energy sustainability, in particular [35]. Similar

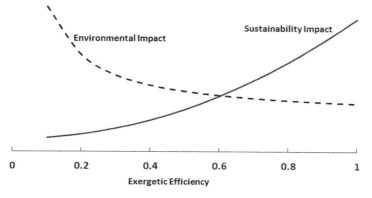

Figure 6.4 Qualitative diagram of the environmental and sustainability impacts of exergetic efficiency.

to their role with ecological and environmental impacts, these concepts, which have their origins in thermodynamics, may assist as a subset of the metrics that offer a thermodynamics perspective in the quest for sustainable development.

Problems

1. A tropical cyclone hits hard a Pacific Island. Identify three probable environmental consequences and three ecological consequences.
2. State three similarities and three differences between ecosystems and industrial ecosystems. Also state in which ways exergy analyses may assist the two.
3. Explain the differences between the natural environment and the model environment used for the calculation of exergy.
4. Explain why an exergoenvironmental analysis cannot be as simple and as precise as an exergy analysis of an engineering system.
5. In your opinion what are five important environmental indicators that would define the state of the environment? How would you measure these indicators?
6. After several design modifications by bright engineers the thermal efficiency of a 600 MW coal power plant is improved from 35.6% to 38.2%. Assuming that the plant operates continuously throughout the year and that all the heat input to the power plant is derived from carbon combustion, what is the annual CO_2 emissions avoidance from this change?
7. Since the 1980s the thermal efficiencies of gas turbines improved from 36% to 64%. For a typical gas turbine with 50 MW rating that uses methane as fuel and operates 70% of the time during a year, what is the annual CO_2 emissions avoidance?
8. Construct the life cycle exergy production and consumption diagram for a 1,000 MW nuclear power plant. List in detail the exergy streams associated with this unit during its entire life.
9. "A life cycle analysis of an energy system unfairly penalizes several systems by assigning phantom environmental costs. For energy conversion systems a simple cost analysis is more appropriate." Comment with a 300 word essay.
10. In reference to Figure 6.4, explain in detail why the environmental impact does not vanish when the exergetic efficiency of processes and systems approaches 100%. Give two examples to support your arguments.
11. In a 500 word essay explain how the discipline of thermodynamics and, in particular, the concept of exergy may be used to achieve some of the sustainability objectives. Connect your answers to the indicators listed in Table 6.2.

References

[1] D. M. Gates, *Energy and Ecology* (Sunderland, MA: Sinauer, 1985).
[2] M. J. Molina and F. S. Rowland, Stratospheric Sink for Chlorofluoromethanes: Chlorine Atom-Catalysed Destruction of Ozone. *Nature*, **249** (1974), 810–12.

[3] E. E. Michaelides, *Energy, the Environment, and Sustainability* (Boca Raton, FL: CRC Press, 2018).

[4] V. Vestreng, G. Myhre, H. Fagerli, S. Reis, and H. Tarrason, Twenty-Five Years of Continuous Sulphur Dioxide Emission Reduction in Europe. *Atmospheric Chemistry and Physics*, **7** (2007), 3663–81.

[5] A. R. Douglas, P. A. Newman, and S. Solomon, The Antarctic Ozone Hole: An Update. *Physics Today*, **67** (2014), 42–8.

[6] International Energy Agency, *Key World Statistics* (Paris: IEA-Chirat, 2018).

[7] S. E. Jorgensen, S. N. Nielsen, and H. Mejer, Emergy, Environ, Exergy and Ecological Modeling. *Ecological Modeling*, **77** (1995), 99–109.

[8] S. E. Jorgensen and J. C. Marques, 2001, Thermodynamics and Ecosystem Theory, Case Studies from Hydrobiology. *Hydrobiologia*, **445**, 1–10.

[9] L. Susani, F. M. Pulselli, S. E. Jorgensen, and S. Bastianoni, Comparison Between Technological and Ecological Exergy. *Ecological Modeling*, **193** (2006), 447–56.

[10] H. Lin, Thermodynamic Entropy Fluxes Reflect Ecosystem Characteristics and Succession. *Ecological Modeling*, **298** (2015), 75–86.

[11] R. J. Holdaway, A. D. Sparrow, and D. A. Coomes, Trends in Entropy Production During Ecosystem Development in the Amazon Basin. *Philosophical Transactions of the Royal Society B*, **365** (2010), 1437–47.

[12] T. Wiedmann and J. Barrett, A Review of the Ecological Footprint Indicator–Perceptions and Methods. *Sustainability*, **2** (2010), 1645–93.

[13] D. Frosh and N. Gallopoulos, *Strategies for Manufacturing*, Scientific American, 261 (1989), 94–102.

[14] R. U. Ayres, and L. W. Ayres, eds. *A Handbook of Industrial Ecology* (Cheltenham, UK: Edward Elgar Publishing, 2002).

[15] Y. Ao, L. Gunnewiek, and M. A. Rosen, Critical Review of Exergy-Based Indicators for the Environmental Impact of Emissions. *International Journal of Green Energy*, **5** (2008), 87–104.

[16] D. Favrat, F. Marechal, and O. Epelly, The Challenge of Introducing an Exergy Indicator in a Local Law on Energy. *Energy*, **33** (2008), 130–36.

[17] C. A. Frangopoulos and Y. C. Caralis, A Method for Taking Into Account Environmental Impacts in the Economic Evaluation of Energy Systems. *Energy Conversion and Management*, **38** (1997), 1751–63.

[18] A. P. Simpson and C. F. Edwards, An Exergy-Based Framework for Evaluating Environmental Impact, *Energy*, **36** (2011), 1442–59.

[19] A. Goldberg, Environmental Exergonomics for Sustainable Design and Analysis of Energy Systems. *Energy*, **88** (2015), 314–21.

[20] S. E. Jorgensen and S. N. Nielsen, Application of Exergy as Thermodynamic Indicator in Ecology. *Energy*, **32** (2007), 673–85.

[21] L. Meyer, G. Tsatsaronis, J. Buchgeister and L. Schebek, Exergoenvironmental Analysis for Evaluation of the Environmental Impact of Energy Conversion Systems. *Energy*, **34** (2015), 75–89.

[22] Y. Lara, F. Petrakopoulou, T. Morosuk, A. Boyano, and G. Tsatsaronis, An Exergy-Based Study on the Relationship between Costs and Environmental Impacts in Power Plants. *Energy*, **138** (2017), 920–28.

[23] M. Goedkoop and R. Spriensma, *The Eco-Indicator 99: A Damage Oriented Method for Life-Cycle Impact Assessment*, 3rd ed., *Methodology Report* (Amersfoort, Netherlands, 2000).

[24] G. Wall, On Exergy and Sustainable Development in Environmental Engineering. *The Open Environmental Engineering Journal*, **3** (2010), 21–32.

[25] D. Maes. and S. Van Passel, Advantages and Limitations of Exergy Indicators to Assess Sustainability of Bioenergy and Biobased Materials. *Environmental Impact Assessment Review*, **45** (2014), 19–29.

[26] R.H. Edgerton, *Available Energy and Environmental Economics* (Toronto: D. C. Heath, 1992).

[27] R. K. Singh, H. R. Murty, S. K. Gupta, and A. K. Dikshit, An Overview of Sustainability Assessment Methodologies. *Ecological Indicators*, **15** (2012), 281–99.

[28] *Ecological Indicators*, Published by Elsevier since 2001.

[29] United Nations, Report of the World Commission on Environment and Development: Our Common Future. www.un-documents.net/our-common-future.pdf, last visited on September 20, 2019.

[30] US Interagency Working Group on Sustainable Development Indicators. Sustainable development in the United States: An experimental set of indicators (Washington, DC, December 1998).

[31] International Institute for Sustainable Development, Compendium of Sustainable Development Indicator Initiatives. www.iisd.org/library/compendium-sustainable-development-indicator-initiatives, last visited February 9, 2019.

[32] T. M. Parris and R.W. Kates, Characterizing and Measuring Sustainable Development. *Annual Review of Environmental Resources*, **28** (2003), 13.1–13.28.

[33] C. J. Romero and P. Linares, Exergy as a Global Energy Sustainability Indicator. A review of the State of the Art. *Renewable and Sustainable Energy Reviews*, **33** (2014), 427–42.

[34] E. Sciubba, From Engineering Economics to Extended Exergy Accounting: A Possible Path from "Monetary" to "Resource-Based" Costing. *Journal of Industrial Ecology*, **8** (2004), 19–40.

[35] N. Lior, Quantifying Sustainability for Energy Development. *Energy Bulletin No. 19*, Intern. Sustainable Energy Development Center under the Auspices of UNESCO, (2015), 8–24.

[36] E. E. Michaelides, Fossil Fuel Substitution with Renewables for Electricity Generation – Effects on Sustainability Goals. *European Journal of Sustainable Development Research*, **4** (1), (2020), em0111.

7 Optimization and Exergoeconomics

Summary

Mathematical optimization has been used since the early part of the 20th century to improve the profitability of systems and processes. The time-value of money that leads to the concepts of net present value, annual worth, and annual cost of capital investment, is very important in the optimization of energy systems that typically operate for very long periods. The method of thermoeconomics (which was formulated in the 1960s) and the similar method of exergoeconomics (which emerged in the 1990s) are two cost-analysis methods extensively used for the optimization of energy systems. Calculus optimization and the Lagrange undetermined multipliers are similarly used mathematical tools. This chapter starts with a succinct exposition of the basic concepts of economics and optimization theory, and continues with the critical examination of the mathematical tools for the optimization of energy conversion systems and processes, including the use of exergy. The uncertainty of the optimum solution, which is an important consideration in all economic analyses, is clarified and an uncertainty analysis for exergy-consuming systems is presented. All exergy optimization processes should be combined with the ingenuity, intuition, and the correct judgment of the engineers who undertake the design of energy systems.

7.1 Mathematical Optimization Models – Duality

Optimization is the method for the determination of the set of conditions that achieve the best result for a given process or system. In a mathematical context optimization implies the minimization or maximization of one or more well-defined performance measures for the process or system. Inherent in all optimization methods are the precise definitions of the system, the processes, and the variables that come in the optimization. In mathematical representation optimization is expressed in terms of a *criterion* to be optimized, which typically is a function of several variables. Thus, for economic and financial optimizations the criterion is defined as the *minimum cost*, the *maximum profit*, or the *maximum revenue* and is expressed as a mathematical function of the optimized variables, the *objective function*. The latter is complemented by a set of conditions or restrictions, which must be satisfied when the best result is achieved, the optimization *constraints*. Examples of such constraints with cost minimization objective functions are: in manufacturing, the production of at least a given number of units to be sold; in transportation, the shipping of at least a given weight of products between two destinations; in the heating of buildings the maintenance of at least a given interior temperature; and in the cooling of buildings the maintenance of at most a given interior temperature. The constraints are expressed as inequalities or equations, such as: "the PV system will produce at least 10 million kWh annually," which is expressed as $W \geq 10 * 10^6$ kWh/yr; or "the interior temperature of the building will not rise above 25°C in the summer," expressed as $T_{int} \leq 25°C$.

The general form of a minimization problem that includes the objective function and k constraints to be satisfied is:

Minimize $F(x_1, x_2, \ldots, x_n)$
Subject to $G_1(x_1, x_2, \ldots, x_n) \geq 0; \quad G_2(x_1, x_2, \ldots, x_n) \geq 0; \quad \ldots G_k(x_1, x_2, \ldots, x_n) \geq 0,$

(7.1)

where $x_1, x_2, x_3, \ldots, x_n$ are variables of the system under consideration that contribute to the optimization and the inequalities or equations $G_1, G_2, G_3, \ldots, G_k$ express the constraints that must be satisfied when the optimum is achieved. The objective function, F, typically expresses a cost in economic/financial problems.

Similarly, the general form of a maximization problem is:

Maximize $F(x_1, x_2, \ldots, x_n)$
Subject to $G_1(x_1, x_2, \ldots, x_n) \leq 0; \quad G_2(x_1, x_2, \ldots, x_n) \leq 0; \quad \ldots G_k(x_1, x_2, \ldots, x_n) \leq 0.$

(7.2)

In this case, the objective function, F, typically expresses revenue or profit.

Solutions to the general optimization problem may be achieved with numerical methods. Classes of optimization problems for which specific solution methods have been developed include:

1. *Linear optimization* – also called *linear programming* – where the functions $F(x_1, x_2, \ldots, x_n)$ and $G_i(x_1, x_2, \ldots, x_n)$ are linear functions of the set of n variables $x_1, x_2, \ldots, x_n$.
2. *Quadratic optimization* – also called *quadratic programming* – where the objective function $F(x_1, x_2, \ldots, x_n)$ is a quadratic function of the variables $x_1, x_2, \ldots, x_n$; and the constraints $G_i(x_1, x_2, \ldots, x_n)$ are linear functions of the variables $x_1, x_2, \ldots, x_n$.
3. *Dynamic programming* relies upon detecting a multistage dependence of the objective function $F(x_1, x_2, \ldots, x_n)$ and achieves an optimum solution by a recursive method.
4. *Geometric programming* applies to optimization problems with objective functions that are posynomials and constraints that are monomials. The method determines the proportions of the component terms of the objective function $F(x_1, x_2, \ldots, x_n)$ for an optimum. This distinctive feature of geometric programming is useful in engineering design applications, particularly applications with circuit design and controls.

5. *The Kuhn–Tucker* method (or theory) applies to all analytic functions for the objective function $F(x_1, x_2, \ldots, x_n)$ and the constraints $G_i(x_1, x_2, \ldots, x_n)$. The corresponding optimization method is a generalization of the Lagrange undetermined multipliers method and applies first-order derivative tests for the determination of the optimum solution.

Of particular interest in optimization studies is *linear optimization* that accounts for a large fraction of applications in management and engineering, principally in the areas of supply management, blending, and power transmission. When the functions $F(x_1, x_2, \ldots, x_n)$ and $G_i(x_1, x_2, \ldots, x_n)$ are linear, the minimization problem of Eq. (7.1) may be written in vectorial form as:

$$\text{Minimize} \quad \vec{c} \bullet \vec{x}$$
$$\text{Subject to} \quad [A] \bullet \vec{x} \geq \vec{b} \quad \text{and} \quad \vec{x} \geq 0, \quad (7.3)$$

where $\vec{x}$ is a vector of n dimensions and comprises the set of the optimization variables $(x_1, x_2, \ldots, x_n)$; the vector $\vec{c}$ is also of n dimensions; the matrix $[A]$ has $k \times n$ dimensions; and the vector $\vec{b}$ is of k dimensions and is related to the k constraints. Equation (7.3) describes a typical cost minimization procedure subject to constraints for the production of at least a specified numbers of units, specified by the vector $\vec{b}$, of each product. The vectors $\vec{x}$ and $\vec{c}$ represent the inputs of the production processes and the cost of each input, which includes energy and labor.

Every linear optimization problem is equivalent to a *dual* problem that represents the inverse operation of the objective function. Minimization problems have duals that are equivalent maximization problems and maximization problems have equivalent dual minimization problems. The dual of the problem described by Eq. (7.3) is:

$$\text{Maximize} \quad \vec{b} \bullet \vec{y}$$
$$\text{Subject to} \quad [A]^T \bullet \vec{y} \leq \vec{c} \quad \text{and} \quad \vec{y} \geq 0. \quad (7.4)$$

The matrix $[A]^T$ is the transpose of $[A]$ and the vectors $\vec{b}$ and $\vec{c}$ are the same as in Eq. (7.3). In this case the dual problem is interpreted as the maximization of the revenue or of the profit, subject to the consumption or usage of at most a given quantity of inputs. The dual of the maximization problem in Eq. (7.4) is the minimization problem of Eq. (7.3). The solutions of the two problems are equivalent and lead to the same optimization of the cost variables $\vec{x}$ and the resource utilization variables $\vec{y}$. The *simplex method/algorithm* – available in several software packages and libraries – generates the solutions to both the primal and its dual optimization problem. Practical linear energy and exergy optimization problems are more easily formulated as cost minimization problems, with the vector $\vec{c}$ being the unit cost of inputs. The constraints, $\vec{b}$, represent the level of output (e.g. number of products) of the process or the system [1].

Example 7.1: Following a somewhat extravagant and exotic party the BΓΔ fraternity members discover that 20 L of tequila, 15 L of triple sec, and 12 L of lemon juice were not consumed. Resisting the temptation to consume the surplus, they decide to market ready-made versions of the classical cocktail "Margarita," using the secret and time-honored fraternity recipes:

	Triple Sec	Tequila	Lemon Juice
Basic Margarita	25%	50%	25%
Dreamy Margarita	40%	30%	30%
Stony Margarita	40%	40%	20%

They can sell all they produce to their neighbors, the ΠΡΣ fraternity, at $30/L for Basic; $20/L for Dreamy; and $25/L for the Stony Margarita. Naturally, the BΓΔ brothers wish to maximize their revenue. Set up this problem and its dual problem as linear optimization problems, determine their solutions, and interpret them.

Solution: The relevant variables of this optimization problem are the quantities of the three products in the optimum solution, in L: x_B, x_D, and x_S. The revenue maximization problem (the primal problem) is the profit maximization from the sale of the products:

Maximize: $30x_B + 20x_D + 25x_S$
subject to: $0.25x_B + 0.40x_D + 0.40x_S \leq 15$
$\quad\quad\quad\quad 0.50x_B + 0.30x_D + 0.40x_S \leq 20$
$\quad\quad\quad\quad 0.25x_B + 0.30x_D + 0.20x_S \leq 12$
and $\quad\quad x_B, x_D, x_S \geq 0$.

The dual problem is:
Minimize: $15y_{TS} + 20y_T + 12y_{LJ}$
subject to: $0.25y_{TS} + 0.50y_T + 0.25y_{LJ} \geq 30$
$\quad\quad\quad\quad 0.40y_{TS} + 0.30y_T + 0.30y_{LJ} \geq 20$
$\quad\quad\quad\quad 0.40y_{TS} + 0.40y_T + 0.20y_{LJ} \geq 25$
and $\quad\quad y_{TS}, y_T, y_{LJ} \geq 0$.

The variables of the dual problem represent the value of the three resources (triple sec, tequila, and lemon juice) to the producers and their units are price units, $/L.

The solution of either the primal or the dual is accomplished using the *simplex method*, which is standard material in books on optimization [1, 2] and is part of most software library packages. The solution to the primal problem is:

$x_B = 25.33$; $x_D = 13.33$; and $x_S = 8.33$, which implies that the production that maximizes revenue is 25.33 L of Basic; 8.33 L of Dreamy; and 13.33 L of Stony Margarita. The total revenue is $1,234.75. One notes that at the optimum solution all their resources (triple sec, tequila, and lemon juice) are used. This does not necessarily happen in all manufacturing optimization problems.

The **dual problem** is interpreted as a minimization of resources (triple sec, tequila, and lemon juice) problem. Its solution is: $y_{TS} = 5.03$; $y_T = 54.99$; and $y_{LJ} = 4.96$ and yields an optimum of $1,234.77 (the $0.02 difference with the primal is due to the rounding of prices to the last cent). These values are the marginal costs (the value) of the three resources to the producers. The producers will buy more of these resources to produce additional units of the products if the market prices of the resources were lower than the marginal cost. For example, the producers would buy additional tequila (to manufacture more of the Basic Margarita) at a price less than $54.99/L; and more triple sec if its price were less than $5.03/L.

This is an example of the general category of optimization problems called *the blending problem* and applies to manufacturing facilities (including refineries) that produce a mix of products using a number of resources as inputs.

7.2 Definitions of Relevant Economic Variables

The following are a few of the important concepts used in economics and management that are helpful in the optimization of energy conversion projects [3–5]. The list (in alphabetical order) is not exhaustive of the concepts and terms used in economic theory.

1. *Average cost:* The total of all *fixed* and *variable* costs for one unit of the product. The average cost is calculated over a period of time, typically one year. For example the average cost of electricity in 2017 was $0.048/kWh; and the average cost of gasoline in 2015 was €1.62/L.
2. *Average revenue:* The total revenue of the product units sold over a period of time, usually one year, divided by the total number of units sold. For example, the average revenue of electricity in 2017 was $0.072/kWh; and the average revenue of cement produced in the XYZ plant during 2016 was ₩3,520/ton.
3. *Average profit:* The difference between average revenue and average cost. For the electric power plant of the above two paragraphs, the average profit of electricity production in 2017 was: $0.072 − $0.048 = $0.024/kWh.
4. *Fixed cost:* All costs that are not affected by the level of business activity or units of the product. Examples are: capital investment, rents, insurance, property taxes, administrative salaries, advertising contracts, and interest on borrowed capital. Fixed costs are contractually determined and are paid regardless of whether or not anything is produced. In energy-related projects the capital investment costs are directly related to the exergetic efficiency of the project: More efficient equipment usually cost more. For example, (all other things being equal) heat pump systems with higher COP (or SEER) typically have higher cost than those with lower COP (or SEER).
5. *Life cycle* or *time horizon:* The time from the inception to the end of the project, including the time for the disposal and storage of equipment and products.
6. *Life-cycle cost:* The sum of all costs – fixed and variable – associated with a project from its inception to its conclusion. The life-cycle cost includes: the planning cost; the capital investment for construction and machinery; rents;

licenses; hiring and salaries of employees; fuel; salvage value; and any abandonment, disposal, and storage costs at the end of the project.

7. *Marginal or incremental cost:* The cost of production of one additional unit of product.
8. *Marginal or incremental revenue:* The revenue of the sale of one additional unit of product.
9. *Opportunity cost:* The monetary equivalent of what is sacrificed, when an investment is pursued. For example, the opportunity cost for a corporation to construct a new power plant is to invest their available capital in 6.8% interest-bearing bonds.
10. *Salvage value:* The price paid by an informed and willing buyer for a plant or business after its operations have terminated. Typically, this is related to the market value of the used equipment, the land, and the buildings. If there are cleaning costs associated with the abandonment or disposal of hazardous materials (e.g. in a nuclear site), the salvage value may be negative.
11. *Sunk cost:* All costs paid in the past, and associated with past activity, that may not be recovered and do not affect any future costs or revenues.
12. *Variable cost:* Any cost associated with the level of business activity or output level including fuel cost, materials cost, labor cost, distribution cost, sales commissions, etc. The variable cost increases monotonically with the number of units produced, but the relationship between units produced and variable cost is not necessarily linear. When the variable cost per unit output decreases with the number of units produced *economies of scale* are realized. In energy-related projects, fuel and materials costs are directly associated with exergy input.

7.3 Time Value of Money – Annualized Cost, Net Present Value

The concepts of the time-value of money and interest are based on the premise that $1 today is worth more than $1 a year from now, the latter is worth more than $1 two years from now, and so on. The time-value of money is determined using one or more of several variables including: the return to capital, the interest rate, the discount rate, the inflation, and the investment risk [6].

While most investments entail the risk of capital – partially or totally – some investments are considered *risk-free*. The interest rate associated with them, r_{rf}, is the lowest interest charged by the capital markets. Short-term USA government securities – the 3-month and 6-month governmental obligations, called *treasury paper* – are considered risk-free investments and typically have very low interest rates. Since the government is a reliable debtor and the investment timeframe is very short, the investors are certain that their capital and interest will be paid in full. The return on the capital expected for other types of debt is significantly higher. Financial institutions and investors lend funds to established corporations at rates 1–10% above the r_{rf} rate and to individuals at higher rates. Certain energy-related activities (e.g. drilling for oil and gas, electricity generation with PV, and thermal solar electricity generation[1]), are considered among the riskier investments and are associated with higher interest rates. The type of the project, capital availability, and market conditions are variables that determine the interest rates charged for all energy projects.

The *compound interest* is used in most commercial transactions, including loans and bonds issued to finance energy projects. With compound interest, effectively, the accrued interest at the end of each period is added to the original capital, P, and generates more interest. The formula for calculating the total amount to be paid in the future with compound interest is:

$$T = P(1+r)^N, \qquad (7.5)$$

where N is the period of the loan, usually measured in years or months. The total interest paid at the end of the N periods is: $I = T - P$. It is apparent that the time period of compounding, N, is as important as the value of the interest rate, r, in the calculation of the interest paid, as Example 7.2 illustrates.

Example 7.2: An entrepreneur borrows $50,000 to be paid after 5 years and has the option to choose either a 12% annual rate or 0.97% monthly rate. Which option should be chosen?

Solution: For both options $P = 50,000$. For the option of the annual rate $N = 5$, $I = 0.12$, and the total payment after 5 years is: $T = \$88,117$. For the option of the monthly rate $N = 60$ (= $12 \ast 5$), $I = 0.0097$ and the total payment after 5 years is: $T = \$89,230$.

The first option of 12% annual rate should be chosen by the entrepreneur.

7.3.1 Equivalence of Funds, Present Value, and Annual Worth

Monetary equivalency of funds implies that an informed investor is indifferent between two sums of funds to be received at different times. For example, if an informed investor is indifferent between $120,000 to be paid in the year 2025 and $140,000 to be paid in 2027, then these 2 sums are *equivalent* to the investor. A consequence of the equivalency principle is that if the same investor is offered $120,000 to be paid in 2025 or $135,000 to be paid in 2027, the investor will unequivocally choose the first option.

The *Present Value*, P_V, of a series of funds paid in the future – or *future returns* – is a method to establish a basis for the comparison of projects. The P_V establishes the

---

[1] Because of this, several governments offer *loan guarantees* for solar, wind, and other renewable energy projects. With governmental guarantees the interest rates charged for renewable energy projects are lower.

currently equivalent monetary value of future returns for informed investors and corporations. The present value, P_V, of an amount, M, to be obtained after N time periods at a *discount rate*, r_d, is:

$$P_V = \frac{M}{(1+r_d)^N}. \tag{7.6}$$

This implies that an amount equal to P_V today is equivalent to a greater amount M after N periods of time – typically after N years. An informed investor is indifferent between a sum of P_V today and a sum M, to be received after N periods of time. A glance at Eqs. (7.5) and (7.6) prove that there is a correspondence between the present value, P_V, and the compound interest, with the discount rate being equivalent to the interest rate: P_V is equivalent to the principal, P, invested today and the future return, M, is equivalent to the total sum, T, expected after N time periods.

Oftentimes there is a constant amount of funds received, expenditures paid, and income derived, that is repeated regularly for several time periods. Examples are mortgages, annuities, and bond payments earned or expended over several years. The constant returns, A, earned or spent in every period of the duration of the project, starting in period 0 (the present), and ending in period N has a present value equal to:

$$P_{VA} = A + \frac{A}{(1+r_d)} + \frac{A}{(1+r_d)^2} + \cdots + \frac{A}{(1+r_d)^N} = A\left[\frac{(1+r_d)^N - 1}{(1+r_d)^N r_d}\right]. \tag{7.7}$$

In several books on economics the present value of the annuity is denoted as (P/A, r_d, N), an expression that specifies the time period and interest rate, two important variables in the calculation of P_{VA}.

Effectively, the concept of the present value is a financial tool that offers an equivalent monetary value for projects during their entire life cycle and bypasses the timeframe of the duration of projects and investments. While the discount rate, r_d, is an important variable for the calculation of the present value, it is neither universal nor time-invariant and has a complex dependence on several other variables, including:

1. The type of corporation that performs the project, public or private. In general, public and nonprofit organizations use lower discount rates.
2. The inherent risk of the project. The riskier the project (e.g. because it uses new and unproven technology), the higher the discount rate is. Building an engine to harvest the waves off the coast of Scotland is a riskier investment than substituting the incandescent bulbs with LEDs in an office building at the heart of Edinburgh.
3. The overall economy and ease of borrowing capital. In general, when the overall economy is recovering, interest rates are lower and so are the discount rates for the corporations.
4. Inflation. Expected future inflation is monotonically correlated with discount rates that rise significantly in years when high inflation is expected.

5. The length of the project. Longer projects are associated with higher economic uncertainty; they are considered to have higher risk; and they bear higher discount rates. Most energy-related projects are long-duration projects: the construction and operation of coal and nuclear power plants; the purchase of a gas turbine for ship propulsion; the addition of a reheater in a power plant; the installation of a GSHP in a building, are all projects expected to last from 20 to 50 years or more. For this reason, they are considered riskier than other investments.

The *Net Present Value (NPV)*, sometimes called *Present Worth (PW)* and *Net Present Worth (NPW)*, is a method used for the comparison of projects and investments. The NPV of a project is the sum of the discounted annual cash flows from the inception of the project to its final disposal. The cash flow, *CF*, in a given period – typically one year – is essentially the sum of all receipts and expenditures related to the project. For a project that commences in year 0 (the present) and is completed in year *N* the NPV is:

$$NPV = CF_0 + \frac{CF_1}{(1+r_d)} + \frac{CF_2}{(1+r_d)^2} + \cdots + \frac{CF_N}{(1+r_d)^N} = \sum_{i=0}^{N} \frac{CF_i}{(1+r_d)^i}. \quad (7.8)$$

The *Annual Worth (AW)* of a project is derived from the NPV. It determines the equivalent annual amount, *AW*, of all the revenues or losses, during the time horizon (life-cycle) of the project. Using Eq. (7.7) the annual worth of a project is:

$$NPV = AW \left[\frac{(1+r_d)^N - 1}{(1+r_d)^N r_d} \right] \Rightarrow AW = NPV \frac{(1+r_d)^N r_d}{(1+r_d)^N - 1}. \quad (7.9)$$

Effectively, the *AW* spreads the NPV of a project in the entire lifecycle of the project as an annuity. The method is frequently used with energy-related projects that may be repeated for long periods of time [1, 4]. The annual worth method may also be applied to the costs of capital investment projects to determine their *annual cost of capital investment*, C_{CI}, which is widely used in exergoeconomic analyses.

Example 7.3: The addition of a regenerator in a gas turbine costs $135,000 and is expected to operate for 20 years. What is the annual cost of this investment if the discount rate of the corporation is 7%, 10%, and 15%?

Solution: Using Eq. (7.9) with the capital cost of $135,000 as the *NPV* we have:

For $r_d = 0.07, C_{CI} = \$12,743$.

For $r_d = 0.10, C_{CI} = \$15,857$.

For $r_d = 0.15, C_{CI} = \$21,658$.

It is observed that a high discount rate causes significant increases in the annual cost of capital investments.

7.4 Thermoeconomics and Exergoeconomics

The decision-making procedure for engineering, and in particular for energy systems, involves several criteria – profitability, environmental stewardship, public perception, social welfare, etc. The criterion of profitability, which is related to the economic parameters of projects, is usually given the highest weight in decision making. Unprofitable projects are not undertaken by corporations in a market-oriented economy. *Thermoeconomics* and *Exergoeconomics* combine principles and concepts from thermal science – thermodynamics, fluid mechanics, and heat transfer – with principles of economics – in particular, cost analysis – to facilitate the decision making procedure of energy systems.

7.4.1 Cost Analysis Methods

The method of thermoeconomics essentially utilizes one or more of the concepts from thermal science – internal energy, enthalpy, heat, work, entropy, and exergy. The method assigns monetary unit costs to these variables and formulates a cost function for the system – the objective function – to be minimized. Thermoeconomics – using energy, work, and heat as the pertinent variables – has been used in early optimization studies to determine the marginal cost of these variables at the several states of thermal systems. The cost minimization information is used to optimize and improve the operation of the energy systems. More recent studies have used either energy or exergy as the pertinent variable that expresses cost [7].

A characteristic of the thermoeconomics method is the combination of all the costs of system components, including fixed costs and capital investment costs, in a single objective function, the *cost function*. The objective function of the cost is expressed for a period of operation, usually one year. The cost function of a single component has the form:

$$C = c_F V_F + C_{CI} + C_R, \tag{7.10}$$

where c_F is the average price of the fuel and V_F is the volume or mass of fuel consumed in a year, with $c_F V_F$ representing the annual cost of the fuel; C_{CI} is the annual cost of capital investment of the component; and C_R is the annual cost of all the other expenses associated with the component. The annual cost of capital investment is obtained from the total capital investment associated with the component, TC_{CI}, using an equation similar to the annual worth, Eq. (7.9):

$$C_{CI} = TC_{CI} \frac{(1+r_d)^N r_d}{(1+r_d)^N - 1}. \tag{7.11}$$

At the optimum operation of the component the cost function, expressed by Eq. (7.10), is minimized.

Alternatively, a *profitability function* may be defined for a component that produces a number of products, P, [7]. The profitability function is to be maximized:

$$J_{prof} = \sum_{i=1}^{P} c_i N_i - (c_F V_F + C_{CI} + C_R), \tag{7.12}$$

where c_i is the average price of the product i; N_i is the number of units of product i generated annually; and P is the total number of products generated by the component. The first term on the r.h.s. represents the annual revenue derived from the products and the terms in parenthesis represent the annual cost for their production, as in Eq. (7.10).

It must be noted – and this is one of the disadvantages of the thermoeconomics method – that the optimization problem based on the profitability function is not the dual of the problem of the cost minimization function. The minimization of Eq. (7.10) is different than the maximization of Eq. (7.12) and the two approaches often lead to different solutions.

Example 7.4: An electricity production corporation considers the investment of $5,600,000 in a new 30 MW gas turbine. The annual fuel cost of the turbine is estimated to be $3,700,000, the maintenance and operation costs are $400,000, and the turbine is expected to operate for 30 years and produce annual revenue $5,200,000. The discount rate of the corporation is 12%. Determine the profitability of the gas turbine.

Solution: The annualized cost of capital investment is obtained from Eq. (7.11) $C_{CI} = \$695,200$.

Therefore: $J_{prof} = 5,200,000 - 3,700,000 - 400,000 - 695,200 = \$404,800$.

The method of *exergoeconomics* is similar to thermoeconomics and is based on the stipulation that exergy is a unique variable – the only variable that assigns realistic monetary costs to the inputs and outputs of thermal systems as well as to the irreversibilities within the systems. All the time-rates of inputs and outputs of systems and components are assigned costs based on exergy rates alone [8, 9]. The exergy rates may be defined for any time period – daily, weekly, monthly, annually – with annual rates being the most convenient parameters to use. The method is applied to every component of a larger system to calculate periodic (most often annual) exergy rate costs at the outlets of the components and to deduce the exergy cost at all the junctures of the system.

For the general schematic diagram of a component with m inlets and n outlets, which is depicted in Figure 7.1, the exergoeconomics method uses the following procedure:

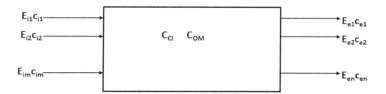

Figure 7.1 Schematic diagram of a component, used for the derivation of the cost balance equation, with the pertinent costs.

1. Assign unit monetary costs, c_{ij} and c_{ej} (e.g. in \$/kJ) to the rates of exergy, E_{ij} and E_{ej}, for all the streams that enter and leave the component.
2. Compute the capital investment costs of the component as a recurring periodic (most often annual) cost, C_{CI}, using an annual worth method.
3. Add all the annual fixed and variable costs, C_{OM}, other than the exergy costs of the inputs.
4. Construct an *exergy cost balance equation* for the component:

$$\sum_{k=1}^{m} c_{ik} E_{ik} + C_{CI} + C_{OM} = \sum_{k=1}^{n} c_{ek} E_{ek}. \qquad (7.13)$$

Typically, components have one or two inlets and outlets [8]. Large energy systems have a number of components, for which a cost balance equation is written. Thus, the number of the cost balance equations is equal to the number of components. For the purpose of the calculations, electric power, work, and heat are considered as streams that enter and exit the components. The unit costs of the exergy inlets at the system level are computed using the market prices of the fuels and other inputs as the annual cost of the streams. For example, the average annual cost of a fuel input per unit exergy of the fuel defines the cost, c_{ij}, with units \$/kJ, €/kJ, £/kJ, etc. The annual cost is calculated by summing the cost of all the fuel used during a year and dividing this cost by the total exergy of the fuel. The unit outlet costs are determined either by closure (auxiliary) equations or by the solution of one of the cost balance equations of the entire system. Several publications and a treatise on exergoeconomics [10] include examples of auxiliary equations and component cost determination.

The outcomes of exergoeconomics analyses are the costs of exergy at all the states of a larger, complex system. After the exergy costs of inputs and outputs are determined, other figures of merit – cost differences, cost ratios, relative costs – may be calculated to assist with the optimization process. Exergetic improvements to the system are accomplished at the components that have the highest exergy costs, or higher relative exergy cost in comparison to other components. Alternatively, if the capital investment and operating costs of a component are very high in comparison to other costs, a reduction of these costs may be recommended for the component at the expense of exergetic efficiency.

One of the drawbacks of thermoeconomics and exergoeconomics analyses is that, because of lack of detailed information, the variable and fixed cost assignments of the several components of a complex system are often allocated subjectively by those performing the optimization and this introduces a personal bias in the calculations, which significantly affects the optimization outcome.

Exergoeconomics analyses, when properly implemented, identify the location and magnitude of exergy destruction in larger energy systems; calculate their monetary costs; and compare the effects of these costs on the product(s) of the system. This information may be used by engineers to identify changes in the system that would reduce the exergetic and monetary cost of the product(s) and would improve the operation of the system. Because the improvements are not always obvious, intuition, experience, and correct judgment are helpful attributes for the engineers who undertake the system improvements.

An extension of the method is the *Extended Exergy Accounting (EEA)* [11, 12], which transforms all the monetary, environmental, and labor costs during the life-cycle of a system to exergy costs (e.g. in kJ or MJ). The exergy-equivalent costs of capital investment, labor, environmental effects, and all other costs are computed as exergy costs, in a general cost-accounting process. Using an exergetic definition for capital (defined as the exergetic equivalent of resources) and labor (the exergetic resource of humans divided by the generated working hours), this method evaluates all the inputs and outputs in terms of exergy, and the objective function (cost minimization) is expressed as exergy destruction minimization. A major problem with the EEA is that the definitions for the labor and capital significantly diverge from the internationally accepted norms and standards for labor and capital, which makes the method incongruent with conventional economic and financial optimization methods. A second limitation is related to the uncertainty associated with the determination of the environmental costs (how clean is clean?), even if this is expressed in exergy terms, as outlined at the end of Section 6.3.1.

In all the optimization studies one has to accommodate capital investment, capital availability, and labor. Unlike the natural resources that are finite and limited, capital (in monetary terms) is created at will by the sovereign governments and central banks. Most sovereign governments run annual budget deficits with funds (capital) injected into the economies by legislative *fiat*. During periods of economic recession, the national deficits grow to create additional capital, which is borrowed by the corporations to undertake projects that stimulate the national economies.[2] Labor, in all its forms and specialties, is also one of the resources that grow with time as the population of the planet increases and has doubled since the 1950s. Because of better educational opportunities, specialized labor (e.g. the number of mechanical engineers and the number of manufacturing technicians) is constantly growing, significantly surpassing the global population growth.

7.4.2 Calculus Optimization

Calculus optimization may be used in thermoeconomics and exergoeconomics analyses when the system to be optimized is simple and the optimization is expressed in terms of a single design variable. Let us consider a simple energy system – a refrigeration plant, a heat pump, a commercial vehicle, or an engine that produces a single product – which satisfies a desired action – keeping a building at the desired temperature range, transporting a number of merchandize at a given distance, producing a number of product units per year, etc. The system encumbers capital cost, paid at the time of its purchase; it operates simply without other fixed costs; and its variable cost is only the exergy consumed. In this case it is convenient to express the capital cost of the system

---

[2] During the 2008 recession, the USA Government and Federal Reserve created and injected $787 billion of capital into the American economy. Many sovereign governments followed suit.

as an annualized cost, using an equation similar to Eq. (7.7) or (7.9) and express the total annual cost as follows:

$$C_T = C_E + C_{CI}, \tag{7.14}$$

where C_T is the total annual cost (expressed in units of a currency per year, e.g. \$/yr); C_E is the annual exergy cost for the operation of the system; and C_{CI} is the annualized capital cost.

Empirical evidence shows that the capital cost of most energy systems is associated with the size of the systems and their exergetic efficiency, η_{II}. Systems with higher exergetic efficiency have higher capital cost. A variable associated with the size of the system is the annual exergy input of the system. A simple equation that expresses this relationship is:

$$C_{CI} = c_1 E_{in} + C_2 \eta_{II}, \tag{7.15}$$

where, E_{in}, is the annual exergy input into the system, c_1 has units of currency per exergy unit per year and C_2, has units currency per year.

Similarly, the annual cost, C_E, may be expressed in terms of the annual exergy input of the system, E_{in}, and the total cost may be expressed in terms of the annual exergy output, E_{out}:

$$\begin{aligned} C_T &= c_T E_{out} \\ C_E &= c_E E_{in} \end{aligned}, \tag{7.16}$$

where c_T and c_E are cost parameters with units currency per unit exergy per year. The ratio E_{out}/E_{in} is the exergetic efficiency of the system, η_{II}. Substituting these variables in Eq. (7.14) the total unit exergy output cost, c_T, is expressed as follows:

$$c_T = (c_E + c_1)\frac{1}{\eta_{II}} + \frac{C_2}{E_{out}}\eta_{II}. \tag{7.17}$$

For fixed cost coefficients and exergy output, the total cost, C_T (or equivalently the unit cost, c_T), of the system is minimized when:

$$\frac{\partial c_T}{\partial \eta_{II}} = 0 \Rightarrow \eta_{II} = \sqrt{\frac{C_2}{E_{out}(c_1 + c_E)}}. \tag{7.18}$$

This is the condition that may be used for the optimum type of system to be purchased. It must be noted that the calculus optimization method is restricted to systems with a single optimization variable, which are, in general, simple energy systems. An inherent assumption of the method is that the cost functions are continuous functions of the variable, η_{II}.

7.4.3 Lagrange Undetermined Multipliers

This method is used for the optimization of complex systems with several design parameters and multiple constraints. Let us consider a large energy system, which has overall exergy input, E_{in}, and overall exergy output, E_{out}. The system is composed of

7.4 Thermoeconomics and Exergoeconomics

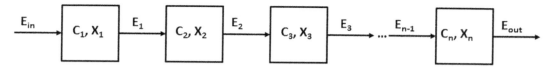

Figure 7.2 A general system with n interacting components to be optimized.

n subsystems (components), each of which is associated with exergy input, E_i; exergy output, E_{i+1}; annual cost for the equipment acquisition and operation C_i; and an inherent design parameter that may be optimized, X_i. Examples of such complex systems are electric power plants, petroleum refineries, and ship engines. The schematic diagram of such a complex system is depicted in Figure 7.2, where the arrows indicate material flows, energy and exergy exchanges. The total annualized cost for the acquisition and operation of the system is equal to the sum of the cost of the exergy input and the annualized cost of the n components:

$$C_T = c_1 E_{in} + C_1 + C_2 + \cdots + C_n. \tag{7.19}$$

This is the objective function of the system to be minimized. For a given product (i.e. a specified exergy output, E_{out}, of the system) one may write the exergy inputs in the components of the larger system as functions of the exergy output of the component and the associated design parameter, X_i:

$$E_{in} = F_1(X_1, E_1), \quad E_1 = F_2(X_2, E_2), \quad E_2 = F_3(X_3, E_3), \ldots, E_{n-1} = F_n(X_n). \tag{7.20}$$

Similarly, the annualized cost function of each unit may be expressed in terms of the exergy output and the design parameter, X_i, associated with component i:

$$C_1 = f_1(X_1, E_1), \quad C_2 = f_2(X_2, E_2), \quad C_3 = f_3(X_3, E_3), \ldots, C_n = f_n(X_n). \tag{7.21}$$

Therefore, the total annual cost for the acquisition and operation of the overall system is:

$$C_T = c_1 E_{in} + f_1(X_1, E_1) + f_2(X_2, E_2) + f_3(X_3, E_3) + \cdots + f_n(X_n). \tag{7.22}$$

The constraints associated with the operation of the entire system with n components may be formulated by rewriting the n equations in Eq. (7.20) as:

$$F_1(X_1, E_1) - E_{in} = 0, \quad F_2(X_2, E_2) - E_1 = 0, \quad F_3(X_3, E_3) - E_2 = 0, \ldots \\ F_n(X_n) - E_{n-1} = 0 \tag{7.23}$$

For the cost minimization of the n components of the entire system, the associated *Lagrangian function* is formed. This function is a linear combination of the objective function, Eq. (7.19), and the constraints of Eq. (7.23):

$$L = c_1 E_{in} + C_1 + C_2 + \cdots + C_n - \sum_{i=1}^{n} \lambda_i [F_i(X_i, E_i) - E_{i-1}], \tag{7.24}$$

where $E_0 = E_{in}$, and the coefficients, λ_i, are the *Lagrange undetermined multipliers*. An inspection of Eq. (7.24) reveals that, in the case of the optimization of an energy system, the Lagrange multipliers have dimensions of currency per unit exergy (e.g. \$/kJ, €/kWh, etc.). The value of λ_i may be interpreted as the marginal cost of exergy at the exit of the *i*th component of the system. The optimization is accomplished when the derivatives of L with respect to the design variables X_i vanish:

$$\frac{\partial L}{\partial X_i} = 0 \Rightarrow \frac{\partial}{\partial X_i}\left\{c_1 E_{in} + \sum_{i=1}^{n}[f_i(X_i, E_i) - \lambda_i[F_i(X_i, E_i) - E_{i-1}]]\right\}. \quad (7.25)$$

The n equations of the last expression and the n constraints of Eq. (7.20) are solved as a system of $2n$ simultaneous equations to determine the n design parameters X_i, and the n multipliers, λ_i, which represent the cost of exergy at the output of every component of the system. The information on exergy costs is used to facilitate improvements of the components and of the entire system: improvements of the exergetic efficiencies of components with the highest exergy cost are prioritized and performed first.

7.4.4 Alternative Optimization Objectives

While the previous sections pertain to the economic optimization of energy systems with objective functions based on cost minimization or profit maximization, it has become apparent in the early 21st century that other optimization objectives are of importance to the community. Such objectives are:

1. Environmental, for example a cleaner environment or minimization of GHGs emitted.
2. Ecological effects, for example the preservation of one or more endangered species.
3. Social, for example diversity and equity among workers in a project.
4. Employment, for example local jobs created by a project.
5. Sustainability, for example accessibility and availability of resources for future generations.
6. Societal risk, for example risk minimization to the population from accidents.
7. Reliability, for example the uninterrupted supply of electricity.
8. Infrastructure development, for example concurrent construction of roads and bridges.
9. Regional economic development, for example creation of new industries and jobs.

The inclusion of additional objectives in optimization studies is desirable [13] but not always feasible. When this becomes feasible it is through the introduction of objective functions based on *ad hoc* assumptions and a great deal of empirical data that are not universal, but apply to specific times and regions.

It is axiomatic that, when more than one objective is considered, the objectives will not be congruent and will not lead to the same solution of the optimization problem. Multiple objectives may be contradictory and conflicting. For example, when the least cost option is considered for a factory and optimization endorses the installation of the least labor-intensive machinery, fewer jobs are created by the factory and local employment suffers. Currently, the minimum economic cost of electric energy production in several countries – including the USA, China, the UK, and France – is by coal and nuclear power plants. However, building more of these power plants is in conflict with environmental, sustainability, and societal risk objectives.

The main limitation in optimization studies with incongruent objectives is the formulation of an objective function that accurately and quantitatively expresses the several objectives. The objective function of these methods rely on a system of *ad hoc* auxiliary equations, environmental indicators, and eco-indicators to assess the environmental impact caused by the multitude of chemical components, but these indicators are largely subjective and vary with time and country (jurisdiction). As a consequence, the application of such optimization methods with incongruent objectives may yield different "optimization results" at different times and at different jurisdictions.

The application of economic optimization with environmental and social aspects, including local employment, was the subject of a study on forest biomass utilization [14]. The environmental and social aspects of optimization are incorporated with the profit from operations into a multi-objective linear programming model that maximizes the NPV of the project, the social benefit of local employment, and the environmental impact of GHG avoidance. The effect of incongruent objectives is apparent in this study with the "trade-offs" between maximizing the NPV (the economic objective) the local employment (the social objective) and the environmental objective. An additional observation is that – in the absence of methodologically researched and generally accepted norms – the assignment of coefficients in the optimization model is highly subjective. As a result, the optimization outcomes are very much influenced by these subjective coefficients. The lack of accurate and widely accepted criteria for the formulation of multi-objective functions is also apparent in [15], which gives an account of sustainability metrics.

7.5 Uncertainty and Other Limitations

Thermodynamics is a branch of Physics, based on physical principles – the laws of thermodynamics – that have been proven to be time-invariant, precise, and accurate. All the pertinent experiments and the entire body of scientific theory of the last two centuries have proven the accuracy of the laws of thermodynamics, on which the concepts of energy and exergy are based. The numerical values of thermodynamic variables are well defined, have very low uncertainty, and are time-invariant: the latent heat of water at 0.1 MPa is $h_{fg} = 2,258.0$ kJ/kg; the uncertainty of this value is less than ± 0.4 kJ/kg ($\pm 0.018\%$); and everyone may count on the fact that this value will be the same after 30 years and after 3 centuries. Similarly, exergy is a variable based on physical laws and the state of the environment. For an environment at $P_0 = 0.1$ MPa and $T_0 = 300$ K, the specific exergy of hexane at $P = 0.5$ MPa and $T = 450$ K is 142.8 kJ/kg with uncertainty ± 0.7 kJ/kg ($\pm 0.49\%$). It is universally accepted that this numerical value will not change outside of its uncertainty limits at any time in the future.

Economics is a social science, based on empirical principles that are derived from contemporary observations of persons and group behaviors. The principles of economics emanate from correlations of data; they are not permanent; and heavily depend on the time of observations and the type of economy they were made (e.g. the price mechanisms are different in free-market economies, regulated economies, and centrally-planned economies). Prices, from which the exergetic costs are derived, reflect the ephemeral scarcity or availability of goods, the markets, financing, and consumer preferences. Unlike internal energy and exergy, the prices of fuels do not reflect any physical laws. As a result, the values of economic variables are time-dependent – actually ephemeral – they exhibit high variability in long time periods, and are laden with significant uncertainty. Economic variables associated with energy (e.g. energy prices and interest rates for energy projects) exhibit higher uncertainty because (in addition to all other effects) the economics of energy are subjected to geopolitical and military events [3, 16]. Figure 7.3 depicts the nominal and real prices of anthracite and gasoline in the USA in the period 1980–2015 [17, 18]. The real prices are expressed as US dollars of 2015. It is observed that both sets of prices exhibit significant fluctuations during this period, which is on the same order of magnitude as the life cycle of an energy project. Table 7.1 shows the time-average of the prices; their standard deviation in the period 1980–2015; and the standard deviation as a percentage of the average.

It is observed in Table 7.1 that the standard deviation of all the prices represents a high fraction of the average. As a consequence, the uncertainty of the gasoline and anthracite prices is significantly higher than the uncertainty of the thermodynamic

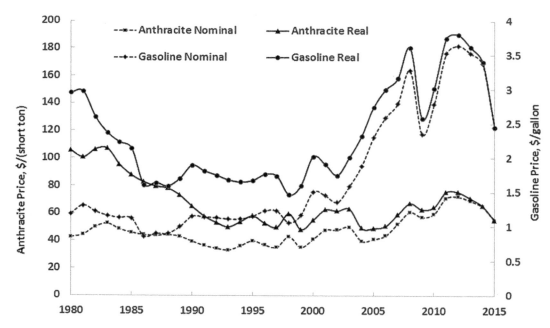

Figure 7.3 Fluctuations of the nominal and real prices of anthracite and gasoline in the USA between 1980 and 2015. Data from [17,18]

7.5 Uncertainty and Other Limitations

Table 7.1 Statistical Factors for the Nominal and Real Prices of Gasoline and Anthracite in the USA. Compilation of data from [17,18].

	Gasoline		Anthracite	
	Nominal ($/gal)	Real ($2015/gal)	Nominal ($/ton)	Real ($2015/ton)
Average	1.74	2.33	47.18	67.95
Standard deviation	0.88	0.71	10.41	17.28
Standard deviation (%)	50.27	30.42	22.07	25.43

variables. If one defines the uncertainty of the prices as 2 standard deviations of the respective variables (95.5% confidence level for normal distributions), then the uncertainty of the real value of the price of gasoline is 60.84% and that of anthracite 50.86%.

All optimization methods make use of the energy prices, in one form or another, to derive realistic costs for the energy and exergy inputs of the systems to be optimized. The inputs in the general system depicted in Figure 7.1 are the annual rates of exergy, E_{ij}, and the cost per unit exergy c_{ij}. The first is derived from calculations of thermodynamic properties and its uncertainty is on the order of 1% (or less for water and air). The second is derived from current prices of energy and its uncertainty during the life cycle of the project is on the order of 50%. The percent uncertainty of the product, $E_{ij} * c_{ij}$, may be obtained using the uncertainty propagation formula:

$$\frac{\delta(E_{ij}c_{ij})}{E_{ij}c_{ij}} \approx \frac{1}{E_{ij}c_{ij}} \sqrt{\left[\frac{\partial(E_{ij}c_{ij})}{\partial E_{ij}}\delta E_{ij}\right]^2 + \left[\frac{\partial(E_{ij}c_{ij})}{\partial c_{ij}}\delta c_{ij}\right]^2}. \quad (7.26)$$

With the percent uncertainty values of 1% and 50% for exergy and unit cost respectively, the estimate for the uncertainty of the input costs in Figure 7.1 is slightly higher than 50%.

This simple estimate of the fuel cost uncertainty demonstrates that the coefficients of the cost minimization objective functions and the cost balance functions in all LCA studies are subject to high uncertainty, much higher than the typical uncertainties encountered with thermodynamic calculations. Because the results of the optimization studies heavily depend on the coefficients of these functions, one must be cognizant that the optimization results are also laden with significant uncertainty.

Other limitations of optimization based on thermoeconomics and exergoeconomics analyses are:

1. A key element of the two analyses is the expression of the capital investment cost as an annual cost, C_{CI}, which is obtained from an equation similar to Eq. (7.11) that includes the discount rate, r_d. While this rate is readily quoted as a constant in academic studies (e.g. 12%, 15%, 20%) the determination of a realistic discount rate for corporations is complex and depends on a variety of economic factors such as: the rate of inflation; the prevailing interest rates; the state of the economy; the availability of capital; the type of corporation; the inherent risk

of the project; and the length of the project. Oftentimes a set of different discount rates is used for the different time-periods of a realistic project.

2. The type of financing of the project adds an extra layer of complexity in the determination of r_d. Oftentimes, large projects are financed by a combination of equity and bonds that bear an interest rate, i, and are sold in the market. The equity capital is made available by the corporation and is subjected to the internal discount rate, r_d, which is significantly higher than the interest rate of the bonds, i. Because the corporation takes a financial risk by issuing the bonds the true discount rate for the bond-financing portion of the capital is higher than i, and is subjectively determined by experienced managers.

3. The life cycles of most energy systems are long and any optimization based on LCA entails predictions long into the future. Nuclear and coal power plants, which were constructed with life cycles of 40 years in the 1950s, have received extensions and now operate for more than 60 years; aircrafts and their engines operate for more than 30 years; industrial boilers operate for more than 40 years; and simple household air-conditioning systems operate for more than 30 years. Technological advances occur during such long periods that are impossible to take into account at the inception of the projects. An example is the remarkable advances in gas turbines between 2010 and 2020 that enabled engineers to manufacture very efficient (close to 65% exergetic efficiency) gas turbine systems. Such high efficiencies were unknown in the 1980s and 1990s, when many of the gas turbines that operated in 2020 were manufactured.

4. The closure (auxiliary) equations that are used for the assignment of costs in the general cost balance equation, Eq. (7.13), are arbitrary and based on *ad hoc* assumptions and transitory data from economics that may easily change over the life cycle of the project. It is not necessary for the closure equations to comply with physical laws, such as the laws of thermodynamics.

5. The inclusion of environmental costs or benefits in the optimization function – quantities that are not standardized, but are highly subjective and differ in the different regions of the globe – significantly increases the uncertainty of the optimization outcome, as it is explained in more detail at the end of Section 6.3.1.

6. Several financial factors that affect the *NPV* of projects are difficult to be expressed as annualized costs. Such factors are chiefly related to taxation procedures and include the allowed depreciation of the capital investment, which affects the annual taxation costs; and governmental investment tax credits and subsidies (several of which expire on a future date) that have become commonplace with renewable energy systems and are accrued for a limited number of years.

7. Most optimization methods are based on the inherent assumption that the variables in the cost and profitability functions to be optimized are independent and continuous, while several of the variables of realistic systems are not. For example, the ratings (hp or kW) of pumps and turbines and the diameters of pipes are not continuous variables. Given that components and systems come in predefined sizes with discrete efficiencies, it is doubtful that components can be found in the market with size and efficiency equal to those

recommended by the optimization results. In such cases, discrete optimization methods must be considered.
8. According to strict mathematical theory, thermoeconomics and exergoeconomics are cost determination methods and not optimization methods. The optimization constraints are often implicit or included in the calculations for exergy. The addition of *ad hoc* auxiliary equations further complicates this matter and makes the entire problem an *ill-posed* problem mathematically that does not have a unique solution. The ill-posedeness was manifested in the so called *CGAM problem* – a 30 MW cogeneration unit – where 4 different teams derived different optimized solutions for the same problem [19]. This notwithstanding, one must note that the four solutions were within the uncertainty of the input data, and that in all four cases engineers with sound judgement would have made good use of the generated solutions.

In summary, the results of thermoeconomics and exergoeconomics evaluations are subject to high uncertainties, primarily because the methods apply concepts, equations, and ephemeral data from the discipline of economics. The uncertainty notwithstanding, one may obtain useful information on the operation of energy systems and may judiciously use this information in the design of energy systems. The experience, operational skill, ingenuity in interpreting results, intuition, and correct judgment are attributes of paramount importance for the engineers who would undertake the design of these systems.

Problems

1. Obtain access in your institution to the computer software that solve the linear optimization problems and learn how to use this software. Then solve the problem of Example 7.1 and its dual using the software.

2. Using your utility bills calculate the average cost of your electric energy consumption in the last six months. You may need to first separate the fixed costs and other fees from your calculations.

3. A corporation offers the following prices and discounts for the purchase of laptop computers:

1–5 computers: $1,400 each.
5–10 computers: $1,200 each.
10–20 computers: $1,100 each.
20–50 computers: $1,000 each.
More than 50 computers: $850 each.

In the month of June 2019 the corporation received separate orders to sell 7, 8, 25, 12, 75, and 35 laptop computers. What is the total revenue and the average revenue during the month?

4. You have an option to use one of 2 credit cards: the first advertises a 12.4% annual rate and the second advertises a simple 1% monthly rate. Which one would you choose?

5. What is the NPV of $10,000 paid annually for 40 years when the discount rate is 5%? How about when the discount rate is 8% and 15%?
6. What is the annual worth for the next 6 years of $50,000 worth today with 8% discount rate?
7. An electric utility considers the investment of $10,000,000 for improvements in an existing nuclear power plant. The annual revenue increase expected is $1,800,000 for 30 years. It is also expected that the annual maintenance and operation costs will increase by $150,000. The discount rate of the corporation is 12%. Determine the NPV and the profitability for this project. For the calculations assume that the investment happens at the beginning (year 0) of the project and all annual revenues and expenses occur at the end of each year.
8. A regenerator is to be added to a gas turbine unit that is expected to be in operation for 15 more years. The cost of the regenerator is $210,000. What is the annual cost of this improvement if the discount rate of the utility is 11%?
9. Explain what are the similarities and differences between thermoeconomics and exergoeconomics.
10. What are the variables that would come into an exergoeconomics analysis of a photovoltaic energy generation system?
11. When multiple objectives are used in an optimization study, the objective function is typically a linear combination of the variables that measure each objective. Prove that such a multiple objective optimization has an infinite set of feasible solutions.
12. Using practical examples explain why the experience, ingenuity, and judgment of engineers are very much needed for the design of energy conversion systems.

References

[1] G. V. Reclaitis, A. Ravindran, and K.M. Ragsdell, *Engineering Optimization – Methods and Applications* (New York: Wiley, 1983).
[2] G. H. Hurlbert, *Linear Optimization* (New York: Springer, 2010).
[3] E. E. Michaelides, *Alternative Energy Sources* (Berlin: Springer, 2012).
[4] C. S. Park, *Contemporary Engineering Economics*, 4th ed. (New Jersey: Pearson, 2007).
[5] W. G. Sullivan, E. M. Wicks, and J. T. Luxhoj, *Engineering Economy*, 12th ed. (New Jersey: Pearson, 2003).
[6] E. E. Michaelides, *Energy, the Environment, and Sustainability* (Boca Raton, FL: CRC Press, 2018).
[7] M. M. El Sayed, *The Thermoeconomics of Energy Conversions* (Oxford: Elsevier, 2003).
[8] G. Tsatsaronis and F. Cziesla, Thermoeconomics. In R. A. Meyers, ed., *Encyclopedia of Physical Science and Technology*, 3rd ed., **16** (Academic Press, 2002), 659–80.
[9] G. Tsatsaronis, Thermoeconomic Analysis and Optimization of Energy Systems. *Progress in Energy and Combustion Science*, **19** (1993), 227–57.
[10] A. Bejan, G. Tsatsaronis, and M. J. Moran, *Thermal Design and Optimization* (New York: Wiley, 1996).
[11] E. Sciubba, Exergo-Economics: Thermodynamic Foundations of a More Rational Resource Use. *International Journal of Energy Research*, **29** (2005), 613–36.

[12] M. V. Rocco, E. Colombo, and E. Sciubba, Advances in Exergy Analysis: A Novel Assessment of the Extended Exergy Accounting Method. *Applied Energy*, **113** (2014), 1405–20.
[13] C. A. Frangopoulos, Recent Developments and Trends in Optimization of Energy Systems. *Energy*, **164** (2018), 1011–20.
[14] C. Cambero and T. Sowlati, Incorporating Social Benefits in Multi-Objective Optimization of Forest-Based Bioenergy and Biofuel Supply Chains. *Applied Energy*, **178** (2016), 721–35.
[15] N. Lior, Quantifying Sustainability for Energy Development. *Energy Bulletin* **19**, International Sustainable Energy Development Center under the Auspices of UNESCO (2015), 8–24.
[16] E. E. Michaelides, A New Model for the Lifetime of Fossil Fuel Resources. *Natural Resources Research*, **26** (2017), 161–75.
[17] Office of Energy Efficiency and Renewable Energy www.energy.gov/eere/vehicles/fact-915-march-7-2016-average-historical-annual-gasoline-pump-price-1929-2015, last visited November 2019.
[18] The U.S. Energy Information Administration www.eia.gov/totalenergy/data/annual/index.php, last visited November 2019.
[19] A. Valero, M. A. Lozano, L. Serra, G. Tsatsaronis, J. Pisa, C. Frangopoulos, and M. R. von Spakovsky, CGAM Problem: Definition and Conventional Solution. *Energy*, **19** (1994), 279–86

Index

abiogenesis, 243
acid rain, 251–52
affinity, 238
aging, 201, 242, 244–46, 250
air-conditioning, 147, 150–51, 155, 172–73, 175–77, 187, 191, 193, 196
algae, 201, 212–13, 248
atmosphere, 27–28, 38, 41, 43–44, 49, 65–67, 69–71, 73
ATP, 202, 204, 216–18, 220–21, 223, 226–30, 233–34, 236, 239–41, 249
average cost, 277
average profit, 277
average revenue, 277

batteries, 184, 194–96
Betz's limit, 53, 79, 141, 143–44
biological system, 214, 234, 238–39, 242
biomass, 201, 205–7, 209, 212–13, 248
boilers, 85
bottoming cycle, 98
Brayton cycle, 19, 95–98, 100, 105
buildings, 147, 149–50, 171–76, 191, 193, 196, 200
burners, 85

CAFE, 183
chemical exergy, 42–45, 48, 56, 68, 70, 74
chlorophyll, 201
coefficient of performance, 22
cogeneration, 79, 101–3, 125
compound interest, 279
compressed air storage, 189
condensers, 83
conduction, 6
conservation, 147–50, 171–72, 179
convection, 6
COP, 176

desalination, 147, 169–70
desired actions, 147–49, 151, 171, 181
discount rate, 280
distillation, 164–65, 167–68, 186
drying, 161–62, 199

dual, 275, 283
dual flashing, 112

eco-exergy, 251, 253–55, 259, 261
eco-indicators, 251, 260, 262
ecosystems, 251–53, 255–59, 261–62, 267, 269–70
efficiencies, 15
electric cars, 184
emergy, 253, 269
energy consumption, 2
energy efficiency, 150
energy storage, 147, 188–91, 193, 196–97, 200
energy substitution, 150, 179
energy supply, 3
entropy, 13
environment, 251–53, 255–62, 264, 266, 268–70
ethanol, 209
exergetic efficiency, 62–63, 71, 86, 103, 123, 126, 132, 135, 141, 144–45, 204–5, 208, 210–13, 217–18, 227–28, 230, 234, 236, 243, 257, 259, 263–65, 269–70, 277, 284, 286, 292
exergoeconomics, 273, 282–85, 291, 293
exergoenvironmental, 251, 259, 270
exergy conservation, 148
exergy destruction, 45, 57, 60–61, 70, 75–76
exergy of closed systems, 32
exergy of open systems, 35

feed-water heaters, 89
first law of thermodynamics, 8
fixed costs, 277
fuel cells, 79, 95, 120–23, 125, 186, 196, 199

gas compression, 153
gasohol, 209
Gemasol, 137
generalized fluxes, 59
generalized forces, 59–60
GHG, 251
Gibbs free energy, 4, 194
glycolysis, 217, 226–28
GSHP, 176

Index

heat exchangers, 79–80, 82, 90
heat reservoirs, 22, 25
heat of combustion, 45
heat pump, 148–50, 167–68, 172, 174–79, 198
heating value, 10
hot rock, 33
hydroelectric, 51
hydrogen storage, 196
hydrosphere, 27–28, 65–67, 69, 73

industrial ecology, 256
intercoolers, 154
irreversible, 5

jet Engines, 105, 107

Lagrange multipliers, 288
lead contamination, 251
life cycle analysis, 264, 270
life cycle cost, 277
lighting, 147–48, 171–73
liquefaction, 147, 156–60
longevity, 214, 242, 245–46
lost work, 57–58, 70, 75
low heating value (LHV), 207

marginal, 278
membranes, 36, 43–44, 46–48, 73
metabolism, 204, 217–18, 221–22, 230, 234–35, 239, 248
methanol, 209
model environment, 30
molten salts, 137
muscular system, 226–28

non-equilibrium thermodynamics, 58, 201, 238
nuclear fuel, 25, 28, 54–56, 63
nutrients, 201, 214–15, 218, 220–22, 230, 234, 236

opportunity cost, 278
optimization, 273–75, 277, 282–86, 288–89, 291–92, 294
optimization objectives, 288
Organic Rankine Cycles, 116
oxidative phosphorylation, 216, 239, 249
ozone depletion, 251

petroleum, 199
phenomenological coefficients, 60
photosynthesis, 201, 203–5, 209, 213, 243, 248
photovoltaic, 127, 130, 146

physical exergy, 37
pinch point, 86, 118
plutonium, 55
present value, PV, 279–80
pumped hydro systems, 189

radiation, 7, 48–50, 65, 126–27, 129, 132, 135–36, 138–40
Rankine cycle, 16, 92–94, 99, 102, 125, 137
reference environment, 68, 70–71
refrigeration, 20, 147, 155–59, 163, 175–76, 193, 197
reheaters, 85
respiratory quotient, 218
reversible, 5

salvage value, 278
second law of thermodynamics, 13
SEER, 176
single-flashing, 110
societal task, 148
solar, 52
solar collectors, 140
Solar Tres, 137
sport, 226
sunk costs, 278
superheaters, 85
sustainability, 251–52, 259, 266–70, 272
sustainability index, 267

terrestrial environment, 26, 28–29, 49, 54, 65–66, 70–71, 73
thermal energy storage, 191
thermic effect, 222, 230, 236
thermoeconomics, 273, 282–85, 291, 293–94
thermoregulation, 222, 235
tides, 52
time horizon, 277
transportation, 147, 157, 179, 181, 183–84, 186–87, 189, 199

uncertainty, 273, 281, 285, 289–93
uncertainty propagation, 291
uranium, 55

variable costs, 278

waves, 52
well-to-wheels efficiency, 184
Wien's Law, 128
wind turbines, 141